The Fluid Dynamics of Cell Motility

Fluid dynamics plays a crucial role in many cellular processes, including the locomotion of cells such as bacteria and spermatozoa. These organisms possess flagella, slender organelles whose time-periodic motion in a fluid environment gives rise to motility. Sitting at the intersection of applied mathematics, physics and biology, the fluid dynamics of cell motility is one of the most successful applications of mathematical tools to the understanding of the biological world.

Based on courses taught over several years, this book details the mathematical modelling necessary to understand cell motility in fluids, covering phenomena ranging from single-cell motion to instabilities in cell populations. Each chapter introduces mathematical models to rationalise experiments, uses physical intuition to interpret mathematical results, highlights the history of the field and discusses notable current research questions. All mathematical derivations are included for students new to the field, and end-of-chapter exercises help to consolidate understanding and practise applying the concepts.

ERIC LAUGA is Professor of Applied Mathematics at the University of Cambridge and a Fellow of Trinity College, Cambridge. He is the author or co-author of over 170 publications in the field of fluid mechanics, biophysics and soft matter. He is a recipient of a CAREER Award from the US National Science Foundation (2008), and of three awards from the American Physical Society: the Andreas Acrivos Dissertation Award in Fluid Dynamics (2006), the François Frenkiel Award for Fluid Mechanics (2015) and the Early Career Award for Soft Matter Research (2018). Eric Lauga is a Fellow of the American Physical Society.

T0179810

Cambridge Texts in Applied Mathematics

All titles listed below can be obtained from good booksellers or from Cambridge University Press. For a complete series listing, visit www.cambridge.org/mathematics.

The Fluid Dynamics of Cell Motility

ERIC LAUGA

University of Cambridge

CAMBRIDGE
UNIVERSITY PRESS

CAMBRIDGE
UNIVERSITY PRESS

University Printing House, Cambridge CB2 8BS, United Kingdom

One Liberty Plaza, 20th Floor, New York, NY 10006, USA

477 Williamstown Road, Port Melbourne, VIC 3207, Australia

314–321, 3rd Floor, Plot 3, Splendor Forum, Jasola District Centre, New Delhi – 110025, India

79 Anson Road, #06–04/06, Singapore 079906

Cambridge University Press is part of the University of Cambridge.

It furthers the University's mission by disseminating knowledge in the pursuit of
education, learning, and research at the highest international levels of excellence.

www.cambridge.org
Information on this title: www.cambridge.org/9781107174658
DOI: 10.1017/9781316796047

First published 2020

Printed in Singapore by Markono Print Media Pte Ltd

A catalogue record for this publication is available from the British Library.

Library of Congress Cataloging-in-Publication Data
Names: Lauga, Eric, author.
Title: The fluid dynamics of cell motility / Eric Lauga.
Description: Cambridge ; New York : Cambridge University Press, 2020. |
    Series: Cambridge texts in applied mathematics | Includes bibliographical references.
Identifiers: LCCN 2020009075 (print) | LCCN 2020009076 (ebook) |
    ISBN 9781107174658 (hardback) | ISBN 9781316626702 (paperback) |
    ISBN 9781316796047 (epub)
Subjects: LCSH: Cells–Motility–Mathematical models. | Fluid dynamics.
Classification: LCC QH647 .L38 2020 (print) | LCC QH647 (ebook) | DDC 571.6/7–dc23
LC record available at https://lccn.loc.gov/2020009075
LC ebook record available at https://lccn.loc.gov/2020009076

ISBN 978-1-107-17465-8 Hardback
ISBN 978-1-316-62670-2 Paperback

To Dominique, Alexis, William and Raphaël

Contents

Preface

Fluid dynamics plays a crucial role in numerous biological processes, from the largest animals to the smallest cells, and research activity at the intersection of fluid dynamics and biology has been steadily growing since the pioneering work of Taylor in the 1950s. Over the last 20 years, fluid mechanics conferences and journals have seen a stark increase in research relevant to, or inspired by, biology. An area of particular interest is the role of low Reynolds number flows in cellular life. Enabled by advances in microscopy and micromanipulation, experimental data have been used to inspire new models and enable hydrodynamic discoveries, while in turn quantitative theories using fluid mechanics have been precisely tested.

One area where this back and forth between theory and experiments has been very successful is the study of cell motility. Exemplified by the swimming of microorganisms such as bacteria and spermatozoa, this is an area where fluid dynamics has not only helped explain many natural phenomena but also one where biology has led to a new understanding of hydrodynamics. Several comprehensive review papers have been written on the subject but often they (including the ones I wrote) do not have all the space required to be fully pedagogical. In this book, I therefore chose to focus on the basic mathematical modelling tools and emphasise the key physical ideas behind active biological fluids.

This book develops, from first principles, the mathematical framework necessary to model and quantify the motion of active microorganisms in viscous fluids and applies these methods to understand a range of phenomena in cellular biophysics, from single-cell motion to instabilities in cell populations. Ideally, this book should be viewed as a case study in biological fluid dynamics, demonstrating how the tools of applied mathematics and continuum mechanics may be harnessed to provide quantitative and physical insights into the biological world. It is my hope that the methodology outlined here will be applicable to many other problems from nature and will inspire students to develop their own approach.

The target audience for this book includes graduate or advanced undergraduate students in either fluid mechanics, applied mathematics or physics who are interested in problems at the intersection of physical and natural sciences. The reader of this book should be familiar with basic hydrodynamics at the undergraduate level, including the derivation of the mass and momentum conservation equations, the Navier–Stokes equations and their boundary conditions and the energy equation. Undergraduate knowledge of mathematical methods, as well as vector and tensor calculus, is also assumed.

Most of the chapters in this book evolved from a set of notes for a course in Part III of the Mathematical Tripos at the University of Cambridge, which had itself developed from lectures given at the University of California, San Diego and at various summer schools. Along with the lecture notes, I have also gathered over 90 relevant mathematical problems, so each chapter includes mathematical exercises to be used for further study or to be assigned as homework or exam questions. It was my goal for the final product to be sufficiently self-contained to be useable as a textbook.

This book has 15 chapters and is organised along three parts, with the roadmap on the next page showing how knowledge from a chapter is essential for a subsequent one. Part I (4 chapters) covers the fundamental biological, mathematical and physical background and presents idealised models of active locomotion; Part II (5 chapters) reviews the mathematical modelling of single-cell locomotion powered by flagella; Part III (6 chapters) is devoted to situations in which cells interact with their environment, either via a background flow, a complex or fluctuating fluid, other cells or boundaries.

The total sum of the work I carried out with many collaborators, postdocs and students has helped me gain a better understanding of the subject, and I am grateful for it. Thank you to P. Katsamba, L. Koens, M. Lisicki, E. Riley, T. Spelman, M. Tătulea-Codrean and K. Wan for giving me feedback on some chapters. Special thanks to A. Chamolly and W. Liao for going through the entire book with a fine and rigorous comb and M. Tătulea-Codrean for her help with the exercises. I am grateful to my colleague R. Goldstein for his feedback and advice and I also thank T. Montenegro-Johnson, D. Das and D. Brumley for their help with some of the figures. Finally, let me express my gratitude to my father, for lending his immense drawing skills to my book, and to my wife, for spending too many hours proofreading it.

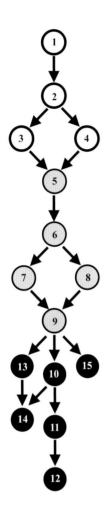

PART ONE

FUNDAMENTALS

1

Biological Background

In this chapter, we give a brief biological introduction necessary not only to understand the context of the mathematical models developed in the following chapters, but also to appreciate the relevance of the biophysical problems addressed. This gives us the opportunity to present a short overview of the role played by fluid dynamics in biology.

1.1 The Biological World

A simple illustration of the taxonomy of the biological world is shown in Fig. 1.1. Organisms are classified as belonging to one of two domains: prokaryotes, which are single cells that lack a nucleus and other important organelles, and eukaryotes, whose cells do contain a nucleus. As we explain below, prokaryotes and eukaryotes also differ fundamentally in the way they move in a fluid. The common ancestor of both domains dates back approximately 3.5 billion years, not long after the start of cellular life on Earth about 4 billion years ago.

Each domain is further subdivided into kingdoms, and for each of them we indicate in Fig. 1.1 the orders of magnitude of the numbers of different species they include. The prokaryotic domain is made up of two kingdoms: bacteria, which represent the majority of prokaryotes and whose locomotion is studied in detail in this book, and the much less studied archaea.

In contrast, four eukaryotic kingdoms exist. The first, protists, contains the simplest eukaryotes, and includes both unicellular and multicellular algae, many of which are self-propelled and are discussed in this book, but also protozoa and slime molds.

Next comes the animal kingdom which, with over a million species, has the largest diversity of species in the entire living world. Groups of animals sharing many common traits include sponges, jellyfish, worms, insects, crustaceans, mollusks and vertebrates. Of course a multitude of organisms move in fluids within

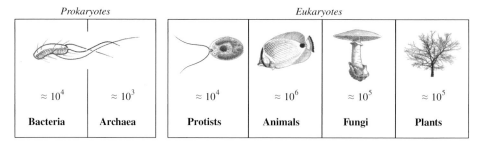

Prokaryotes   *Eukaryotes*

$\approx 10^4$	$\approx 10^3$	$\approx 10^4$	$\approx 10^6$	$\approx 10^5$	$\approx 10^5$
Bacteria	**Archaea**	**Protists**	**Animals**	**Fungi**	**Plants**

Figure 1.1 Taxonomy of the biological world. Life is divided into two domains, prokaryotes (cells lacking a nucleus) and eukaryotes (cells with a nucleus). Prokaryotes are further divided into two kingdoms (bacteria and archaea) while eukaryotes are typically divided into four kingdoms (protists, animals, fungi and plants). The order of magnitude of the numbers of different species is given for each kingdom. Drawings courtesy of Jacques Lauga.

this kingdom, from self-propelled cells to small insects and large whales. In this book, we characterise in particular the locomotion of spermatozoa, male gametes involved in the sexual reproduction of many animals.

The last two kingdoms involve living organisms that are in general not capable of self-propulsion and are not discussed in this book. Fungi are spore-producing organisms whose digestion takes place outside their body (they include molds, yeasts, truffles and mushrooms), while plants use photosynthesis to produce their nutrients (they include both vascular and non-vascular plants, seed plants, angiosperms and monocots).

1.2 Fluid Dynamics in Biology

Fluids are relevant to all forms of life, from cells to higher organisms, and are involved in virtually all biological processes. Fluids in movement also play important biological roles, in the case of both gases (e.g. air) and liquids (e.g. water). The classical book by Vogel (1996) gives a comprehensive overview of the impact of moving fluids in biology, and it has long been a tradition of fluid dynamicists to develop mathematical models to capture these effects (Lighthill, 1975).

Broadly speaking, biological fluid dynamics problems may be separated into two categories: internal and external problems. Internal fluid dynamics is concerned with fluid motion inside an organism. The two most famous examples of internal problems are blood circulation, which encompasses problems with a range of length and timescales (Pedley, 2000; Popel and Johnson, 2005), and the respiratory system, which includes gas exchange and multiphase fluid physics (Grotberg, 1994; Heil and Hazel, 2011). Another important example at the cellular scale is the cytoplasmic streaming inside cells (see Goldstein (2015) in the case of plant cells), while other internal fluids relevant to human health include synovial (joints) and cerebrospinal (brain and spinal cord) fluids.

In contrast, external problems are concerned with situations in which a fluid is located outside an organism of interest or its appendages and usually involve some degree of activity or deformation. The classical external problem is swimming (Childress, 1981), where the moving limbs of a human, or the fins of a fish, induce a flow in the surrounding fluid resulting in hydrodynamic stresses which, when integrated over the entire surface of the body, lead to a net propulsive force and locomotion. Another classical example is the impact of wind on plant growth (De Langre, 2008; Gardiner et al., 2016).

Clearly, the distinction between internal and external problems is not very precise. For example, the flow around a moving red blood cell is internal from the point of view of the organism whose blood is being examined, but external from the point of view of the cell. However, this distinction between two broad categories of problems has proven useful in grouping together biological systems involving fluid mechanics with similar dynamical regimes (Lighthill, 1975).

1.3 Biological Locomotion

The section of biofluid dynamics at the heart of this book is locomotion: How do living organisms explore a surrounding fluid and how are they constrained by it? Notably, the physical hydrodynamics of biological locomotion encompasses a large variety of problems with length scales ranging over seven orders of magnitude – from a few microns for cellular locomotion to tens of metres for the largest mammal. Accordingly, the dynamical regimes appropriate to capture all of biology include all of incompressible fluid dynamics, from the Stokes flow regime to turbulent boundary layers.

With this in mind, an alternative taxonomy is often proposed by fluid physicists to separate biological problems according to their relevant fluid dynamical regime (Childress, 1981). Consider a swimmer of size L propelling itself at speed U in a viscous, incompressible fluid of density ρ and dynamic viscosity μ. Ignoring unsteady effects, two timescales govern the motion of the fluid surrounding the swimmer. The first timescale, $t_1 = L/U$, is the typical time for a perturbation to the quiescent fluid to be advected along the swimmer. The second timescale, $t_2 = \rho L^2/\mu$, is the relevant diffusive timescale for a perturbation to the fluid to be dissipated away by the action of viscosity. Flows for which $t_1 \ll t_2$ are therefore dominated by advective transport whereas those where $t_1 \gg t_2$ are controlled by viscous effects.

Since both t_1 and t_2 have dimension of time, their ratio t_2/t_1 is dimensionless, and is in fact the relevant steady Reynolds number, Re, for the swimmer,

$$\text{Re} \equiv \frac{\rho L U}{\mu}. \tag{1.1}$$

Locomotion problems tend to be understood according to the magnitude of their Reynolds number. Well-studied high Reynolds number locomotion problems include fish swimming, flying birds and insects. While very few organisms undergo locomotion in the intermediate regime, Re $\sim$ 1 to 10, another world exists where Re $\ll$ 1, the world of swimming microorganisms to which this book is devoted.

1.4 Locomotion at Low Reynolds Number

Microorganisms represent the bulk of the world's biomass, with over 10^{30} living cells, most of them prokaryotes. The ability to self-propel in a fluid, called motility, is a property of many cellular organisms, both prokaryotes and eukaryotes, and both unicellular and multicellular (Bray, 2000). The importance of local fluid motion to these organisms was long recognised but first modelled only in the 1950s, with the seminal work of Taylor and Lighthill at the University of Cambridge, and has since been at the centre of many studies at the intersection of applied mathematics, physics and biology.

The field has now reached a point where we understand the hydrodynamic framework required to model cell locomotion, and in particular the fluid dynamics of bacteria, plankton, spermatozoa and mammalian reproduction. In this book we cover the fundamental fluid dynamical modelling approaches for cell motility from the ground up. By focusing on a few model organisms, and showing systematically how a simple model may be constructed and solved, we are able to answer a series of biophysical questions of increasing complexity. Some model organisms considered repeatedly in this book are illustrated in Fig. 1.2, in a picture drawn to approximate relative scale. In order of increasing length scales of the cell body we tackle the propulsion of bacteria, spermatozoa, algae and ciliates.

1.5 Organelles that Confer Cell Motility

Before focusing on hydrodynamic aspects, we need to understand one important biological feature, namely the exact manner in which self-propelled cells create their swimming motion. All organisms studied in this book use slender whip-like organelles called flagella (singular, flagellum) in order to generate time-varying motion within the surrounding fluid. How do these organisms deform or move their flagella? Prokaryotes and eukaryotes have evolved fundamentally different solutions to this question, and while the same term of 'flagella' is used to denote

PROKARYOTES   Bacterium

EUKARYOTES

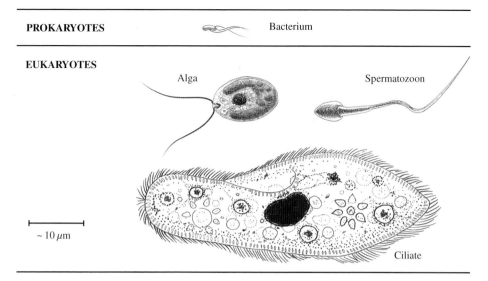

                    Alga   Spermatozoon

~ 10 μm

                                                            Ciliate

Figure 1.2 An illustration of the four model microorganisms that are studied in this book, drawn to relative scale; scale bar: 10 μm (approximate). Prokaryote: flagellated bacterium (model organism: *Escherichia coli*). Eukaryotes: flagellated spermatozoon (model invertebrate: sea urchin spermatozoon; model vertebrate: human spermatozoon); biflagellate alga (model organism: genus *Chlamydomonas*); ciliate (model organism: genus *Paramecium*). Drawings courtesy of Jacques Lauga.

the appendages of both, they are radically different in the manner in which they are molecularly actuated (Alberts et al., 2007). Both mechanisms, which were elucidated in the 1970s, are introduced briefly here and a more precise mathematical approach is offered in Chapter 5.

1.5.1 Prokaryotes

Prokaryotic cells (which we restrict here to mean bacteria) swim by rotating slender polymeric helices called flagellar filaments. The rotation of the filaments is driven by an intricate piece of molecular machinery called the bacterial rotary motor, a reversible stepper motor driven by ion fluxes, about 45 nm in diameter and able to generate rotation speeds of up to 300 Hz and torques of up to 5×10^3 pN nm (Berg, 2004).

The rotation of the motor is transmitted to the helices through a short flexible rod called the flagellar hook, which acts as a universal joint (see illustration in Fig. 1.3). Each helical filament is linked to a hook, which is actuated by an individual motor. The term 'flagella' then refers to the 'motor + hook + filament' apparatus. While some cells only have one motor, and thus actuate only one filament, many bacteria

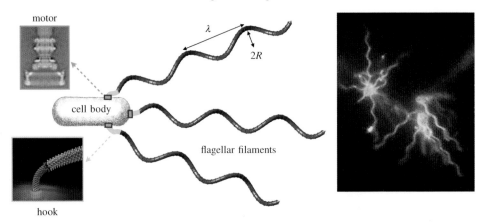

Figure 1.3 Left: flagellated bacterium where rotary motors embedded in the cell wall each use a short flexible filament called a hook. The hook acts as a universal joint to rotate passively a helical flagellar filament of pitch λ, which is on the order of microns, and radius R, which is typically hundreds of nanometres in size. Right: fluorescence microscopy picture of two *E. coli* bacteria displaying multiple identical flagellar filaments. Motor image reproduced with permission from Thomas et al. (2001). Hook picture courtesy of K. Namba, Osaka University. *E. coli* picture courtesy of Howard Berg, Harvard University.

employ multiple motors, each of them rotating an individual helical filament, as is the case for the oft-studied Escherichia coli (*E. coli*, Berg 2004). In all cases, locomotion of the cell results from the overall force and torque balance on the collection of rotating and interacting helices in the fluid.

The flagellar filaments, which are polymers made of a single protein conveniently called flagellin and all about 20 nm in thickness, are able to take one of 11 polymorphic forms depending on the chemical conditions and the behaviour of the cell body they are attached to (Macnab and Ornston, 1977). One of these forms is termed 'normal' and is used for forward locomotion by cells in their natural environment (wild-type cells); it has pitch $\lambda \approx 2.3 \, \mu$m (i.e. the wavelength measured along the axis of the helix) and radius $R \approx 200$ nm (i.e. the radius of the cylinder on which the helix is coiled; see Fig. 1.3). During normal swimming, the normal flagellar filaments rotate at about 100 Hz with respect to the background quiescent fluid.

Importantly, from the point of view of the flagellar filaments the propulsion mechanism employed by bacteria is passive. During locomotion, each flagellar filament undergoes rigid-body rotation from an actuation point localised at one end (the motor). If one helical filament were to be cut in two, the half that is no longer connected to the hook would stop rotating and as such would no longer induce

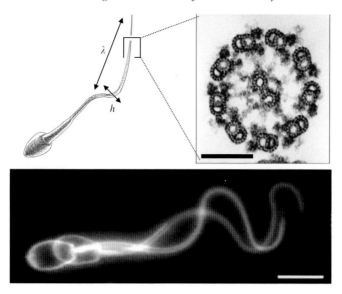

Figure 1.4 Top left: model spermatozoon swimming using active deformation of a flexible flagellum of wavelength λ, which is on the order of tens of microns, and amplitude h, which is typically a few microns. Drawings courtesy of Jacques Lauga. Top right: cross section of the internal structure of a flagellum (axoneme) showing the 9+2 structure of polymeric filaments (microtubule doublets); scale bar: 100 nm. Picture reproduced from Alberts et al. (2007). Bottom: picture of a ram spermatozoon at two times separated by half a beating cycle (0.28 s); scale bar: 10 μm. Picture used with permission of John Wiley and Sons from Woolley (2010); permission conveyed through Copyright Clearance Center, Inc.

locomotion. This passive mechanism for bacteria is to be contrasted with the active mechanism at the heart of eukaryotic swimming.

1.5.2 Eukaryotes

All swimming eukaryotes generate flows and propulsive forces from the actuation of flexible flagella. Despite carrying the same name as the prokaryotic apparatus, the 100 nm-thick eukaryotic flagella are instead similar to muscle fibres. Along an internal flagellar structure called the axoneme, long polymeric filaments (microtubule doublets) arranged in a 9+2 conformation as shown in Fig. 1.4 are made to slide past each other under the action of ATP-driven molecular motors (dynein). This sliding is converted geometrically into bending waves which propagate along flagella with wavelengths of tens of microns and frequencies on the order of tens of hertz.

While a range of flagellar arrangements exists, we address three prototypical situations in this book, as illustrated in Fig 1.2. The first is a model spermatozoon (see also Fig. 1.4), where a cell body is propelled by a single flagellum undergoing approximately planar waving motion and pushing the cell forward. The second case is a biflagellate such as the green algae genus *Chlamydomonas*. These cells are pulled from the front by two flagella that often beat with mirror-image symmetry (Goldstein, 2015). The third example encompasses larger organisms called ciliates such as the protozoon *Paramecium*. Ciliates possess a large number of short flagella (given in this situation the appellation of cilia), whose asymmetric back-and-forth deformation strokes are coordinated in patterns of metachronal waves, allowing the cell to undergo three-dimensional helical motion.

It is significant to stress that, in contrast to prokaryotes, the actuation mechanism of eukaryotic flagella is active. The forcing from the molecular motors is distributed spatially along the entire flagellum. The molecular fuel, ATP, is produced by specialised organelles located in the thick portion of the flagellum, called the mid-piece, at the junction between the cell body and the rest of the flagellum, and diffuses along the axoneme. If a eukaryotic flagellum were to be cut in two, both halves would continue to beat, and self-propel, as long as they have access to ATP. Eventually, the half of the flagellum no longer connected to the cell body (and thus to the mid-piece) would run out of ATP and then stop beating. An important consequence of this active actuation mechanism is the time-varying deformation of the flagella. While bacterial flagellar filaments undergo rigid-body rotation, the shape of a eukaryotic flagellum changes in time, leading to a more complex fundamental balance of forces.

1.6 Cellular Locomotion as a Case Study in Modelling

Equipped with this biological background, we develop in this book mathematical models to capture the fluid-based locomotion of microorganisms and their interactions with relevant environments. We should keep in mind that this specialised topic represents but a small subset of all biofluid locomotion problems, which are themselves only a small portion of all external fluid dynamics relevant to biology, itself only a small part of the field of biological fluid dynamics.

By the end of this book, the reader will have learned how to exploit various modelling approaches to capture biological behaviour – from idealised two-dimensional models to flow singularities; from far-field interactions to near-field details. There remain vast and unexplored areas outside the scope of this book, and we hope that readers will adopt a similar attitude to model and understand many of these new problems.

Further Reading

A comprehensive introduction to biology is offered in a textbook format in the classical volume by Campbell and Reece (2014). For a focus on molecular biology at the cellular level, the reader is referred to Alberts et al. (2007), while cell motility is given a biological overview by Bray (2000).

A number of specialised review articles written over the last 50 years have shed light on many aspects of the fluid dynamics of cell locomotion and could be used as companion pieces to this book. Notable reviews include those on the locomotion of protozoa (Jahn and Votta, 1972), the general hydrodynamics of cell locomotion (Brennen and Winet, 1977; Lauga and Powers, 2009), the fluid dynamics of bacteria (Lauga, 2016), the motion of aquatic microorganisms (Pedley and Kessler, 1992; Guasto et al., 2012), collective effects in cell suspensions (Pedley and Kessler, 1992; Koch and Subramanian, 2011), and the fluid mechanics of spermatozoa (Gaffney et al., 2011) and mammalian reproduction (Fauci and Dillon, 2006).

On larger length scales, a variety of books and review articles provide good entry points to the physics and biology of flying and swimming, in particular fish locomotion (Lighthill, 1975; Triantafyllou et al., 2000; Fish and Lauder, 2006; Videler, 2012), flying birds (Alexander, 2002) and insects (Ellington, 1984; Dudley, 2002; Wang, 2005). Finally, the reader interested in a pedagogical overview of the physics at play in a broad class of biological systems might enjoy some of the popular books by Steve Vogel (Vogel, 1988; Vogel and Calvert, 1993).

2

The Fluid Dynamics of Microscopic Locomotion

This chapter is devoted to an overview of the fundamental physical concepts and mathematical equations required to construct models of cell locomotion. We start from the general setup of locomotion in fluids at arbitrary scales before focusing on the low Reynolds number limit. We show in particular how the Stokes limit of hydrodynamics can be exploited along with symmetry arguments to derive general properties of microscopic locomotion.

2.1 Dynamics of Locomotion

A prototypical fluid-based locomotion problem is illustrated in Fig. 2.1. An organism of length L, immersed in a viscous fluid, varies its shape in time with an average frequency ω, resulting in its locomotion. The deformation kinematics of the organism may be described in a Lagrangian fashion by representing the shape of the swimmer as the set of material points, $\mathbf{r}_s(\mathbf{r}_0, t)$, denoting the instantaneous location at time t of the material point located at $\mathbf{r}_0$ at $t = 0$. Knowledge of the field $\mathbf{r}_s(\mathbf{r}_0, t)$ for all values of $\mathbf{r}_0$ and all times is then sufficient to characterise the shape during the entire swimming motion.

As the organism deforms its shape, it sets up a flow in the surrounding fluid that imparts hydrodynamic stresses and leads to locomotion. Denoting by σ the stress tensor in the fluid, $\mathbf{u}(\mathbf{r}, t)$ the fluid velocity and ρ the mass density, conservation of linear momentum is written in a continuum fashion as Cauchy's incompressible momentum equations,

$$\rho \frac{\partial \mathbf{u}}{\partial t} + \rho(\mathbf{u} \cdot \nabla)\mathbf{u} = \nabla \cdot \sigma, \quad \nabla \cdot \mathbf{u} = 0, \tag{2.1}$$

where we ignore the presence of any external force density in the fluid such as gravity (Batchelor, 1967). Fluids that are said to be Newtonian are characterised by a constant dynamic viscosity μ and have a symmetric stress tensor, which is given by $\sigma = -p\mathbf{1} + 2\mu\mathbf{E}$, where p is the dynamic pressure, $\mathbf{1}$ is the identity

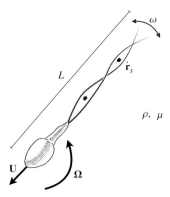

Figure 2.1 Schematic setup for a swimmer moving by surface deformation in a Newtonian fluid of constant density ρ and dynamic viscosity μ. The typical length scale of the swimmer is denoted L and the typical frequency at which its shape is deforming is ω. The surface of the swimmer is described by a Lagrangian field $\mathbf{r}_s(\mathbf{r}_0, t)$ deforming with velocity $\dot{\mathbf{r}}_s$ everywhere along the swimmer ('the swimming gait'). The instantaneous swimming kinematics are denoted $\mathbf{U}$ and $\boldsymbol{\Omega}$, which are in general functions of time. Drawing courtesy of Jacques Lauga.

tensor and $\mathbf{E}$ is the rate of strain tensor, equal to the symmetric part of velocity gradient, $\mathbf{E} = (\nabla\mathbf{u} + \nabla\mathbf{u}^T)/2$. In that case, Eq. (2.1) becomes the incompressible Navier–Stokes equations,

$$\rho\frac{\partial\mathbf{u}}{\partial t} + \rho(\mathbf{u} \cdot \nabla)\mathbf{u} = -\nabla p + \mu\nabla^2\mathbf{u}, \quad \nabla \cdot \mathbf{u} = 0. \tag{2.2}$$

In order to solve Eq. (2.2), we need to provide boundary conditions for the flow. Let us denote the instantaneous linear and angular velocities of the organism by $\mathbf{U}(t)$ and $\boldsymbol{\Omega}(t)$, respectively. The instantaneous velocity of any point $\mathbf{r}_s$ on the surface of the swimmer is given by the superposition of the deformation of the organism in the body-fixed frame, i.e. $\dot{\mathbf{r}}_s \equiv \partial\mathbf{r}_s/\partial t$, and the velocity at which that frame translates and rotates, $\mathbf{U}$ and $\boldsymbol{\Omega}$. The lab frame velocity boundary conditions for Eq. (2.2) are therefore written

$$\mathbf{u}(\mathbf{r}_s, t) = \dot{\mathbf{r}}_s + \mathbf{U}(t) + \boldsymbol{\Omega}(t) \times \mathbf{r}_s, \tag{2.3}$$

where the velocity $\dot{\mathbf{r}}_s$ is often referred to as the 'swimming gait'. The additional boundary conditions far from the swimmer depend on the specific setup but are typically the no-slip boundary conditions either on the boundaries of the domain (swimming in confinement or near a surface) or at infinity (swimming in an unbounded fluid).

In order to close this system, we need additional equations since both $\mathbf{U}(t)$ and $\mathbf{\Omega}(t)$ are still unknown. These are given by Newton's second law for the organism, written in the absence of external forces as

$$\frac{\mathrm{d}}{\mathrm{d}t}\left(m(t)\mathbf{U}(t)\right) = \mathbf{F}(t), \tag{2.4a}$$

$$\frac{\mathrm{d}}{\mathrm{d}t}\left(\mathbf{I}(t)\cdot\mathbf{\Omega}(t)\right) = \mathbf{T}(t), \tag{2.4b}$$

where $m(t)$ is the mass of the organism and $\mathbf{I}(t)$ is its instantaneous moment-of-inertia tensor, defined as

$$\mathbf{I}(t) \equiv \iiint_V \rho\left[(\mathbf{r}\cdot\mathbf{r})\mathbf{1} - \mathbf{rr}\right]\,\mathrm{d}V, \tag{2.5}$$

both of which are allowed to vary in time to account for biological processes and shape changes. In Eq. (2.4), $\mathbf{F}(t)$ and $\mathbf{T}(t)$ are the instantaneous hydrodynamic forces and torques, given by surface integrals over the shape of the organism:

$$\mathbf{F}(t) = \iint_S \sigma\cdot\mathbf{n}\,\mathrm{d}S, \quad \mathbf{T}(t) = \iint_S \mathbf{r}\times(\sigma\cdot\mathbf{n})\,\mathrm{d}S, \tag{2.6}$$

with $\mathbf{n}$ the unit normal to the surface S of the organism pointing into the fluid. We note that Eq. (2.4) provides the correct number of scalar equations, namely six, in order to determine the six degrees of freedom quantified by the three components of $\mathbf{U}$ and the three components of $\mathbf{\Omega}$. Importantly, the flow and the swimming motion are fully coupled since the swimming kinematics appear as boundary conditions for the flow problem (Eq. (2.3)) while the fluid stresses are required to determine the swimming kinematics (Eq. (2.4)).

2.2 Reynolds Numbers

A major simplification arises for small organisms, such as the swimming cells at the heart of this book, by investigating the relative magnitude of each term in the equations of motion. Using L and ω^{-1} as characteristic length and timescales, and a typical swimming velocity U as a characteristic velocity scale, the Navier–Stokes equations, Eq. (2.2), may be non-dimensionalised, leading to

$$\mathrm{Re}_\omega\frac{\partial\mathbf{u}}{\partial t} + \mathrm{Re}(\mathbf{u}\cdot\nabla)\mathbf{u} = -\nabla p + \nabla^2\mathbf{u}, \tag{2.7}$$

where the scale of the pressure is taken to be the viscous scale, $p \sim \mu U/L$, and where we have used the same symbols as the dimensional equation for notation

		Length	Velocity	Frequency	Re	Re$_\omega$
Bacterium		10 μm	10 μm/s	100 Hz	10^{-4}	10^{-2}
Spermatozoon		100 μm	100 μm/s	10 Hz	10^{-2}	10^{-1}
Ciliate	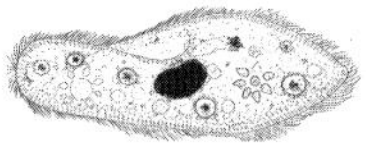	100 μm	1 mm/s	10 Hz	10^{-1}	10^{-1}
Tadpole		1 cm	10 cm/s	10 Hz	10^{3}	10^{3}
Small fish	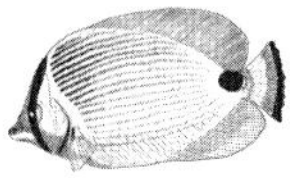	10 cm	10 cm/s	10 Hz	10^{4}	10^{5}
Penguin		1 m	1 m/s	1 Hz	10^{6}	10^{6}
Sperm whale	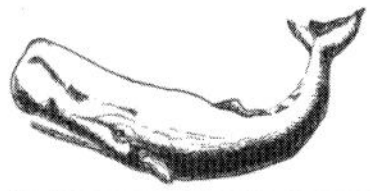	10 m	1 m/s	0.1 Hz	10^{7}	10^{7}

Figure 2.2 Characteristic Reynolds numbers for the locomotion of seven representative organisms in water ($\mu/\rho \approx 10^{-6}$ m^2/s). Three organisms belong to the low Reynolds number world (bacterium, spermatozoon and ciliate) and four swimming at high Reynolds number (tadpole, small fish, penguin and sperm whale) with orders of magnitude for their typical length (L), maximum velocity (U) and typical frequency ($\omega/2\pi$) scales. Drawings courtesy of Jacques Lauga.

convenience. Two dimensionless numbers appear in Eq. (2.7), namely

$$\text{Re} \equiv \frac{\rho L U}{\mu}, \quad \text{Re}_\omega \equiv \frac{\rho L^2 \omega}{\mu}. \tag{2.8}$$

The steady Reynolds number, Re, has already been discussed in the previous chapter. In contrast, Re$_\omega$ is the oscillatory Reynolds number, i.e. a Reynolds number for which the velocity scale is the typical local fluid velocity induced by the oscillatory motion of the shape, namely $L\omega$. Note that Re and Re$_\omega$ do not necessarily have the same magnitude; for example, an insect hovering in air at a fixed position would have a small Re but a very large Re$_\omega$ due to the high-frequency motion of its wings.

Focusing on locomotion in water, seven typical organisms and the order of magnitude of their characteristic scales (typical length, maximum velocity and typical frequency) and corresponding values of Re and Re$_\omega$ are displayed in Fig. 2.2. Three

of these organisms belong to the low Reynolds number world (bacterium, spermatozoon and ciliate) while the other four swim at high Reynolds numbers (tadpole,
small fish, penguin and sperm whale). Note that ciliates are at the limit of what
would be considered low Reynolds number locomotion, and in fact are known to
be able to take advantage of inertia for escape purposes (Hamel et al., 2011).

2.3 The Stokes Equations

For small cellular organisms we see that both Re and Re_ω are small, and thus
the Navier–Stokes equations may be approximated by the incompressible Stokes
equations,

$$\nabla p = \mu \nabla^2 \mathbf{u}, \quad \nabla \cdot \mathbf{u} = 0. \tag{2.9}$$

Numerous exact and approximate solutions of the Stokes equations exist in reference monographs (Happel and Brenner, 1965; Kim and Karrila, 1991; Leal, 2007)
and we will draw from this extensive literature to characterise the flows relevant to
swimming cells.

In anticipation of the rest of this book, four mathematical consequences of
Eq. (2.9) should be noted at this point. First, it is possible to show rigorously that
solutions to the Stokes equations with prescribed velocity boundary conditions are
unique (Kim and Karrila, 1991). As a consequence, if a flow solution for a particular setup is obtained using any mathematical method (or guesswork), then it is the
unique solution to the problem.

Second, since the velocity field is divergence-free, we may compute the divergence of Eq. (2.9) and obtain

$$\nabla^2 p = 0. \tag{2.10}$$

The pressure in a Stokes flow is therefore harmonic. As a result, computing the
Laplacian of the Stokes equations leads to

$$\nabla^4 \mathbf{u} \equiv \nabla^2 \nabla^2 \mathbf{u} = \mathbf{0}, \tag{2.11}$$

and the velocity field is said to be biharmonic (i.e. its Laplacian is harmonic).
Further, since the curl of a gradient is always zero, taking the curl of Eq. (2.9)
leads to an equation for the flow vorticity, $\omega = \nabla \times \mathbf{u}$, namely

$$\nabla^2 \omega = \mathbf{0}, \tag{2.12}$$

and the vorticity of a Stokes flow is also harmonic.

Finally, it is important to note that the momentum equation in Eq. (2.9) may be
rewritten conveniently as an equation for the stress tensor as

$$\nabla \cdot \sigma = \mathbf{0}, \tag{2.13}$$

a result that is in fact true for any continuum material in the absence of inertia. Integrating Eq. (2.13) on a volume of fluid and using the divergence theorem then lead to the result that the net hydrodynamic force (or torque) acting on any closed parcel of fluid is identically zero.

2.4 Low Reynolds Number Dynamics

Assuming a small Reynolds number for the flow has many consequences for the dynamics of a swimming organism. Let us consider the relative magnitude of the terms on both sides of Newton's law for linear momentum, Eq. (2.4a). The results we derive also hold for the conservation of angular momentum, Eq. (2.4b). The typical order of magnitude of the change of momentum for the swimmer of mass m is mU/τ, where τ is a relevant acceleration or deceleration timescale. In contrast, the characteristic hydrodynamic force is $F \sim \sigma L^2$, where σ is the order of magnitude of a hydrodynamic stress. At low Reynolds number, stresses in response to fluid motion at speed U scale as $\sigma \sim \mu U/L$ since U/L is the relevant shear rate in the fluid. A low Reynolds number hydrodynamic force thus scales as $F \sim \mu UL$. Balancing the change in linear momentum with the magnitude of the force leads to an estimate for the relevant timescale τ, namely

$$m\frac{U}{\tau} \sim \mu UL \rightarrow \tau \sim \frac{m}{\mu L}. \tag{2.14}$$

How does this acceleration/deceleration timescale τ compare to the two relevant timescales for the motion of the organism, namely the swimming timescale, L/U, and the body oscillation period, ω^{-1}? Using the fact that the mass of the organism scales as $m \sim \rho_s L^3$, where ρ_s is its density, it is straightforward to obtain the ratios of τ with these timescales as dimensionless groups, namely

$$\frac{\tau}{L/U} \sim \frac{\rho_s UL}{\mu} \quad \text{and} \quad \frac{\tau}{\omega^{-1}} \sim \frac{\rho_s L^2 \omega}{\mu}. \tag{2.15}$$

Comparing Eq. (2.15) with Eq. (2.8), we recognise the appearance of the Reynolds numbers. Specifically we can rewrite the ratio of timescales in Eq. (2.15) as

$$\frac{\tau}{L/U} \sim \frac{\rho_s}{\rho}\text{Re} \quad \text{and} \quad \frac{\tau}{\omega^{-1}} \sim \frac{\rho_s}{\rho}\text{Re}_\omega. \tag{2.16}$$

For small organisms moving in fluids such that $\text{Re} \ll 1$ and $\text{Re}_\omega \ll 1$ we therefore obtain that $\tau \ll L/U$ and $\tau \ll \omega^{-1}$. On any of the timescales relevant to the motion of the organism in the fluid, the inertial accelerations and decelerations are thus instantaneous (Purcell, 1977). Swimming at low Reynolds number is therefore a quasi-steady problem and to a good approximation the left-hand-side terms in Eq. (2.4) may be neglected. Swimming in a Stokes flow occurs thus in such a way that the total instantaneous hydrodynamic force and torque are zero for all times, i.e.

$$\mathbf{F}(t) = \mathbf{0}, \quad \mathbf{T}(t) = \mathbf{0}. \tag{2.17}$$

Low Reynolds number locomotion may thus be understood as an eigenvalue problem where the values of $\mathbf{U}(t)$ and $\mathbf{\Omega}(t)$ are the unique instantaneous solutions to Eq. (2.17). We note that removing the inertial terms in Eq. (2.4) means that all frames are now inertial frames, allowing us to solve low Reynolds number swimming problems in any convenient frame of reference, for example a time-varying swimming frame.

In the rest of this book, we refer to the type of locomotion where the fluid satisfies the Stokes equations, Eqs. (2.9), as 'low Reynolds number locomotion', with the understanding that this appellation actually assumes the Reynolds numbers to be exactly zero. In turn, the fact that the swimming organism is force- and torque-free, Eq. (2.17), is referred to as the 'free-swimming' constraint.

2.5 Rate of Work and Dissipation

The energy budget for low Reynolds number swimming takes a simple form because all transport terms usually present at finite Reynolds numbers disappear in the Stokes limit. The relevant quantity to compute is the rate of work done by the swimming organism on the fluid, $\dot{W}$, given by

$$\dot{W} = -\iint_S \mathbf{u} \cdot \sigma \cdot \mathbf{n} \, dS, \tag{2.18}$$

where the minus sign arises from our convention to have the normal $\mathbf{n}$ to the surface of the swimmer pointing into the fluid. Using summation convention and the divergence theorem (which requires the normal in the opposite direction) we can write the integral as an integral on the volume V of the surrounding fluid as

$$\dot{W} = -\iint_S u_i \sigma_{ij} n_j \, dS = \iiint_V \frac{\partial}{\partial x_j} (u_i \sigma_{ij}) \, dV, \tag{2.19}$$

where we have assumed that far from the swimmer the flows and stresses decay sufficiently fast in order to neglect the contribution of the surface at infinity. Using the fact that the stress tensor is divergence-free, Eq. (2.13), that the flow is incompressible and that the rate of strain tensor is symmetric, we obtain

$$\dot{W} = \iiint_V \frac{\partial u_i}{\partial x_j} \sigma_{ij} \, dV = 2\mu \iiint_V \frac{\partial u_i}{\partial x_j} E_{ij} \, dV = 2\mu \iiint_V E_{ij} E_{ij} \, dV. \tag{2.20}$$

The last term on the right-hand side of Eq. (2.20) is the rate of viscous dissipation in the fluid. Since this is a sum of squares, we obtain that $\dot{W}$ is always positive, and is in fact positive definite if the flow far from the swimmer decays to zero (the only flows for which $E_{ij} = 0$ everywhere in the fluid are solid-body motions).

Physically, the result in Eq. (2.20) indicates that all work done by the swimmer on the fluid is dissipated instantaneously in the viscous fluid.

2.6 Forced vs. Force-Free Motion

2.6.1 Separation Between Thrust and Drag

A long-standing approach of scientists working on fluid-based locomotion is to understand the motion of the organism as a balance between thrust and drag forces (Childress, 1981). Thrust can be thought of as being produced intrinsically by the shape-changing organism, which then moves at such a speed that the drag exerted by the moving fluid exactly balances this thrust. For problems with finite Reynolds numbers, this decomposition is problematic due to the nonlinear response of the fluid to an applied deformation. In other words, the drag and thrust understood as separate problems do not produce the distribution of hydrodynamic stresses obtained during locomotion. In contrast, locomotion in a Stokes flow is a situation where thrust and drag can be separated rigorously. This is because the Stokes equations, Eq. (2.9), are linear in their boundary conditions and thus each term in Eq. (2.3) may be tackled separately, as we now do.

Consider first the problem where the boundary conditions on the surface of the swimmer are given by the swimming gait,

$$\mathbf{u}(\mathbf{r}_s, t) = \dot{\mathbf{r}}_s, \tag{2.21}$$

together with the prescribed, problem-specific boundary conditions at infinity. As a result of the boundary conditions, Eq. (2.21), and the fact that the organism is not allowed to swim in this first problem, a net force and torque are exerted by the flow on the swimmer. We refer to these as the thrust force, $\mathbf{F}_{th}(t)$, and torque, $\mathbf{T}_{th}(t)$. In an experiment, this first problem could thus be realised only if an external force and torque are used to cancel out the thrust and prevent locomotion. Solving for the values of this force and torque requires solving for the Stokes flow around the boundary of the organism, which, for an arbitrary shape, is in general not doable analytically. However, because the Stokes equations are linear, they can at least be written formally as surface integrals of the body deformation as

$$\mathbf{F}_{th}(t) = \iint_S \tilde{\mathbf{F}}(\mathbf{r}_s) \cdot \dot{\mathbf{r}}_s \, dS, \tag{2.22a}$$

$$\mathbf{T}_{th}(t) = \iint_S \tilde{\mathbf{G}}(\mathbf{r}_s) \cdot \dot{\mathbf{r}}_s \, dS, \tag{2.22b}$$

where the tensors $\tilde{\mathbf{F}}$ and $\tilde{\mathbf{G}}$ are functions of the instantaneous shape of the swimmer and may be obtained numerically, for example using a boundary integral formulation of the flow equations (Pozrikidis, 1992) (see also Chapter 6, Exercises).

Consider now as a second problem the situation in which the swimmer, with its instantaneous shape S, is frozen and is only allowed to move as a rigid body with linear and angular velocities denoted by $\mathbf{U}(t)$ and $\mathbf{\Omega}(t)$ (here t is merely used as a parameter to indicate that the shape is frozen at this particular value of time). The boundary condition in that case may be written for any point on the surface as

$$\mathbf{u}(\mathbf{r}_s, t) = \mathbf{U}(t) + \mathbf{\Omega}(t) \times \mathbf{r}_s. \tag{2.23}$$

This rigid-body motion leads to hydrodynamic drag forces, $\mathbf{F}_d(t)$, and torques, $\mathbf{T}_d(t)$, exerted on the body. By linearity of the Stokes equations, these are related linearly to $\mathbf{U}(t)$ and $\mathbf{\Omega}(t)$ through an instantaneous resistance matrix $\mathbf{R}(t)$ as (Happel and Brenner, 1965)

$$\begin{pmatrix} \mathbf{F}_d(t) \\ \mathbf{T}_d(t) \end{pmatrix} = -\mathbf{R}(t) \cdot \begin{pmatrix} \mathbf{U}(t) \\ \mathbf{\Omega}(t) \end{pmatrix}. \tag{2.24}$$

It is a classical result that the resistance matrix is symmetric (a consequence of the reciprocal theorem for Stokes flows, discussed in the next section) and positive definite (a consequence of the energy equation, Eq. (2.20)).

By linearity, the thrust (Eq. (2.22)) and drag problems (Eq. (2.24)) may be added and swimming occurs when the total instantaneous forces and torques acting on the body cancel out,

$$\mathbf{F}_{th}(t) + \mathbf{F}_d(t) = \mathbf{0}, \tag{2.25a}$$
$$\mathbf{T}_{th}(t) + \mathbf{T}_d(t) = \mathbf{0}. \tag{2.25b}$$

Since the tensor $\mathbf{R}$ in Eq. (2.24) is symmetric and positive definite it can always be inverted, and the swimming kinematics may be written formally as

$$\begin{pmatrix} \mathbf{U}(t) \\ \mathbf{\Omega}(t) \end{pmatrix} = \mathbf{R}^{-1}(t) \cdot \begin{pmatrix} \mathbf{F}_{th}(t) \\ \mathbf{T}_{th}(t) \end{pmatrix}. \tag{2.26}$$

An organism swimming at low Reynolds number is thus one that moves with velocities such that the drag experienced at these velocities balances exactly the thrust forces that would be produced if the swimmer could not move. In contrast to problems at finite Reynolds numbers, this is an exact mathematical statement in the Stokes limit.

2.6.2 Integral Theorem

An alternative approach to the separation between thrust and drag was proposed by Stone and Samuel (1996) who derived an integral theorem using the principle of virtual work, which for Stokes flows is referred to as the reciprocal theorem (see proof in Chapter 6, Exercises). This theorem states that for two incompressible Stokes flows $\mathbf{u}$ and $\hat{\mathbf{u}}$ with associated stress fields σ and $\hat{\sigma}$ and subject to no body forces inside the fluid, the virtual rates of work of one flow on the other are equal, i.e.

$$\iint_S \mathbf{u} \cdot \hat{\sigma} \cdot \mathbf{n}\, dS = \iint_S \hat{\mathbf{u}} \cdot \sigma \cdot \mathbf{n}\, dS. \tag{2.27}$$

The theorem written in this form is valid provided that the contribution of the 'surface at infinity' may be neglected, and follows in an elementary fashion from an application of the divergence theorem, similar to what we did to derive the energy equation, Eq. (2.20).

We now apply this theorem to the problem of locomotion. To do so, we have to pick two solutions to the Stokes equations with boundary conditions acting on the same instantaneous geometry. For the first solution $(\mathbf{u}, \sigma)$, let us choose the free-swimming problem characterised by the boundary conditions in Eq. (2.3), i.e.

$$\mathbf{u}(\mathbf{r}_s) = \dot{\mathbf{r}}_s + \mathbf{U}(t) + \mathbf{\Omega}(t) \times \mathbf{r}_s. \tag{2.28}$$

The relevant geometry is therefore the shape of the swimmer at time t.

For the second solution $(\hat{\mathbf{u}}, \hat{\sigma})$, let us instead pick the same solution as the one arising in the 'drag' problem in the previous section, i.e. the swimmer at time t with its instantaneous shape frozen and only allowed to move as a rigid body. The rigid-body motion occurs with velocities denoted by $\hat{\mathbf{U}}(t)$ and $\hat{\mathbf{\Omega}}(t)$. The boundary conditions on the swimmer for this solution are therefore

$$\hat{\mathbf{u}}(\mathbf{r}_s) = \hat{\mathbf{U}}(t) + \hat{\mathbf{\Omega}}(t) \times \mathbf{r}_s. \tag{2.29}$$

As a result of rigid-body motion, the swimmer experiences net drag forces and torques given by

$$\iint_S \hat{\sigma} \cdot \mathbf{n}\, dS = \hat{\mathbf{F}}(t), \quad \iint_S \mathbf{r}_s \times (\hat{\sigma} \cdot \mathbf{n})\, dS = \hat{\mathbf{T}}(t). \tag{2.30}$$

The right-hand side of Eq. (2.27) may be simplified since the rigid-body kinematics from Eq. (2.29) are constant and can be factored out of the integral, so we obtain

$$\iint_S \hat{\mathbf{u}} \cdot \sigma \cdot \mathbf{n}\, dS = \hat{\mathbf{U}}(t) \cdot \iint_S \sigma \cdot \mathbf{n}\, dS + \hat{\mathbf{\Omega}}(t) \cdot \iint_S \mathbf{r}_s \times (\sigma \cdot \mathbf{n})\, dS. \tag{2.31}$$

In the free-swimming problem, locomotion occurs with no net force or torque (see Eq. (2.17)), and thus both integrals on the right-hand side of Eq. (2.31) are zero.

In order to evaluate the left-hand side of Eq. (2.27), we use the boundary conditions in Eq. (2.28) that separate the integrals into three terms, so that Eq. (2.27) now becomes

$$\iint_S \dot{\mathbf{r}}_s \cdot \hat{\boldsymbol{\sigma}} \cdot \mathbf{n} \, dS + \mathbf{U}(t) \cdot \iint_S \hat{\boldsymbol{\sigma}} \cdot \mathbf{n} \, dS + \boldsymbol{\Omega}(t) \cdot \iint_S \mathbf{r}_s \times (\hat{\boldsymbol{\sigma}} \cdot \mathbf{n}) \, dS = 0. \quad (2.32)$$

The second and third integrals in Eq. (2.32) are the forces and torques applied in the rigid-body motion problem, Eq. (2.30), so that the reciprocal theorem leads finally to

$$\mathbf{U}(t) \cdot \hat{\mathbf{F}}(t) + \boldsymbol{\Omega}(t) \cdot \hat{\mathbf{T}}(t) = - \iint_S \dot{\mathbf{r}}_s \cdot \hat{\boldsymbol{\sigma}} \cdot \mathbf{n} \, dS. \quad (2.33)$$

Since the magnitudes and directions of the force and torque $\hat{\mathbf{F}}$ and $\hat{\mathbf{T}}$ in Eq. (2.33) are arbitrary, this integral formulation leads effectively to six equations allowing us to formally determine all components of both $\mathbf{U}(t)$ and $\boldsymbol{\Omega}(t)$ as a function of the swimming gait. Instead of separating thrust and drag, the integral theorem shows that both forced and force-free motion are linked intimately and that one may be exploited to characterise the other.

As an illustration, consider the case of a spherical organism, and let us use the integral theorem to compute the value of its swimming velocity. Since the flow around a sphere in solid-body translation is the classical solution due to Stokes (Happel and Brenner, 1965), the swimming velocity may be found exactly in this case. Denoting the radius of the sphere by a, the hydrodynamic force density in the second solution $(\hat{\mathbf{u}}, \hat{\boldsymbol{\sigma}})$ is constant on the surface of the sphere, given by

$$\hat{\boldsymbol{\sigma}} \cdot \mathbf{n} = -\frac{3\mu}{2a} \hat{\mathbf{U}}(t). \quad (2.34)$$

Using the Stokes formula for the drag of a sphere, $\hat{\mathbf{F}}(t) = -6\pi\mu a \hat{\mathbf{U}}(t)$, Eq. (2.33) becomes

$$6\pi\mu a \hat{\mathbf{U}}(t) \cdot \mathbf{U}(t) = -\frac{3\mu}{2a} \hat{\mathbf{U}}(t) \cdot \iint_S \dot{\mathbf{r}}_s \, dS. \quad (2.35)$$

Since this equality is true for any value of $\hat{\mathbf{U}}(t)$, we obtain necessarily

$$\mathbf{U}(t) = -\frac{1}{4\pi a^2} \iint_S \dot{\mathbf{r}}_s \, dS. \quad (2.36)$$

A spherical organism therefore swims with velocity equal to minus the surface average of the velocity imposed by the swimming gait. Using similar algebra, the rotation speed of the sphere may also be obtained as a surface average, as shown by Stone and Samuel (1996) (see also Exercises).

2.7 Properties of Low Reynolds Number Locomotion

The linearity of the Stokes equations has already been exploited a few times in this chapter. We now demonstrate how it can also be used to derive two properties valid for any organism self-propelling at low Reynolds numbers.

2.7.1 Rate Independence

The first property can be stated as follows: The distance travelled by a low Reynolds number swimmer as its surface changes from one configuration to another does not depend on the rate at which the surface deformation occurs, but only on its geometry. For any periodic deformation of a swimmer, the total distance travelled per period is therefore only a function of the sequence of shapes displayed by the swimmer during one period.

To demonstrate this, let us consider for simplicity a case where a swimmer translates but does not rotate. Using Eqs. (2.22) and (2.26), or alternatively Eq. (2.33), we can always write formally the instantaneous swimming velocity of the organism as an integral on the shape deformation,

$$\mathbf{U}(t) = \iint_S \tilde{\mathbf{H}}(\mathbf{r}_s) \cdot \dot{\mathbf{r}}_s \, dS, \qquad (2.37)$$

where the tensor $\tilde{\mathbf{H}}$ is a function of the whole instantaneous shape of the swimmer, but nothing else. The distance travelled by the organism between two times t_0 and t_1, denoted $\mathbf{\Delta}$, is then written as the time integral

$$\mathbf{\Delta} = \int_{t_0}^{t_1} \mathbf{U}(t) \, dt. \qquad (2.38)$$

Consider now two different time-parameterisations of the same sequence of shapes, as illustrated in Fig. 2.3. For the first parametrisation, the time t is increasing between t_0 and t_1 while a second parametrisation is measured by a different time, t', increasing between t'_0 and t'_1. Since the sequences of shapes are the same, they can be mapped from one to the other and we have $\mathbf{r}'_s(t') = \mathbf{r}_s(t)$. We next compute the distance travelled along the parametrisation t', denoted by $\mathbf{\Delta}'$, as

$$\mathbf{\Delta}' = \int_{t'_0}^{t'_1} \mathbf{U}'(t') \, dt' = \int_{t'_0}^{t'_1} \left[\iint_{S'} \tilde{\mathbf{H}}(\mathbf{r}'_s) \cdot \frac{d\mathbf{r}'_s}{dt'} \, dS' \right] dt'. \qquad (2.39)$$

We then do a change of variable $t' \to t$ in the integral. Since the tensor $\tilde{\mathbf{H}}$ only depends on the current shape, we have $\mathbf{H}(\mathbf{r}'_s(t')) = \mathbf{H}(\mathbf{r}_s(t))$. Using the chain rule we can also write

$$\frac{d\mathbf{r}_s}{dt} = \frac{d\mathbf{r}'_s}{dt'} \frac{dt'}{dt}, \qquad (2.40)$$

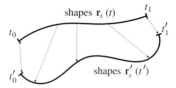

Figure 2.3 Rate independence of locomotion: swimming through the same sequence of shapes but at different rates. A mapping between the increasing
time-parameterisations $t \in [t_0, t_1]$ and $t' \in [t'_0, t'_1]$ is such that the shapes $\mathbf{r}'_s(t')$
are identical to the shapes $\mathbf{r}_s(t)$. The net distances travelled during the two intervals are then identical.

and thus the total distance travelled becomes

$$\Delta' = \int_{t_0}^{t_1} \left[\iint_{S'} \tilde{\mathbf{H}}(\mathbf{r}'_s) \cdot \frac{\mathrm{d}\mathbf{r}'_s}{\mathrm{d}t'} \, \mathrm{d}S' \right] \frac{\mathrm{d}t'}{\mathrm{d}t} \, \mathrm{d}t = \int_{t_0}^{t_1} \left[\iint_{S} \tilde{\mathbf{H}}(\mathbf{r}_s) \cdot \frac{\mathrm{d}\mathbf{r}_s}{\mathrm{d}t} \, \mathrm{d}S \right] \mathrm{d}t = \Delta.$$

$$(2.41)$$

For any parametrisation going through the same sequence of shapes, we obtain
$\Delta' = \Delta$, and the swimming kinematics are therefore rate-independent.

An alternative way to see this result, often used to introduce the geometrical
description of low Reynolds number swimming, consists in combining Eqs. (2.39)
and (2.40) to write the total displacement over one stroke as

$$\Delta = \int_{t_0}^{t_1} \iint_{S} \tilde{\mathbf{H}}(\mathbf{r}_s) \cdot \dot{\mathbf{r}}_s \, \mathrm{d}S \, \mathrm{d}t. \qquad (2.42)$$

We then note in this integral that $\dot{\mathbf{r}}_s \mathrm{d}t \equiv \mathrm{d}\mathbf{r}_s$ is the exact differential element along
the shape of deformation and therefore Eq. (2.42) can be recast formally as an
integral over deformation space,

$$\Delta = \int_{\mathbf{r}_s} \left(\iint_{S} \tilde{\mathbf{H}}(\mathbf{r}_s) \mathrm{d}S \right) \cdot \mathrm{d}\mathbf{r}_s, \qquad (2.43)$$

which depends only on the sequence of shapes $(\mathbf{r}_s)$ and is independent of the rate
of deformation.

Note that during this discussion we have not said anything about the boundary
conditions away from the swimmer. Clearly if the boundary conditions were to depend on the flow itself, this result would no longer be valid, for example in the case
of locomotion near a flexible surface (Trouilloud et al., 2008). The property of rate
independence thus remains valid in all situations where the boundary conditions
remain unaffected by the swimming motion, for example in an unbounded fluid or
near rigid surfaces.

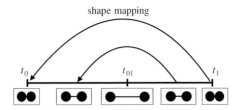

Figure 2.4 Reciprocal swimming: time-periodic motion with a sequence of shapes that are identical under time-reversal. The periodic sequence of shapes between t_0 and t_1 may be split into two intervals, $[t_0, t_{01}]$ and $[t_{01}, t_1]$, and the sequence of shapes between t_{01} and t_1 can be mapped directly onto those between t_0 and t_{01}, as illustrated schematically with a two-sphere swimmer.

2.7.2 Scallop Theorem

One of the most famous consequences of the property of rate independence is the scallop theorem, formalised originally by Purcell (1977). In his influential article, Purcell identifies shape kinematics, called reciprocal, which are never able to induce swimming at low Reynolds number. Formally, the scallop theorem states that if a sequence of shapes displayed by a swimmer deforming in a time-periodic fashion is identical when viewed under a time-reversal transformation, the swimmer cannot move on average.

We now present an intuitive demonstration for the scallop theorem, while a formal proof may be found in Ishimoto and Yamada (2012). Consider a swimmer undergoing a periodic shape deformation on a time interval $[t_0, t_1]$. If the sequence of shapes of the swimmer is identical when viewed under a time-reversal transformation, then the time interval can be split into two portions as dictated by the mapping of identical shapes unto themselves. The mapping is illustrated in Fig. 2.4 together with the shapes of a two-sphere swimmer at five times during its periodic deformation, for illustrative purposes. The interval $[t_0, t_1]$ is split into $[t_0, t_{01}] \cup [t_{01}, t_1]$ and all shapes between $[t_0, t_{01}]$ are mapped into $[t_{01}, t_1]$, as shown schematically. Let us now compute the total displacement travelled by the swimmer, $\mathbf{\Delta}_{t_0 \to t_1}$. By construction we have $\mathbf{\Delta}_{t_0 \to t_1} = \mathbf{\Delta}_{t_0 \to t_{01}} + \mathbf{\Delta}_{t_{01} \to t_1}$. By reversibility, it is clear that $\mathbf{\Delta}_{t_{01} \to t_1} = -\mathbf{\Delta}_{t_1 \to t_{01}}$. But from the rate-independence property, we also have equality of the displacement, $\mathbf{\Delta}_{t_1 \to t_{01}} = \mathbf{\Delta}_{t_0 \to t_{01}}$, since they both go through the same sequence of shapes. Consequently $\mathbf{\Delta}_{t_0 \to t_1} = \mathbf{\Delta}_{t_0 \to t_{01}} - \mathbf{\Delta}_{t_0 \to t_{01}} = \mathbf{0}$. The total displacement travelled by a swimmer undergoing reciprocal kinematics is thus zero. In other words, a reciprocal swimmer is a non-swimmer.

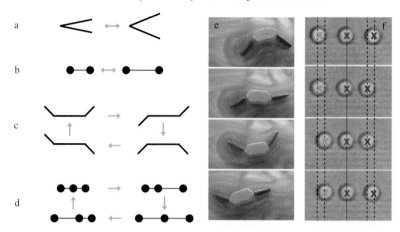

Figure 2.5 Low Reynolds number locomotion with small degrees of freedom. Left: theoretical illustration; Right: experimental implementation. Swimmers with one degree of freedom cannot swim (a: Purcell's scallop; b: two-sphere dumb-bell) but two degrees of freedom may be used to induce locomotion under non-reciprocal deformation (c: Purcell's two-hinge swimmer; e: macroscale experimental realisation (Chan (2009) reproduced with permission)); d: three-sphere swimmer; f: microscale experimental implementation using optical tweez-ers, used with permission of the Royal Society of Chemistry from Leoni et al. (2009); permission conveyed through Copyright Clearance Center, Inc.

2.7.3 Consequences

An important consequence of the scallop theorem is that swimmers with a single degree of freedom cannot swim at low Reynolds numbers since one-degree-of-freedom kinematics are necessarily reciprocal. This can be verified by noting that for a swimmer with a single degree of freedom, denoted θ, the general form of the velocity, Eq. (2.37), can be simply written as $\mathbf{U}(t) = \hat{\mathbf{D}}[\theta(t)]\dot{\theta}$ for some function $\hat{\mathbf{D}}$. As a result, the net displacement travelled, $\mathbf{\Delta}$, is

$$\mathbf{\Delta} = \int_{t_0}^{t_1} \hat{\mathbf{D}}[\theta(t)]\frac{\mathrm{d}\theta}{\mathrm{d}t}\,\mathrm{d}t = \int_{\theta_0}^{\theta_1} \hat{\mathbf{D}}[\theta]\,\mathrm{d}\theta = \mathbf{0}, \qquad (2.44)$$

since periodicity implies that $\theta_0 = \theta_1$.

Examples of one-degree-of-freedom non-swimmers are illustrated in Fig. 2.5: Purcell's one-hinged scallop (Fig. 2.5(a)) and a two-sphere dumbbell (Fig. 2.5(b)). At high Reynolds numbers, motion with a single degree of freedom is effective, e.g. the back-and-forth motion of scuba-diving fins or the flapping of insect wings (Dudley, 2002), but these are all inertial propulsive effects. Similarly, reciprocal kinematics can lead to locomotion in non-Newtonian fluids (see Chapter 15).

By showing us what can never swim, the scallop theorem therefore suggests ways to design effective swimmers. For shape changes to lead to swimming, it is a necessary condition to undergo non-reciprocal motion. For simple swimmers, we therefore need at least two degrees of freedom and to actuate them in a non-reciprocal way. Examples of such swimmers include Purcell's two-hinged swimmer (illustrated schematically in Fig. 2.5(c) with a macroscale experimental implementation (Chan, 2009) shown in Fig. 2.5(e)) or three aligned spheres (sketch in Fig. 2.5(d) and microscale experimental implementation using optical tweezers (Leoni et al., 2009) in Fig. 2.5(f)). Many other examples exist; see e.g. Lauga (2011b) and references therein.

Instead of having only a few degrees of freedom, biological swimmers are able to deform their shapes continuously. How should they do this in order to swim effectively? Purcell's scallop theorem shows that in order to induce locomotion, a clear direction of time needs to be embedded in the shape deformations. The prototypical deformations with this property are waves, either longitudinal or transverse, and indeed waves are ubiquitous in the biological world. In the next chapter we analyse the most elementary wave, namely that of an infinite two-dimensional sheet, and we extend this to other propulsive waves later in this book.

Further Reading

A comprehensive overview of fluid dynamics with both mathematical rigour and an emphasis on physical interpretation is offered in the textbooks by Batchelor (1967) and Guyon et al. (2001). The mathematical treatment of Stokes flows can be found in the books by Happel and Brenner (1965) and Kim and Karrila (1991), while their role in transport phenomena is reviewed by Leal (2007). Some further insight on the impact of fluid dynamics on locomotion can be found in the classical monographs by Lighthill (1975) and Childress (1981).

Exercises

1. How fast would an adult human swimming at the pool have to move her arms in order to induce flow with the same Reynolds number as a swimming bacterium?

2. A spherical organism swims at low Reynolds number along a straight line with constant speed, when it instantly stops its shape deformation. Considering Newton's second law for the organism, determine its subsequent trajectory and the maximum distance travelled. What is the typical value of this distance travelled in the case of a swimming bacterium?

3. How do the scaling results for acceleration and deceleration timescales in Section 2.4 change at high Reynolds number?

4. Show that rigid bodies cannot swim at low Reynolds number.

5. According to the scallop theorem, it is necessary for bodies deforming period-ically in a Stokes flow to undergo reciprocal deformations. Show that the re-quirement of non-reciprocal deformation is, however, an insufficient condition by constructing two counter-examples: (a) a body undergoing non-reciprocal deformation with a swimming velocity that is not instantaneously zero but is zero when time-averaged; (b) a body undergoing non-reciprocal deformation with a swimming velocity identically zero for all times.

6. The Stokes flow due to a rigid sphere of radius a rotating with angular velocity $\hat{\boldsymbol{\Omega}}$ is given by $\hat{\mathbf{u}} = \left(a^3/r^3\right) \hat{\boldsymbol{\Omega}} \times \mathbf{r}$. Use this solution to derive the rotational ver-sion of Eq. (2.36) giving the angular velocity, $\boldsymbol{\Omega}$, of a torque-free sphere with an imposed surface velocity $\dot{\mathbf{r}}_s$ as

$$\boldsymbol{\Omega}(t) = -\frac{3}{8\pi a^3} \int_S \mathbf{n} \times \dot{\mathbf{r}}_s \, dS.$$

7. A dumbbell is composed of two rigid spheres of radii a_1 and a_2 linked by a mechanism with negligible hydrodynamic effect (see illustration in Fig. 2.5(b)). The distance $\ell(t)$ between the two spheres varies in a prescribed periodic fash-ion. Assuming that the dumbbell is force-free at all times, show by exploiting the linearity of Stokes flow that it cannot swim on average.

8. The radii of the spheres in the dumbbell from Exercise 7 now vary periodically in time, i.e. $a_1(t)$ and $a_2(t)$. Ignoring hydrodynamic interactions between the spheres, and thus assuming that the force on each sphere is independent of the other one, derive an equation for the mean dumbbell swimming speed as an average involving the two radii and the distance between the spheres. This is called the *pushme-pullyou* swimmer (Avron et al., 2005).

9. Consider the three-sphere swimmer illustrated in Fig. 2.5(d), with three aligned spheres of radii a_1, a_2 and a_3. The distances between the pairs of spheres vary periodically in time in a prescribed fashion. Show that, if we ignore hydrody-namic interactions between the spheres, force-free motion implies that no net locomotion is possible.

3

The Waving Sheet Model

In this chapter we introduce the historically important model of swimming at low Reynolds numbers originally proposed by Taylor (1951), now considered a classical case. In his paper, Taylor set out to investigate the possibility of swimming in a fluid without inertia at all, a possibility that was at odds with physical intuition at the time. Since waves are the fundamental non-reciprocal kinematics, and since microorganisms were observed to deform their flagella in a wave-like fashion, he focused on the simplest setup possible, namely that of a flexible two-dimensional sheet deforming as a travelling wave of transverse displacements. In this chapter, considering waves with both transverse and longitudinal motion, we show that indeed inertia-less swimming is possible, and that the sheet motion can be used to model both swimming using flagella and pumping using cilia. By computing the rate of work of the wave on the fluid, and its optimisation, we then illustrate how this simple two-dimensional model can be exploited to interpret the two modes of deformation of cilia arrays that are observed experimentally.

3.1 Biological Motivation

The waving sheet model is illustrated in Fig. 3.1. The sheet is infinite, two-dimensional and undergoes prescribed waving motion in a viscous fluid whose dynamics we solve for in the Stokes flow limit. Mathematically, this setup is simple enough to be solved analytically in the asymptotic limit of long wavelengths.

From a biological standpoint, an important property of the model is that it allows us to capture two biophysical situations of interest at once. If the sheet is free to move it can model a free swimmer several wavelengths long at low Reynolds number – for example, the flagellar propulsion of eukaryotic spermatozoa. This was indeed the original motivation in Taylor (1951). In that case the sheet is assumed to be inextensible, the waving motion is mostly transverse to the direction of

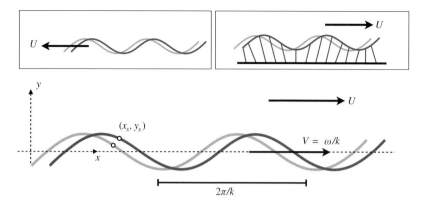

Figure 3.1 Top: two situations that can be modelled by a two-dimensional waving sheet: an active flagellum (left) and the envelope of cilia arrays (right). Bottom: waving sheet of wavelength $2\pi/k$, frequency ω and wave speed $V = \omega/k$. Working in a frame in which the sheet centreline is stationary, x and y are the lab frame coordinates while x_s and y_s denote the material points on the sheet. The steady flow induced at infinity is denoted by $U\mathbf{e}_x$.

propagation of the wave and the quantity of interest is the value of its swimming speed (Fig. 3.1, top left).

    If instead the sheet cannot move on average, then it induces a flow in the bulk that can be seen as a model for the fluid motion induced by large cilia arrays. Indeed, if cilia are present at high density on a biological tissue, we can draw at each instant a continuous surface as the envelope of the cilia tips, and then solve for the dynamics of this envelope as a proxy for the collective dynamics of the many cilia (Blake, 1971c). Since cilia arrays deform in a coordinated fashion as metachronal waves (see more details in Section 3.6), the continuous envelope also undergoes waving deformation but it can stretch and include both transverse and longitudinal motion. The quantity of interest in that case is the value of the flow induced at infinity in the direction parallel to the direction of propagation of the wave (Fig. 3.1, top right).

    Remarkably, in the frame translating with the swimmer, the free-swimming problem is mathematically equivalent to the cilia envelope problem. There the sheet does not move on average but it induces a flow at infinity equal to minus its swimming speed. As a consequence, solving for the fluid flow induced by the waving sheet and computing the velocity induced far from it allow us to address these two different biophysical situations at the same time. As shown in later chapters, the waving sheet model can also be used to tackle problems involving interactions, including swimming near boundaries (Chapter 11), synchronisation between swimmers (Chapter 12) and locomotion in complex fluids (Chapter 15).

3.2 Setup

Our notation for the model is indicated in Fig. 3.1 (bottom). The wave propagates along the x direction with speed V. Its infinite shape is periodic in space with wavenumber k and in time with frequency ω such that $V = \omega/k$. The lab frame coordinate system is denoted (x, y).

The kinematics of material points on the sheet, (x_s, y_s), are assumed to be described by the periodic waving motion

$$x_s = x + A \cos(kx - \omega t - \phi), \tag{3.1a}$$

$$y_s = B \sin(kx - \omega t), \tag{3.1b}$$

where A is the amplitude of the longitudinal wave, B is the transverse amplitude and ϕ is a phase difference between the two deformation modes. With this formulation, the trajectories of material points are ellipses in (x, y) space.

We solve for the fluid motion above the sheet in the region $y > y_s$ and the flow is described by the Stokes equations, Eq. (2.9). Since the kinematics of the sheet are prescribed, the no-slip boundary condition sets the value of the flow velocity on the sheet, namely

$$\mathbf{u}(x = x_s, y = y_s) = \frac{\partial}{\partial t}(x_s, y_s). \tag{3.2}$$

Far from the sheet, we prescribe that no net force or pressure gradient be induced and we solve for the unknown velocity U induced along the direction of propagation of the wave as

$$\lim_{y \to \infty} \mathbf{u} = U\mathbf{e}_x. \tag{3.3}$$

The resulting value of U is then interpreted as either the opposite of a locomotion speed when viewing the sheet as a model for a free swimmer with many wavelengths or as the value of a pumping speed when viewing the sheet as a model for the motion of large cilia arrays. Note that, since the wave is infinite, a shift in time $t \to t + \delta t$ is equivalent to a shift in space $x \to x + V\delta t$ and therefore the value of U does not depend on time.

In order to solve the fluid dynamics problem, we non-dimensionalise it using k^{-1} and ω^{-1} as characteristic length and timescales, respectively (the pressure scale is thus $\mu\omega$). Expecting to solve the problem asymptotically in the long-wavelength limit $A, B \ll k^{-1}$, we define the small parameter ϵ as

$$\epsilon^2 \equiv (Ak)^2 + (Bk)^2, \tag{3.4}$$

which in turn sets the values of the dimensionless longitudinal and transverse amplitudes, a and b, as

$$a \equiv \frac{Ak}{[(Ak)^2 + (Bk)^2]^{1/2}}, \quad b \equiv \frac{Bk}{[(Ak)^2 + (Bk)^2]^{1/2}}. \tag{3.5}$$

Using the same symbols for notation convenience, the dimensionless kinematics of the sheets become

$$x_s = x + \epsilon a \cos(x - t - \phi), \tag{3.6a}$$

$$y_s = \epsilon b \sin(x - t), \tag{3.6b}$$

and we aim to determine the asymptotic behaviour of U in the limit $\epsilon \to 0$.

3.3 Asymptotic Solution

3.3.1 Governing Equations

Since the hydrodynamic problem is two-dimensional, we can introduce a streamfunction ψ defined such that the velocity components are given by

$$u_x = \frac{\partial \psi}{\partial y}, \quad u_y = -\frac{\partial \psi}{\partial x}. \tag{3.7}$$

The use of the streamfunction enforces the incompressibility condition automatically. Taking then the curl of the Stokes equations, we obtain that the vorticity is $\omega = \nabla \times \mathbf{u} = -(\nabla^2 \psi)\mathbf{e}_z$, where z is the direction perpendicular to (x, y). Since the vorticity is harmonic, Eq. (2.12), the streamfunction satisfies the biharmonic equation (Batchelor, 1967)

$$\nabla^2 \nabla^2 \psi = \nabla^4 \psi = 0. \tag{3.8}$$

The boundary conditions for the flow on the sheet, Eq. (3.2), become in a dimensionless form

$$u_x(x_s, y_s) = \left.\frac{\partial \psi}{\partial y}\right|_{(x_s, y_s)} = \epsilon a \sin(x - t - \phi), \tag{3.9a}$$

$$u_y(x_s, y_s) = -\left.\frac{\partial \psi}{\partial x}\right|_{(x_s, y_s)} = -\epsilon b \cos(x - t). \tag{3.9b}$$

Defining $\phi_x \equiv x - t - \phi$ and $\phi_y \equiv x - t$ for notation convenience, Eq. (3.9) can be written for ψ as

$$\left.\frac{\partial \psi}{\partial y}\right|_{(x + \epsilon a \cos\phi_x, \epsilon b \sin\phi_y)} = \epsilon a \sin\phi_x, \tag{3.10a}$$

$$\left.\frac{\partial \psi}{\partial x}\right|_{(x + \epsilon a \cos\phi_x, \epsilon b \sin\phi_y)} = \epsilon b \cos\phi_y. \tag{3.10b}$$

Finally, the boundary condition at infinity in Eq. (3.3) becomes the two conditions for the streamfunction,

$$\lim_{y\to\infty} \frac{\partial\psi}{\partial y} = U, \tag{3.11a}$$

$$\lim_{y\to\infty} \frac{\partial\psi}{\partial x} = 0, \tag{3.11b}$$

which can be written in a condensed fashion as

$$\lim_{y\to\infty} \nabla\psi = U\mathbf{e}_y, \tag{3.12}$$

and the goal of the problem is to solve for U for given values of a, b and ϕ asymptotically as $\epsilon \ll 1$.

It is apparent that the boundary conditions in Eq. (3.10) are nonlinear in the small parameter ϵ. They make the problem impossible to solve by hand for all values of ϵ, justifying the need for an asymptotic approach. We then look for a solution as a regular perturbation expansion in the small parameter ϵ. Since the boundary conditions scale as ϵ, there is no flow at order ϵ^0 and the first non-trivial fluid motion occurs at order ϵ, so we write

$$\psi = \epsilon\psi_1 + \epsilon^2\psi_2 + \cdots, \tag{3.13a}$$

$$U = \epsilon U_1 + \epsilon^2 U_2 + \cdots. \tag{3.13b}$$

Anticipating the need to expand the boundary conditions in Eq. (3.10), we rewrite them as

$$\frac{\partial}{\partial y}(\epsilon\psi_1 + \epsilon^2\psi_2 + \cdots)\Big|_{(x+\epsilon a\cos\phi_x, \epsilon b\sin\phi_y)} = \epsilon a\sin\phi_x, \tag{3.14a}$$

$$\frac{\partial}{\partial x}(\epsilon\psi_1 + \epsilon^2\psi_2 + \cdots)\Big|_{(x+\epsilon a\cos\phi_x, \epsilon b\sin\phi_y)} = \epsilon b\cos\phi_y, \tag{3.14b}$$

which, using a Taylor expansion, become, up to $O(\epsilon^2)$,

$$\epsilon\frac{\partial\psi_1}{\partial y}\Big|_{(x,0)} + \epsilon^2\frac{\partial\psi_2}{\partial y}\Big|_{(x,0)} + \epsilon^2 b\sin\phi_y\frac{\partial^2\psi_1}{\partial y^2}\Big|_{(x,0)}$$
$$+\epsilon^2 a\cos\phi_x\frac{\partial^2\psi_1}{\partial x\partial y}\Big|_{(x,0)} = \epsilon a\sin\phi_x, \tag{3.15a}$$

$$\epsilon\frac{\partial\psi_1}{\partial x}\Big|_{(x,0)} + \epsilon^2\frac{\partial\psi_2}{\partial x}\Big|_{(x,0)} + \epsilon^2 b\sin\phi_y\frac{\partial^2\psi_1}{\partial x\partial y}\Big|_{(x,0)}$$
$$+\epsilon^2 a\cos\phi_x\frac{\partial^2\psi_1}{\partial x^2}\Big|_{(x,0)} = \epsilon b\cos\phi_y. \tag{3.15b}$$

We now proceed to solve the problem order by order until we obtain the leading-order value of U.

3.3.2 Solution at Order ϵ

The streamfunction at order ϵ, ψ_1, satisfies Eq. (3.8) subject to the boundary conditions

$$\left.\frac{\partial \psi_1}{\partial y}\right|_{(x,0)} = a \sin \phi_x = \mathcal{R}\left\{-aie^{-i\phi}e^{i(x-t)}\right\}, \tag{3.16a}$$

$$\left.\frac{\partial \psi_1}{\partial x}\right|_{(x,0)} = b \cos \phi_y = \mathcal{R}\left\{be^{i(x-t)}\right\}, \tag{3.16b}$$

$$\lim_{y\to\infty} \nabla\psi_1 = U_1\mathbf{e}_y, \tag{3.16c}$$

where we have used $\mathcal{R}$ to denote the real part of a complex number.

The general unit-speed 2π-periodic solution to Eq. (3.8) may be found in the form of a Fourier series, with terms of the form $\psi \sim f(y)e^{im(x-t)}$, and we obtain

$$\psi(x,y,t) = \bar{\psi} + Ky + Gy^2 + Hy^3 \tag{3.17}$$
$$+ \sum_{n\geqslant 1}\left[(C_n + yD_n)e^{-ny} + (E_n + yF_n)e^{ny}\right]e^{in(x-t)},$$

where capital letters denote undetermined constants and where we imply taking real parts in each instance where we use complex notation (as allowed for linear operations).

The value of $\bar{\psi}$ in Eq. (3.17) is irrelevant since a streamfunction is only defined up to a constant. For the solution at first order, ψ_1, both series of coefficients E_n and F_n have to be set to zero, as otherwise the flow increases exponentially away from the sheet. Next, we see that the constant H sets a quadratic flow $\psi \sim y^3$, which is associated with a constant pressure gradient. Assuming that no net pressure gradient is maintained along the sheet leads to $H = 0$. Similarly, the constant G sets a linear shear flow $\psi \sim y^2$ and thus it has to be set to zero to ensure that no net force is induced on the sheet. Finally, given the form of the boundary conditions, Eq. (3.16), we see that only the Fourier mode $n = 1$ is relevant at the boundary and therefore the solution at order ϵ takes the form

$$\psi_1 = K_1y + (C_1 + yD_1)e^{-y}e^{i(x-t)}, \tag{3.18}$$

where K has been renamed K_1 to identify it with the first-order solution.

In Eq. (3.18), three constants remain to be determined, which are found using the boundary conditions in Eq. (3.16). First, from the condition at infinity in Eq. (3.16c) we can identify K_1 with the swimming speed at first order, U_1. Second, using Eq. (3.16a) we obtain

$$(D_1 - C_1)e^{i(x-t)} + U_1 = -aie^{-i\phi}e^{i(x-t)}, \tag{3.19}$$

and therefore $U_1 = 0$ and $C_1 - D_1 = aie^{-i\phi}$. Finally, Eq. (3.16b) leads to $iC_1 = b$, which allows us to fully determine the solution in complex notation as

$$\psi_1 = -ib(1 + y)e^{-y}e^{i(x-t)} - iaye^{-y}e^{i(x-t-\phi)}. \tag{3.20}$$

Alternatively, written in real space it is given by

$$\psi_1 = b(1 + y)e^{-y}\sin(x - t) + aye^{-y}\sin(x - t - \phi). \tag{3.21}$$

Importantly, we obtained $U_1 = 0$ and therefore there is no swimming at order ϵ. We could in fact have predicted that, by symmetry, the flow speed induced at infinity is necessarily even in ϵ. A change of sign $\epsilon \to -\epsilon$ is equivalent to a translation along the x direction by half a wavelength. But since the waving setup is translationally invariant, the swimming speed should remain unchanged by this translation and is therefore even in ϵ. The smallest order at which we could expect the swimming speed to be nonzero is thus $O(\epsilon^2)$.

3.3.3 Solution at Order ϵ^2

At order ϵ^2, the streamfunction ψ_2 satisfies the biharmonic equation, Eq. (3.8), with boundary conditions on the sheet obtained from the Taylor expansion shown in Eq. (3.15) and with the condition at infinity, Eq. (3.12), leading to

$$\left.\frac{\partial\psi_2}{\partial y}\right|_{(x,0)} = -b\sin\phi_y\left.\frac{\partial^2\psi_1}{\partial y^2}\right|_{(x,0)} - a\cos\phi_x\left.\frac{\partial^2\psi_1}{\partial x\partial y}\right|_{(x,0)}, \tag{3.22a}$$

$$\left.\frac{\partial\psi_2}{\partial x}\right|_{(x,0)} = -b\sin\phi_y\left.\frac{\partial^2\psi_1}{\partial x\partial y}\right|_{(x,0)} - a\cos\phi_x\left.\frac{\partial^2\psi_1}{\partial x^2}\right|_{(x,0)}, \tag{3.22b}$$

$$\lim_{y\to\infty}\nabla\psi_2 = U_2\mathbf{e}_y. \tag{3.22c}$$

The solution to Eq. (3.8) for ψ_2 is still of the general form shown in Eq. (3.17), where here again the coefficients $\bar{\psi}$, G, H, E_n and F_n are identically zero. Furthermore, by inspection of Eqs. (3.22a)–(3.22b), we see that the boundary condition for ψ_2 on the sheet involves products of sines and cosines, and thus only terms with no x dependence or with $n = 2$ in the series from Eq. (3.17) are part of the solution. The streamfunction ψ_2 takes therefore the form

$$\psi_2 = K_2 y + (C_2 + yD_2)e^{-2y}e^{2i(x-t)}. \tag{3.23}$$

By considering the boundary condition far from the sheet, Eq. (3.22c), we see directly that $K_2 = U_2$. To conclude the calculation and compute the value of U_2, we can bypass solving for C_2 and D_2 by noting that Eq. (3.23) is periodic in space (and time) and therefore an average of $\partial\psi_2/\partial y$ over a wavelength of the sheet leads to the value of U_2, i.e.

$$\left\langle \frac{\partial\psi_2}{\partial y} \right\rangle = U_2. \tag{3.24}$$

We then note that the value of $\partial\psi_2/\partial y$ is given directly in Eq. (3.22a). Using the first-order result in Eq. (3.21), we can evaluate all required derivatives as

$$\frac{\partial\psi_1}{\partial x} = b(1+y)e^{-y}\cos(x-t) + aye^{-y}\cos(x-t-\phi), \tag{3.25}$$

$$\left.\frac{\partial^2\psi_1}{\partial x \partial y}\right|_{(x,0)} = a\cos(x-t-\phi), \tag{3.26}$$

$$\frac{\partial\psi_1}{\partial y} = -bye^{-y}\sin(x-t) + a(1-y)e^{-y}\sin(x-t-\phi), \tag{3.27}$$

$$\left.\frac{\partial^2\psi_1}{\partial y^2}\right|_{(x,0)} = -b\sin(x-t) - 2a\sin(x-t-\phi). \tag{3.28}$$

Substituting these values into Eq. (3.22a) and averaging over one wavelength of the sheet leads to the two averages

$$\langle a\cos\phi_x a\cos\phi_x \rangle = \frac{1}{2}a^2, \tag{3.29a}$$

$$\langle b\sin\phi_y(-b\sin\phi_y - 2a\sin\phi_x) \rangle = -\frac{1}{2}b^2 - ab\cos\phi, \tag{3.29b}$$

and therefore to the final result,

$$U_2 = \frac{1}{2}(b^2 + 2ab\cos\phi - a^2). \tag{3.30}$$

3.3.4 Physical Interpretation

The leading-order value of the flow induced at infinity occurs therefore at $O(\epsilon^2)$ and is given by Eq. (3.30). Returning to dimensional quantities, one may then write the velocity induced at infinity in the fluid as $U\mathbf{e}_x$, where

$$U = \frac{\omega k}{2}(B^2 + 2AB\cos\phi - A^2), \tag{3.31}$$

with next-order terms that are $O(A^2k^2)$, $O(B^2k^2)$ and $O(ABk^2)$ smaller. Since the solution for U is zero at order ϵ^3 by symmetry, the next nonzero terms appear at $O(\epsilon^4)$ (Taylor, 1951; Childress, 1981; Vélez-Cordero and Lauga, 2013).

When interpreting Eq. (3.31) in the context of swimming and flagellar motion, we have to first consider the consequences of the inextensibility of the flagellum. If material points on the flagellum undergo transverse displacements with amplitude ϵ, it can be shown that inextensibility contributes only to an $O(\epsilon^2)$ longitudinal motion of material points on the sheet (and thus $A = 0$ for the leading-order motion; see Exercises), leading to a contribution of inextensibility to the swimming velocity only at $O(\epsilon^4)$. Beyond that order, the calculation of all terms in the expansion of U in powers of ϵ can be performed numerically (computer-extended series) and all coefficients are rational numbers. While the resulting power series has a radius of convergence of slightly less than 1, its singularity can be removed by an Euler transformation to make it converge for all values of ϵ (Sauzade et al., 2011).

In the case of swimming ($A = 0$), the leading-order solution in Eq. (3.31) is always positive and thus the model predicts that locomotion takes place at velocity $-U\mathbf{e}_x$, which is in the direction opposite to that of the wave propagation. This result is in agreement with experiments where, for example, a spermatozoon is seen to send bending waves propagating from the cell body towards the free (distal) end of the flagellum, leading to locomotion head first (Gaffney et al., 2011).

When viewing the sheet as an envelope model for the coordinated motion of cilia tips, the sign of Eq. (3.31) is undetermined. While purely transverse modes ($A = 0$) lead to $U > 0$, pure longitudinal waving ($B = 0$) predicts $U < 0$, and a combination of longitudinal and transverse motion leads to a flow whose sign depends on their relative phase, ϕ, and on the relative amplitudes between the transverse and the longitudinal motions.

Interestingly, we note that the viscosity of the fluid does not affect the value of the flow induced by the sheet. This is best interpreted using dimensional analysis. Given the deformation of the sheet and the properties of the fluid, the flow created in the low Reynolds number limit depends on the parameters $U = f(A, B, k, \omega, \mu, \phi)$. As the viscosity of the fluid is the only parameter that includes dimensions of mass, it drops out of the problem and we have

$$ U = \frac{\omega}{k} f(Ak, Bk, \phi), \tag{3.32} $$

which is consistent with the result of Eq. (3.31). For the fluid viscosity to affect the value of the swimming speed, the sheet should deform not by imposing its kinematics but by forcing dynamical quantities, for example, internal active forces or bending moments.

3.4 Rate of Work

The rate of work of the waving sheet on the fluid is defined in Eq. (2.18), where the integral is evaluated along the instantaneous outline of the sheet. Since this is a quadratic quantity and each term in the integrand occurs at order ϵ or above, the leading-order rate of work occurs at $O(\epsilon^2)$ and is given per unit length in the z direction by

$$\dot{W}_2 = -\int_0^{2\pi} \mathbf{u}_1 \cdot \sigma_1 \cdot \mathbf{n}\big|_{y=0}\, dx, \qquad (3.33)$$

where the integral is evaluated at $y = 0$, which is the location of the sheet at order ϵ^0 (evaluating the integral on the actual position of the sheet gives a subdominant correction in the limit of small ϵ).

Splitting the integral along the components of the velocity, we obtain per unit z length

$$\dot{W}_2 = -\int_0^{2\pi} u_{1,x}\sigma_{1,xy}\big|_{y=0}\, dx - \int_0^{2\pi} u_{1,y}\sigma_{1,yy}\big|_{y=0}\, dx. \qquad (3.34)$$

Using the derivatives in Eqs. (3.25) and (3.27), we obtain the flow components as

$$u_{1,x} = -bye^{-y}\sin(x-t) + a(1-y)e^{-y}\sin(x-t-\phi), \qquad (3.35a)$$
$$u_{1,y} = -b(1+y)e^{-y}\cos(x-t) - aye^{-y}\cos(x-t-\phi). \qquad (3.35b)$$

From these equations we can deduce directly the non-dimensional shear stress as

$$\sigma_{1,xy}\big|_{(y=0)} = \left(\frac{\partial u_{1,x}}{\partial y} + \frac{\partial u_{1,y}}{\partial x}\right)\Big|_{y=0} = -2a\sin(x-t-\phi), \qquad (3.36)$$

and thus obtain directly that the first integral in Eq. (3.34) is equal to $2\pi a^2$.

In order to compute the second term in Eq. (3.34), we need to calculate the normal stress and therefore the pressure. This is obtained by evaluating the x component of the Stokes equations,

$$\frac{\partial p_1}{\partial x} = \frac{\partial}{\partial y}\nabla^2\psi_1 = 2be^{-y}\sin(x-t) + 2ae^{-y}\sin(x-t-\phi). \qquad (3.37)$$

This is easily integrated as

$$p_1 = -2be^{-y}\cos(x-t) - 2ae^{-y}\cos(x-t-\phi), \qquad (3.38)$$

leading to the normal stress

$$\sigma_{1,yy}\big|_{(y=0)} = \left(-p_1 + 2\frac{\partial u_{1,y}}{\partial y}\right)\Big|_{y=0} = 2b\cos(x-t), \qquad (3.39)$$

and thus the second term in Eq. (3.34) integrates to $2\pi b^2$.

We therefore obtain that the leading-order rate of work on the fluid is given, per wavelength along x and per unit length in the z direction, by

$$\frac{\dot{W}_2}{2\pi} = a^2 + b^2. \tag{3.40}$$

Note that in dimensional units, the leading-order statement becomes, per unit area of the sheet,

$$\frac{\dot{W}}{\text{area}} = \mu\omega^2 k (A^2 + B^2), \tag{3.41}$$

with next-order terms that are $O(A^2k^2)$ and $O(B^2k^2)$ smaller.

3.5 Wave Optimisation

With the fluid dynamics and rate of work of the waving sheet solved at leading order, we can now enquire about its optimal modes of deformation. Viewing the waving sheet as a model for the deformation of large ciliary array, we can derive formally the parameter values of the waving motion that lead to the largest induced flow for a fixed amount of energy expended by the cilia (Brennen, 1974; Brennen and Winet, 1977). This derivation assumes of course that the majority of the work done by the deforming cilia is dissipated in the fluid above the ciliary array.

There are three degrees of freedom that affect the magnitude of the flow in the far field and the rate of work, namely the two amplitudes, a and b, and their relative phase, $-\pi < \phi \leqslant \pi$. Given the result in Eq. (3.40), we see that fixing the value of the average rate of work is equivalent to fixing the value of $a^2 + b^2$. Solving the optimal sheet problem then reduces to finding the optimal values of a, b and ϕ that maximise the magnitude of U_2 in Eq. (3.30) for a constant value of $a^2 + b^2$.

Defining $R^2 \equiv a^2 + b^2$, we may write $a = R\sin\theta$ and $b = R\cos\theta$ with the parameter θ allowed to take any value between 0 and $\pi/2$ without loss of generality since both amplitudes are positive. Using Eq. (3.30) we then see that the flow induced by the sheet in the far field has magnitude

$$U_2 = \frac{1}{2}R^2(\cos^2\theta - \sin^2\theta + 2\cos\theta\sin\theta\cos\phi) \equiv \frac{1}{2}R^2 f(\theta, \phi), \tag{3.42}$$

where the function f is therefore given by $f(\theta, \phi) \equiv \cos 2\theta + \sin 2\theta \cos \phi$.

For the swimming speed to be maximal in magnitude, the angles ϕ and θ need to be stationary points of f and thus we need to solve

$$\frac{\partial f}{\partial \phi} = \frac{\partial f}{\partial \theta} = 0. \tag{3.43}$$

The partial derivative in ϕ leads to

$$\frac{\partial f}{\partial \phi} = -\sin\phi \sin 2\theta, \tag{3.44}$$

while the partial derivative in the other direction is given by

$$\frac{\partial f}{\partial \theta} = 2(\cos 2\theta \cos\phi - \sin 2\theta). \tag{3.45}$$

Setting Eq. (3.44) to be zero we see that either $\sin 2\theta = 0$ or $\sin\phi = 0$. If $\sin 2\theta = 0$ then since either $\sin\theta$ or $\cos\theta$ is zero, the motion is either purely vertical ($\theta = 0$, $a = 0$) or purely tangential ($\theta = \pi/2$, $b = 0$) with associated velocity

$$U_2 = \pm\frac{1}{2}R^2. \tag{3.46}$$

Note that in order for Eq. (3.45) to then be zero, we need $\cos\phi = 0$, i.e. $\phi = \pm\pi/2$, but in fact the value of the phase does not affect the value of the swimming speed in that case.

In order to investigate the nature of this first stationary point, we need to compute the matrix of second derivatives, or Hessian H, as

$$H \equiv \begin{pmatrix} \frac{\partial^2 f}{\partial\theta^2} & \frac{\partial^2 f}{\partial\phi\partial\theta} \\ \frac{\partial^2 f}{\partial\phi\partial\theta} & \frac{\partial^2 f}{\partial\phi^2} \end{pmatrix} = \begin{pmatrix} -4(\sin 2\theta \cos\phi + \cos 2\theta) & -2\sin\phi \cos 2\theta \\ -2\sin\phi \cos 2\theta & -\cos\phi \sin 2\theta \end{pmatrix}. \tag{3.47}$$

At the points $(\theta, \phi) = (0, \pm\pi/2)$ and $(\theta, \phi) = (\pi/2, \pm\pi/2)$ corresponding to purely vertical and purely horizontal waving motion, we have a Hessian

$$H = \pm\begin{pmatrix} -4 & \mp 2 \\ \mp 2 & 0 \end{pmatrix}, \tag{3.48}$$

which always has one positive and one negative eigenvalue since it is a symmetric matrix with negative determinant. When the waving motion of the sheet is either purely longitudinal or purely transverse, the flow speed obtained in Eq. (3.46) as a stationary point of the optimisation procedure is therefore a saddle point.

The second possibility for the derivative in Eq. (3.44) to be zero is that $\sin\phi = 0$, which means that either $\phi = 0$ or $\phi = \pi$. If $\phi = 0$, having zero derivative in Eq. (3.45) leads to $\tan 2\theta = 1$ and thus $\theta = \pi/8$. This corresponds to the situation where the amplitudes are proportional to each other as $a = (\sqrt{2} - 1)b$. In that case, the velocity is given by $2U_2 = R^2(\cos 2\theta + \sin 2\theta) = \sqrt{2}R^2$ and thus

$$U_2 = \frac{\sqrt{2}}{2}R^2. \tag{3.49}$$

At that point, $(\theta, \phi) = (\pi/8, 0)$, the Hessian matrix is

$$H \equiv \begin{pmatrix} -4\sqrt{2} & 0 \\ 0 & -\sqrt{2}/2 \end{pmatrix}, \tag{3.50}$$

which has two negative eigenvalues. In that case, the flow induced at infinity with magnitude in Eq. (3.49) is the maximal possible value for all θ and ϕ.

The third and final possibility is $\phi = \pi$; then Eq. (3.45) implies that $\tan 2\theta = -1$ and therefore $2\theta = 3\pi/4$, corresponding to waving with relative amplitudes $a = (\sqrt{2}+1)b$. As a result, the velocity is $2U_2 = R^2(\cos 2\theta - \sin 2\theta) = -\sqrt{2}R^2$ and thus

$$U_2 = -\frac{\sqrt{2}}{2}R^2. \tag{3.51}$$

The Hessian in the case where $(\theta, \phi) = (3\pi/8, \pi)$ is

$$H \equiv \begin{pmatrix} 4\sqrt{2} & 0 \\ 0 & \sqrt{2}/2 \end{pmatrix}, \tag{3.52}$$

which has two positive eigenvalues. The corresponding flow velocity in Eq. (3.51) is therefore the minimal possible value. Note however that the flows in Eqs. (3.49) and (3.51) have the same magnitude but opposite sign.

3.6 Comparison with Experiments: Metachronal Waves of Cilia

The optimisation procedure in the previous section revealed two energetically optimal solutions, which create a flow with the same speed but opposite velocity. One of the optimal solutions induces a flow in the same direction as the propagating wave ($V > 0, U_2 > 0$) while the other optimum induces a flow in the direction opposite to the wave ($V > 0, U_2 < 0$). Both optima create a flow about 41% greater in magnitude than that induced by either purely tangential or purely longitudinal modes.

The biological interpretation of the existence of these two solutions is the two known types of metachronal waves of cilia observed in nature. Individual cilia on the surface of an organism, or on a biological tissue, are known to deform in an asymmetric, non-reciprocal way. In order to create a net flow and escape the constraints of the scallop theorem, an individual cilium deforms periodically using a two-stroke motion. During one half of its motion, the cilium undergoes an effective stroke creating a large force on the fluid; it is followed during the second half of the motion by a different recovery stroke where the cilium bends to decrease its hydrodynamic drag. Examples of these effective-recovery strokes are illustrated in Fig. 3.2 (left) for cilia on the surface of the protozoa *Opalina* (Fig. 3.2(a)) and

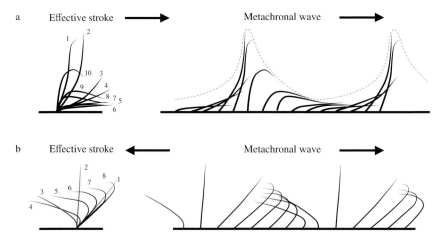

Figure 3.2 The two types of metachronal waves of cilia arrays. (a) Symplectic waves of cilia with wave and effective strokes in the same direction (here from left to right), observed for the protozoon *Opalina* (b) Antiplectic waves where the wave travels in the direction opposite to the effective stroke, as observed for *Paramecium*. Cilia redrawn from Brennen and Winet (1977), courtesy of Jacques Lauga.

Paramecium (Fig. 3.2(b)), where the numbering in the figure labels the sequence of shapes during the full periodic stroke. More details on the hydrodynamics of a wall-bounded cilium are given in Section 11.5.

In a dense array where cilia are closely packed, the coordination between different cilia allows their collective actuation on the fluid to possess additional degrees of freedom even if each cilium deforms in an identical manner. It is observed experimentally that such coordination in ciliary arrays takes the form of metachronal waves where each cilium deforms as its neighbours but with a small phase difference, resulting in a collective envelope of waves of deformation. This is akin to the Mexican waves created by spectators in a football stadium. A model of hydrodynamic interactions between cilia and of metachronal waves is further discussed in Section 12.1.

Experimental observations reveal that two types of metachronal waves exist, similar to our mathematical optimisation problem having two solutions. In the first type of metachronal wave, termed symplectic, the effective stroke of an individual cilium (and thus the net force it exerts on the fluid) is in the same direction as the wave propagation. This is the type of wave displayed by *Opalina*, as illustrated in Fig. 3.2(a), and it is analogous to the first optimal wave derived previously ($V > 0$, $U_2 > 0$). In the second type of metachronal wave, called antiplectic, the effective stroke acts in the direction opposite to that of the wave. This is the case for *Paramecium* (Fig. 3.2(b)) and is analogous to the second optimum ($V > 0$, $U_2 < 0$). The solution to the mathematical optimisation problem has therefore allowed us to provide an interpretation to the existence of two types of ciliary waves as modes of flow transport with maximal energy efficiency.

Further Reading

While the waving sheet model was originally analysed in the Stokes limit, the impact of fluid inertia has been addressed by a few studies (Reynolds, 1965; Tuck, 1968; Brennen, 1974). In particular, inertia leads to a systematic decrease in the swimming speed down to half its Stokes value at high Reynolds numbers (see Exercises). The initial-value problem of starting a waving sheet from rest and the importance of transient effects was analysed by Pak and Lauga (2010), while the question of optimal large-amplitude waveforms was addressed computationally by Montenegro-Johnson and Lauga (2014).

Exercises

1. An infinite two-dimensional waving sheet is swimming using two travelling-wave modes operating at different frequencies and wavelengths. At small amplitude in the waving motion, the dimensionless location of material points (x_s, y_s) in the frame swimming with the sheet is given by $x_s = x + \epsilon a \sin[3(x - t)]$, $y_s = \epsilon b \sin(x - t)$, using notation similar to that in the chapter. Show that the swimming speed is zero for a specific value of the ratio b/a.

2. Many microorganisms self-propel in complex fluids, for example, in mucus (see Chapter 15). A model of mucus is a fluid with a shear-dependent viscosity for which the stress tensor is given by $\sigma = -p\mathbf{1} + 2\mu(\mathbf{E})\mathbf{E}$. The viscosity μ is a known function of the rate of strain tensor $\mu(\mathbf{E}) = \mu_0 f(E_{ij}E_{ij})$, where μ_0 has dimensions of viscosity and f is a well-behaved dimensionless function. Assuming that Taylor's waving sheet is swimming in such a fluid, show that at leading order in ϵ the swimming speed is identical to the Newtonian speed.

3. An inextensible waving sheet undergoes transverse waving motion with the location of its material points in the swimming frame given by $x_s = x$, $y_s = \epsilon \sin(x - t)$ at $O(\epsilon)$. Show that the first effect of inextensibility is an $O(\epsilon^2)$ velocity boundary condition along x. Show that this higher-order term does not influence the leading-order value of the swimming speed.

4. Following Exercise 3, compute the dimensionless swimming speed at $O(\epsilon^4)$ of an inextensible sheet deforming as a transverse travelling wave.

5. By considering the full Navier–Stokes equations with Reynolds number Re $=$ $\omega/(k^2\nu)$, show that the swimming speed at $O(\epsilon^2)$ of the transversely waving sheet differs from the Stokes speed as

$$\frac{U(\text{Re})}{U(\text{Re}=0)} = \frac{1+F}{2F}, \quad F = \left[\frac{1+(1+\text{Re}^2)^{1/2}}{2}\right]^{1/2}.$$

Derive a similar expression for the ratio between the rate of work at $O(\epsilon^2)$ of the transversely waving sheet at finite and zero Reynolds numbers.

4

The Squirmer Model

This chapter introduces the second canonical model of low Reynolds number swimming, namely that of the squirming sphere. Originally proposed by Lighthill (1952), and later extended by Blake (1971b), this model is a variation of the waving sheet adapted to a finite-size swimmer. Here, the surface of a spherical body undergoes periodic small-amplitude deformation of its surface, leading to instantaneous velocity boundary conditions applied on an effective spherical frame. Since Lighthill's original paper, the squirmer model has been used and extended in a variety of setups, in particular to provide an alternative envelope model of the metachronal waves of cilia for finite-size organisms. We first derive the classical swimming squirmer model for a translating swimmer before discussing its extension to rotational motion. We then show how to link the motion of points on the surface of a deformable sphere described in a Lagrangian fashion to the squirmer model written naturally in an Eulerian framework. We finish by comparing the results of the model with flow measurements around the flagellated green alga *Volvox*.

4.1 Axisymmetric Squirmer

4.1.1 Setup

We start the analysis by focusing on the simplest setup possible, namely the swimming of a spherical body of radius a. We assume that the spherical surface, located in an unbounded fluid, remains undeformed but applies velocity boundary conditions to the surrounding fluid instantaneously. These effective boundary conditions are related to the actual deformation of a continuous surface in Section 4.4.

The setup is illustrated in Fig. 4.1. We use spherical coordinates with the origin at the centre of the sphere in order to describe the flow. We assume that the flow is axisymmetric and thus has no dependence on the azimuthal angle ϕ. The velocity boundary conditions, denoted by $u_r(a, \theta, t)$ and $u_\theta(a, \theta, t)$, are imposed in

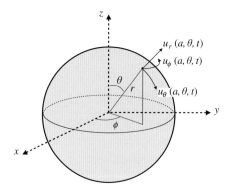

Figure 4.1 Spherical squirmer of radius a described using spherical coordinates (r, θ, ϕ). Velocity boundary conditions are imposed along the radial (r) and polar (θ) directions at the surface of the spherical body with no dependence on ϕ, leading to axisymmetric flow and to swimming along the symmetry axis (z) (calculation in Sections 4.1–4.2). The rotation of the squirmer can also be modelled by adding an azimuthal (ϕ) component of the velocity (Section 4.3).

the swimming frame and are allowed to be arbitrary. They result in swimming of the spherical body with velocity $U(t)\mathbf{e}_z$ so that in the swimming frame the flow at infinity is

$$\mathbf{u}(r \to \infty) = -U(t)\cos\theta\mathbf{e}_r + U(t)\sin\theta\mathbf{e}_\theta. \tag{4.1}$$

In addition to translation, rotation along the z axis may be induced by also imposing an azimuthal component of the velocity, $u_\phi(a, \theta, t)$. This is analysed in Section 4.3.

4.1.2 Form of the Solution

In order to derive the general form of the solution, we follow here the intuitive method presented by Lighthill (1952). We aim to solve the Stokes equations, Eq. (2.9), with prescribed boundary conditions $u_r(a, \theta, t)$ and $u_\theta(a, \theta, t)$.

As seen in Chapter 2, the velocity field in a Stokes flow is biharmonic and satisfies Eq. (2.11). It is then straightforward to show that, due to the incompressibility of the velocity field, $\nabla \cdot \mathbf{u} = 0$, Eq. (2.11) implies that

$$\nabla^2\nabla^2(\mathbf{r} \cdot \mathbf{u}) = 0, \tag{4.2}$$

where $\mathbf{r}$ is the position vector, i.e. $\mathbf{r} = r\mathbf{e}_r$. The result in Eq. (4.2) is a self-contained equation for the radial component of the velocity, u_r. Once u_r is determined, the second velocity component, u_θ, may be obtained using incompressibility, i.e. by integrating explicitly

$$\frac{1}{r^2}\frac{\partial}{\partial r}\left(r^2 u_r\right) + \frac{1}{r\sin\theta}\frac{\partial}{\partial\theta}(u_\theta\sin\theta) = 0. \tag{4.3}$$

In order to find all possible solutions to Eq. (4.2) we use the classical result that the general axisymmetric solutions to $\nabla^2 \Phi = 0$ are $\Phi \sim r^n P_n(\cos\theta)$ and $\Phi \sim r^{-1-n} P_n(\cos\theta)$, where P_n is the Legendre polynomial of order n (Riley et al., 1999). Finding the solutions to Eq. (4.2) means we therefore have to solve

$$\nabla^2(r u_r) \sim r^n P_n(\cos\theta) \quad \text{or} \quad \sim r^{-1-n} P_n(\cos\theta). \tag{4.4}$$

Given the form of the right-hand side of Eq. (4.4), we look for separable solutions of the form $u_r(r,\theta) = f(r)P_n(\cos\theta)$. The decaying solutions to the homogeneous part of Eq. (4.4), denoted by $\tilde{u}_n$, are given by

$$\tilde{u}_{n,r} = \frac{1}{r^{n+2}} P_n(\cos\theta), \quad \tilde{u}_{n,\theta} = \frac{n}{2r^{n+2}} V_n(\theta), \quad n \geqslant 0, \tag{4.5}$$

where the function V_n is defined to be

$$V_n(\theta) = \frac{2}{\sin\theta} \int_{\cos\theta}^{1} P_n(u)\, du = \frac{2}{n(n+1)} P_n'(\cos\theta) \sin\theta, \tag{4.6}$$

and where we use primes to denote derivatives with respect to the argument in the Legendre polynomials. Note that the last equality in Eq. (4.6) is a consequence of the fact that the Legendre polynomial satisfies the Legendre differential equation, given by

$$\frac{d}{dx}\left((1-x^2)\frac{d}{dx} P_n(x) \right) + n(n+1)P_n(x) = 0. \tag{4.7}$$

Up to multiplicative constants, which are irrelevant due to linearity, the particular solutions to Eq. (4.4), which we denote by $\bar{u}_n$, have components

$$\bar{u}_{n,r} = \frac{1}{r^n} P_n(\cos\theta), \quad \bar{u}_{n,\theta} = \frac{1}{r^n}\left(\frac{n}{2} - 1\right) V_n(\theta), \quad n \geqslant 1. \tag{4.8}$$

Finally, in addition to the decaying solutions in Eqs. (4.5)–(4.8) the constant velocity $\hat{u} = e_z$ is also a solution, with components

$$\hat{u}_r = \cos\theta, \quad \hat{u}_\theta = -\sin\theta. \tag{4.9}$$

4.1.3 Full Solution

With the form of the solution obtained in the previous section, we are now in a position to compute the solution to the Stokes flow problem exactly. Given the appearance of the functions P_n and V_n for the radial and polar components of the velocity, these provide the natural basis on which to decompose the boundary conditions. We thus write

$$u_r(a,\theta,t) = \sum_{n\geqslant 0} A_n(t)P_n(\cos\theta), \quad u_\theta(a,\theta,t) = \sum_{n\geqslant 1} B_n(t)V_n(\theta), \tag{4.10}$$

where the coefficients $A_n(t)$ and $B_n(t)$ can vary in time.

If the angular dependence of the boundary conditions is given in a different form from that in Eq. (4.10), we can invert Eq. (4.10) by taking advantage of two standard orthogonality formulae for Legendre polynomials,

$$\int_{-1}^{1} P_n(u) P_m(u) \, du = \frac{2\delta_{nm}}{2n+1}, \tag{4.11a}$$

$$\int_{-1}^{1} (1-u^2) P'_n(u) P'_m(u) \, du = \frac{2n(n+1)\delta_{nm}}{2n+1}, \tag{4.11b}$$

and find explicit expressions for the coefficients

$$A_n(t) = \frac{2n+1}{2} \int_0^\pi u_r(a,\theta,t) P_n(\cos\theta) \sin\theta \, d\theta, \tag{4.12a}$$

$$B_n(t) = \frac{1}{8} n(n+1)(2n+1) \int_0^\pi u_\theta(a,\theta,t) V_n(\theta) \sin\theta \, d\theta. \tag{4.12b}$$

Given the general form of the solution obtained in Section 4.1.2, the flow in the squirming problem is a linear superposition of the three flow fields derived in Eqs. (4.5), (4.8) and (4.9) as

$$\mathbf{u} = \sum_{n\geqslant 0} \alpha_n a^{n+2} \tilde{\mathbf{u}}_n + \sum_{n\geqslant 1} \beta_n a^n \bar{\mathbf{u}}_n + \Gamma \hat{\mathbf{u}}. \tag{4.13}$$

The coefficients α_n, β_n and Γ depend generally on time but in what follows we leave all time dependence implicit to simplify notation. Applying the boundary conditions in Eq. (4.10) allows us to compute the values of the coefficients in Eq. (4.13) as

$$\Gamma = -U, \tag{4.14a}$$

$$\alpha_0 = A_0, \tag{4.14b}$$

$$\alpha_1 = \frac{1}{2}(2B_1 + A_1 - U), \tag{4.14c}$$

$$\beta_1 = \frac{1}{2}(A_1 - 2B_1 + 3U), \tag{4.14d}$$

$$\alpha_n = \left(1 - \frac{n}{2}\right) A_n + B_n, \quad (n \geqslant 2), \tag{4.14e}$$

$$\beta_n = \frac{n}{2} A_n - B_n, \quad (n \geqslant 2). \tag{4.14f}$$

Using $P_0 = 1$, $P_1(\cos\theta) = \cos\theta$ and $V_1(\theta) = \sin\theta$, the full solution for the instantaneous flow components in the swimming frame is given by

$$u_r(r,\theta) = -UP_1 + A_0\left(\frac{a}{r}\right)^2$$ (4.15)

$$+ \frac{1}{2}(2B_1 + A_1 - U)\left(\frac{a}{r}\right)^3 P_1 + \frac{1}{2}(A_1 - 2B_1 + 3U)\frac{a}{r}P_1$$

$$+ \sum_{n\geq 2} A_n\left[\frac{n}{2}\left(\frac{a}{r}\right)^n + \left(1 - \frac{n}{2}\right)\left(\frac{a}{r}\right)^{n+2}\right]P_n$$

$$+ \sum_{n\geq 2} B_n\left[\left(\frac{a}{r}\right)^{n+2} - \left(\frac{a}{r}\right)^n\right]P_n,$$

$$u_\theta(r,\theta) = UV_1$$ (4.16)

$$+ \frac{1}{4}(2B_1 + A_1 - U)\left(\frac{a}{r}\right)^3 V_1 - \frac{1}{4}(A_1 - 2B_1 + 3U)\frac{a}{r}V_1$$

$$+ \sum_{n\geq 2} \frac{n}{2}\left(\frac{n}{2} - 1\right)A_n\left[\left(\frac{a}{r}\right)^n - \left(\frac{a}{r}\right)^{n+2}\right]V_n$$

$$+ \sum_{n\geq 2} B_n\left[\frac{n}{2}\left(\frac{a}{r}\right)^{n+2} + \left(1 - \frac{n}{2}\right)\left(\frac{a}{r}\right)^n\right]V_n,$$

where we have used P_n and V_n to denote the angular functions $P_n(\cos\theta)$, and $V_n(\theta)$, respectively. The flow associated with a nonzero value of A_0 in Eq. (4.15) is a net instantaneous source or sink of fluid located inside the sphere. In what follows, we set A_0 to zero and thus assume volume-conserving motion.

4.2 Free-Swimmer Squirmer

4.2.1 Swimming Velocity and Flow

The solution in Eqs. (4.15)–(4.16) satisfies the Stokes equations with all boundary conditions, but the value of the swimming speed, U, is still undetermined. In the lab frame, the slowest-decaying term in Eqs. (4.15)–(4.16) varies spatially as $1/r$, which is the signature of a net force being applied to the fluid. Specifically, a point force of strength $F\mathbf{e}_z$ located at the origin creates instantaneously a flow called the stokeslet (see more details in Chapters 6 and 9) whose components are given by (Happel and Brenner, 1965)

$$u_r = \frac{2F}{r}P_1, \quad u_\theta = -\frac{F}{r}V_1,$$ (4.17)

and are consistent with the $1/r$ components in Eqs. (4.15)–(4.16). To ensure that the squirmer remains force-free for all times (see Eq. (2.17)) we need to set this coefficient to zero, i.e. $2\beta_1 = A_1 - 2B_1 + 3U = 0$, and the swimming speed is thus given by

$$U = \frac{1}{3}(2B_1 - A_1). \tag{4.18}$$

Note that as an alternative argument we may require the flows in Eqs. (4.15)–(4.16) to have finite total kinetic energy, which also requires setting the $1/r$ mode to zero (Lighthill, 1952; Blake, 1971b).

As a result of Eq. (4.18), the final flow solution for the instantaneous free-swimming squirmer is given by

$$u_r(r, \theta) = -\frac{1}{3}(2B_1 - A_1)P_1 + \frac{2}{3}(A_1 + B_1)\left(\frac{a}{r}\right)^3 P_1 \tag{4.19a}$$

$$+ \sum_{n \geqslant 2} A_n \left[\frac{n}{2}\left(\frac{a}{r}\right)^n + \left(1 - \frac{n}{2}\right)\left(\frac{a}{r}\right)^{n+2}\right] P_n$$

$$+ \sum_{n \geqslant 2} B_n \left[\left(\frac{a}{r}\right)^{n+2} - \left(\frac{a}{r}\right)^n\right] P_n,$$

$$u_\theta(r, \theta) = \frac{1}{3}(2B_1 - A_1)V_1 + \frac{1}{3}(A_1 + B_1)\left(\frac{a}{r}\right)^3 V_1 \tag{4.19b}$$

$$+ \sum_{n \geqslant 2} \frac{n}{2}\left(\frac{n}{2} - 1\right) A_n \left[\left(\frac{a}{r}\right)^n - \left(\frac{a}{r}\right)^{n+2}\right] V_n$$

$$+ \sum_{n \geqslant 2} B_n \left[\frac{n}{2}\left(\frac{a}{r}\right)^{n+2} + \left(1 - \frac{n}{2}\right)\left(\frac{a}{r}\right)^n\right] V_n.$$

Inspecting this solution reveals three important points. First, the value of the swimming speed only depends on the modes A_1 and B_1 of the boundary conditions. All the other surface modes create flows and contribute to the total dissipation but do not lead to swimming. The most efficient squirmer, i.e. the one moving with highest speed for a given amount of dissipation in the fluid, is the one for which $A_n = B_n = 0$ for all $n \geqslant 2$, with a resulting flow that is irrotational and decays spatially as $1/r^3$ (see Exercises).

Second, we note that it is theoretically possible to design a stealth swimmer that moves without perturbing the flow surrounding it. In order to achieve this, first consider the irrotational swimmer where $A_n = B_n = 0$ for all $n \geqslant 2$. Next, choose the values of A_1 and B_1 to satisfy $A_1 + B_1 = 0$. In that case, the flow in Eq. (4.19) is zero in the lab frame while the swimming speed is given by

$$U = \frac{1}{3}(2B_1 - A_1) = B_1. \tag{4.20}$$

A simple interpretation of this particular swimming motion can be made by evaluating its boundary conditions, which are given by

$$u_r(a, \theta) = -B_1 P_1 = -B_1 \cos\theta, \quad u_\theta(a, \theta) = B_1 V_1 = B_1 \sin\theta, \tag{4.21}$$

and can thus be written as $\mathbf{u}(a, \theta, t) = -B_1 \mathbf{e}_z$. This corresponds to a swimmer through which fluid penetrates uniformly along the z direction with velocity $-B_1$ and thus translates with velocity B_1 in the lab frame (it is therefore associated with an internal flow). Note that in order to achieve stealth swimming, both radial and tangential velocities are required (i.e. both A_1 and B_1). Hence a tangential squirmer that is only able to induce angular boundary conditions cannot swim without perturbing the surrounding fluid.

Finally, we note that, in the lab frame of reference, the slowest-decaying terms in Eq. (4.19) are the non-swimming A_2 and B_2 terms. The associated $1/r^2$-decaying flows are interpreted physically in Chapter 9 as force dipoles (or stresslets).

4.2.2 Velocity from Reciprocal Theorem

It is also possible to compute the value of the swimming speed, Eq. (4.18), using an alternative method that bypasses the flow calculation. To do so, we apply the integral result obtained in Chapter 2 using the reciprocal theorem of Stokes flows, Eq. (2.36), which links the swimming velocity with the surface integral of the flow perturbation applied in the swimming frame. By symmetry, the swimming velocity is expected to be directed along z, so we project Eq. (2.36) along the z axis and obtain the axisymmetric integral for the swimming speed, U, as

$$U = -\frac{1}{2} \int_0^\pi \mathbf{e}_z \cdot \mathbf{u}(a, \theta) \sin \theta \, d\theta, \qquad (4.22)$$

where the surface velocity, $\mathbf{u}$, is given by the boundary conditions in Eq. (4.10). Since $\mathbf{e}_z \cdot \mathbf{u} = u_r \cos \theta - u_\theta \sin \theta$, Eqs. (4.10) and (4.22) lead to

$$U = \frac{1}{2} \sum_{n \geqslant 1} B_n \int_0^\pi \sin^2 \theta V_n(\theta) \, d\theta - \frac{1}{2} \sum_{n \geqslant 0} A_n \int_0^\pi \sin \theta \cos \theta P_n(\cos \theta) \, d\theta. \qquad (4.23)$$

Using the orthogonality conditions in Eq. (4.11) along with Eq. (4.6) and the fact that $\cos \theta = P_1$, we see that all integrals in Eq. (4.23) disappear except for the $n = 1$ terms, leading to

$$U = \frac{1}{2} B_1 \int_0^\pi \sin^3 \theta \, d\theta - \frac{1}{2} A_1 \int_0^\pi \sin \theta \cos^2 \theta \, d\theta = \frac{1}{3}(2B_1 - A_1), \qquad (4.24)$$

which agrees with the result in Eq. (4.18) obtained by the full calculation.

4.3 Rotating Squirmer

The analysis so far has led to swimming of the organism along a straight line with no rotation. The axisymmetric model may however be modified to include rotational motion. For example, the green alga *Volvox* rotates around its posterior-anterior axis in a manner that allows it to adjust its response to light signals and to bias its locomotion towards light sources and phototaxis (Drescher et al., 2010b).

In addition to radial and polar velocity boundary conditions, we thus now include an axisymmetric flow component in the azimuthal direction, $u_\phi(a, \theta, t)$, imposed in the swimming frame. This leads to rotation along the symmetry axis, with rotational velocity Ωe_z to be determined. In the frame rotating with the squirmer, the boundary condition at infinity is therefore given by

$$\mathbf{u}(r \to \infty) = -\Omega \mathbf{e}_z \times \mathbf{r}. \tag{4.25}$$

4.3.1 Flow

Since the azimuthal boundary condition does not depend on ϕ, the flow solution is axisymmetric, and there is no azimuthal pressure gradient. The ϕ component of the Stokes equations is thus a solution to

$$\mathbf{e}_\phi \cdot \nabla^2 \mathbf{u} = 0. \tag{4.26}$$

Developing Eq. (4.26) and using the fact that neither u_r nor u_θ depend on ϕ, we obtain that the azimuthal flow component, u_ϕ, satisfies

$$\nabla^2 u_\phi = \frac{u_\phi}{r^2 \sin^2 \theta}. \tag{4.27}$$

For axisymmetric flow, the flow component u_ϕ does not enter the incompressibility condition, and thus, given Eq. (4.27), we get the interesting result that the azimuthal flow evolves independently of the radial and polar problem studied in the previous section.

The general solution to Eq. (4.27) may be found using separation of variables. Looking for the azimuthal velocity as $u_\phi = r^\lambda g_\lambda(\theta)$, we obtain that g_λ verifies the differential equation

$$\lambda(\lambda + 1)g_\lambda + g''_\lambda + \frac{g'_\lambda}{\tan \theta} = \frac{g_\lambda}{\sin^2 \theta}. \tag{4.28}$$

The solution to Eq. (4.28) is closely related to Legendre polynomials. Indeed, consider the function $f_n(\theta) = P_n(\cos \theta)$. Given that P_n satisfies the Legendre differential equation, it is straightforward to see that f_n is a solution to

$$n(n + 1)f_n + f''_n + \frac{f'_n}{\tan \theta} = 0. \tag{4.29}$$

Differentiating Eq. (4.29) once leads to Eq. (4.28) provided that we identify $g \equiv f'$ and $n(n+1) \equiv \lambda(\lambda+1)$. We thus have

$$g_\lambda(\theta) = \frac{\mathrm{d} f_n}{\mathrm{d}\theta} = -\sin\theta P_n'(\cos\theta), \qquad (4.30)$$

with $\lambda = n$ or $\lambda = -1-n$. From Eq. (4.6) we see that, up to a rescaling, this is the $V_n(\theta)$ solution appearing in the polar component of the flow. Expressed in the moving frame and therefore keeping all terms increasing at most like r in the far field, the general solution to Eq. (4.27) is of the form

$$u_\phi(r, \theta, t) = \Lambda(t)\frac{r}{a}V_1(\theta) + \sum_{n \geqslant 1} \gamma_n(t) \left(\frac{a}{r}\right)^{n+1} V_n(\theta). \qquad (4.31)$$

Given the form of Eq. (4.31), we can decompose the azimuthal boundary condition into

$$u_\phi(a, \theta, t) = \sum_{n \geqslant 1} C_n(t) V_n(\theta), \qquad (4.32)$$

and may use the same projection as in Eq. (4.12) to obtain the coefficients C_n as surface integrals of $u_\phi(a, \theta, t)$. Keeping all time dependence implicit for simplicity, the boundary conditions at infinity and on the sphere lead to

$$\Lambda = -\Omega a, \qquad (4.33\text{a})$$
$$\gamma_1 = C_1 + \Omega a, \qquad (4.33\text{b})$$
$$\gamma_n = C_n, \quad (n \geqslant 2), \qquad (4.33\text{c})$$

and therefore the final solution for the azimuthal component of the instantaneous flow is given by

$$u_\phi(r, \theta) = (\Omega a + C_1)\left(\frac{a}{r}\right)^2 V_1 - \Omega r V_1 + \sum_{n \geqslant 2} C_n \left(\frac{a}{r}\right)^{n+1} V_n. \qquad (4.34)$$

4.3.2 Rotational Velocity of Free Squirmer

Similar to the swimming speed resulting in Section 4.2 from the requirement of force-free motion, the value of the rotational velocity is a consequence of the absence of net hydrodynamic torque in free-swimming motion (Eq. (2.17)). Indeed, as is further developed in Chapter 9, the flow induced by point torque $L\mathbf{e}_z$ located at the origin is given by $\mathbf{u} = u_\phi \mathbf{e}_\phi$, with

$$u_\phi = \frac{L\sin\theta}{8\pi\mu r^2}. \qquad (4.35)$$

Comparing this solution with the $1/r^2$ flow component in Eq. (4.34) we see that the total torque acting on the surrounding fluid is zero if and only if the rotational velocity satisfies

$$\Omega = -\frac{C_1}{a},$$
(4.36)

with a resulting final solution for the instantaneous azimuthal component of a freely rotating squirmer:

$$u_\phi(r,\theta) = C_1 \frac{r}{a} V_1 + \sum_{n \geqslant 2} C_n \left(\frac{a}{r}\right)^{n+1} V_n.$$
(4.37)

An important difference between the flow induced by the translational motion of the squirmer, Eq. (4.19), and that created by its rotational motion, Eq. (4.37), must be noted. Specifically, the flow induced by the translating squirmer expressed in the lab frame includes in general a $1/r^3$ decaying term proportional to the swimming modes A_1 and B_1. In contrast, the velocity induced by the rotational part of the squirmer, Eq. (4.37), when expressed in the lab frame is simply given by

$$u_\phi(r,\theta) = \sum_{n \geqslant 2} C_n \left(\frac{a}{r}\right)^{n+1} V_n,$$
(4.38)

which has lost all trace of the C_1 mode responsible for the rotation. In contrast to purely tangential squirmers, which cannot be stealth swimmers for translational motion and always stir the fluid as a result of their swimming, we see that it is possible to have a stealth rotating squirmer by setting $C_n = 0$ for $n \geqslant 2$ so that the resulting flow contains no signature of its rotational motion.

4.3.3 Rotational Velocity from Reciprocal Theorem

In the same fashion as for the swimming velocity in Section 4.2.2, the rotational velocity of the squirmer may be derived directly from the azimuthal boundary conditions. We make use of the rotational version of the reciprocal identity in Eq. (2.36) (see Exercises in Chapter 2) linking the instantaneous rotational velocity, $\Omega(t)$, to the velocity boundary conditions on the surface of the sphere, $\mathbf{u}$, as the integral

$$\Omega(t) = -\frac{3}{8\pi a^3} \iint_S \mathbf{n} \times \mathbf{u} \, dS.$$
(4.39)

Since the problem is axisymmetric, the rotational velocity only has a z component, denoted by Ω. Projecting Eq. (4.39) in the z direction and carrying out the ϕ component of the integration, we obtain the integral

$$\Omega = -\frac{3}{4a} \int_0^\pi (\mathbf{e}_r \times \mathbf{u}) \cdot \mathbf{e}_z \sin\theta \, d\theta = -\frac{3}{4a} \int_0^\pi u_\phi \sin^2\theta \, d\theta.$$
(4.40)

With the azimuthal boundary conditions given by Eq. (4.32), the rotational velocity is given by the sum

$$\Omega = -\frac{3}{4a} \sum_{n \geqslant 1} C_n \int_0^\pi \sin^2 \theta V_n(\theta) \, d\theta. \tag{4.41}$$

Invoking orthogonality of the Legendre polynomials as in Section 4.2.2, we conclude that all integrals in Eq. (4.41) are zero except the one for $n = 1$. Hence the only term contributing to the value of Ω is C_1, leading to

$$\Omega = -\frac{3}{4a} C_1 \int_0^\pi \sin^3 \theta \, d\theta = -\frac{C_1}{a}, \tag{4.42}$$

which is in agreement with the direct result of Eq. (4.36).

4.4 Envelope Model

The analysis so far has assumed that the velocity boundary conditions were applied on the surface of an effective sphere of radius a. We now return to the original 'envelope' motivation for this problem and consider the periodic, small-amplitude deformation of a flexible surface embedded on a sphere. The goal of this section is to derive the mathematical relationship between the motion of points on the surface of the deformable body, described naturally by tracking points in a Lagrangian fashion, and the Eulerian squirmer formalism from the previous sections. For simplicity, we assume that all deformations remain axisymmetric, and refer to the work of Pedley et al. (2016) for the inclusion of azimuthal deformation and the derivation of a rotating envelope model.

The setup is illustrated in Fig. 4.2. We describe the deformation of the flexible surface in a Lagrangian fashion by tracking the motion of material points on the surface using spherical coordinates $(r = R(\theta_0, t), \theta = \theta(\theta_0, t), \phi)$, where θ_0 is the polar angle for a reference Lagrangian point located at a position (R, θ, ϕ) at time t. Since the motion is axisymmetric, we can ignore the azimuthal angle ϕ in what follows.

Assuming that the surface motion has a small dimensionless amplitude ϵ away from a sphere of radius a, we decompose the deformation along the modes that appeared naturally as part of the squirming analysis, i.e.

$$R(\theta_0, t) = a \left[1 + \epsilon \sum_{n \geqslant 2} \alpha_n(t) P_n(\cos \theta_0) \right], \tag{4.43a}$$

$$\theta(\theta_0, t) = \theta_0 + \epsilon \sum_{n \geqslant 1} \beta_n(t) V_n(\theta_0). \tag{4.43b}$$

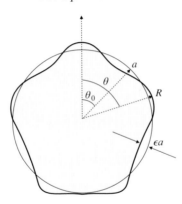

Figure 4.2 Envelope model: A flexible spherical surface of radius a undergoes axisymmetric deformation of small dimensionless amplitude ϵ. The material points
originating from a reference angle θ_0 are located at time t at a radial distance
$R(\theta_0, t)$ and have a polar angle $\theta(\theta_0, t)$.

Note that, in keeping with Blake (1971b), the sum in Eq. (4.43a) only starts at
$n = 2$ because the mode with $n = 0$ does not conserve volume (although it is easy
to relax this hypothesis; see Pedley et al., 2016) and $n = 1$ merely shifts the centre
of the sphere. The sum in Eq. (4.43b) starts at $n = 1$ because V_1 is the first nonzero
mode.

From the Lagrangian motion in Eq. (4.43), we can then invoke the no-slip boundary condition in order to derive the instantaneous velocity boundary conditions
evaluated at the current location of material points (R, θ) at time t. In the radial
direction we obtain

$$u_r(R, \theta, t) = \dot{R} = a\epsilon \sum_{n \geq 2} \dot{\alpha}_n(t) P_n(\cos \theta_0), \qquad (4.44)$$

while the polar velocity component is given by

$$u_\theta(R, \theta, t) = R\dot{\theta} = a\epsilon \left[1 + \epsilon \sum_{n \geq 2} \alpha_n(t) P_n(\cos \theta_0) \right] \sum_{n \geq 1} \dot{\beta}_n(t) V_n(\theta_0). \qquad (4.45)$$

Looking for solutions of the flow problem and the swimming velocity as power
series expansions in $\epsilon \ll 1$,

$$\mathbf{u} = \epsilon \mathbf{u}_1 + \epsilon^2 \mathbf{u}_2 + \cdots, \qquad (4.46a)$$

$$U = \epsilon U_1 + \epsilon^2 U_2 + \cdots, \qquad (4.46b)$$

we proceed by carrying out Taylor expansions of Eqs. (4.44)–(4.45). This provides
the flow problem at each order with boundary conditions applied on the effective
spherical surface of radius a (so-called domain perturbation).

At order ϵ, the linearisation of Eqs. (4.44)–(4.45) leads to the boundary conditions

$$u_{1,r}(a,\theta_0,t) = a \sum_{n\geqslant 2} \dot{\alpha}_n(t) P_n(\cos\theta_0), \tag{4.47a}$$

$$u_{1,\theta}(a,\theta_0,t) = a \sum_{n\geqslant 1} \dot{\beta}_n(t) V_n(\theta_0). \tag{4.47b}$$

Since these boundary conditions take exactly the form assumed in the analysis of the spherical squirmer (see Eq. (4.10)), the full free-squirming solution of Eq. (4.19) is, at that order, given by

$$A_0 = A_1 = 0, \tag{4.48a}$$

$$A_{1,n}(t) = a\dot{\alpha}_n, \quad n \geqslant 2, \tag{4.48b}$$

$$B_{1,n}(t) = a\dot{\beta}_n(t), \quad n \geqslant 1. \tag{4.48c}$$

In particular, the swimming speed at first order is given by

$$U_1(t) = \frac{2}{3} a\dot{\beta}_1(t), \tag{4.49}$$

which is an exact derivative and thus averages to zero by the periodicity of β_1. Similar to what was obtained for the swimming of a waving sheet in Chapter 3, net locomotion only appears with a quadratic dependence on the deformation amplitude ϵ.

In order to derive the Eulerian boundary conditions at order ϵ^2, we use the fact that for any quantity of interest $g = \epsilon g_1 + \epsilon^2 g_2 + \cdots$ defined on the deforming surface (R,θ), a Taylor expansion about (a,θ_0) up to order ϵ^2 is given by

$$g|_{(R,\theta)} = \epsilon g_1|_{(a,\theta_0)} + \epsilon^2 g_2|_{(a,\theta_0)} + \epsilon(R-a)\frac{\partial g_1}{\partial r}\bigg|_{(a,\theta_0)} + \epsilon(\theta-\theta_0)\frac{\partial g_1}{\partial \theta}\bigg|_{(a,\theta_0)}, \tag{4.50}$$

with the last three terms all contributing to order ϵ^2 since both $R-a$ and $\theta-\theta_0$ are of order ϵ. Since the order-ϵ solution has been derived analytically, we can compute all the required derivatives exactly, allowing us to write the second-order boundary conditions as the product of infinite series

$$\frac{u_{2,r}(a,\theta_0,t)}{a} = \left(\sum_{n\geqslant 2}\alpha_n P_n\right)\left[2\dot{\beta}_1 P_1 + \sum_{n\geqslant 2}2(\dot{\alpha}_n + \dot{\beta}_n)P_n\right]$$

$$- \left(\sum_{n\geqslant 1}\beta_n V_n\right)\left[\sum_{n\geqslant 2}\dot{\alpha}_n\frac{\mathrm{d}}{\mathrm{d}\theta_0}P_n\right], \tag{4.51}$$

$$\frac{u_{2,\theta}(a,\theta_0,t)}{a} = \left(\sum_{n\geqslant 2}\alpha_n P_n\right)\left[2\dot{\beta}_1 V_1 + \sum_{n\geqslant 2}\left\{n\left(1-\frac{n}{2}\right)\dot{\alpha}_n + (2n+1)\dot{\beta}_n\right\}V_n\right]$$

$$- \left(\sum_{n\geqslant 1}\beta_n V_n\right)\left[\sum_{n\geqslant 1}\dot{\beta}_n\frac{\mathrm{d}}{\mathrm{d}\theta_0}V_n\right], \tag{4.52}$$

where P_n are functions of the reference argument $\cos\theta_0$, V_n are functions of θ_0 and where the coefficients α_n and β_n are all implicit functions of time.

The result in Eqs. (4.51)–(4.52) provides the link between the Lagrangian description of the surface, Eq. (4.43), and the Eulerian squirmer description from the previous section. The amplitudes of the second-order squirming modes, $A_{2,n}$ and $B_{2,n}$, may be found by exploiting the projection formula, Eq. (4.12). Since these integrals involve triple products of Legendre polynomials, they are related to so-called Gaunt coefficients (Gaunt, 1929) for which fast numerical evaluation algorithms can be used (Xu, 1996). The only coefficients relevant for the locomotion of the deforming interface are the first modes, i.e. the values of $A_{2,1}$ and $B_{2,1}$, since $U_2 = \frac{1}{3}(2B_{2,1} + A_{2,1})$, and they can be evaluated analytically in principle. The instantaneous swimming speed of the spherical envelope model was computed heroically by Blake (1971b), revealing that, at leading order, all pairs of adjacent modes of the surface deformation contribute to the swimming speed.

4.5 Comparison with Experiments: *Volvox* Locomotion

We close this chapter by comparing the predictions of the squirmer model with experiments. We consider the green alga *Volvox carteri*, a multicellular organism roughly spherical in shape and used as a model to study a range of problems in biological hydrodynamics (Goldstein, 2015). Distributed on the surface of *Volvox carteri* are a few thousand cells much smaller in size than the organism and each equipped with two beating flagella protruding into the surrounding fluid (see Fig. 4.3(a) for a picture and Fig. 4.3(b) for a schematic representation). The flagella deform in time to create large-scale flows and induce locomotion of the organism.

In Fig. 4.3(c) we reproduce measurements of the time-averaged flow, obtained using particle-image velocimetry, induced by the multiple flagella when the cell is held in place by a pipette (Brumley et al., 2015). This flow is well captured by the

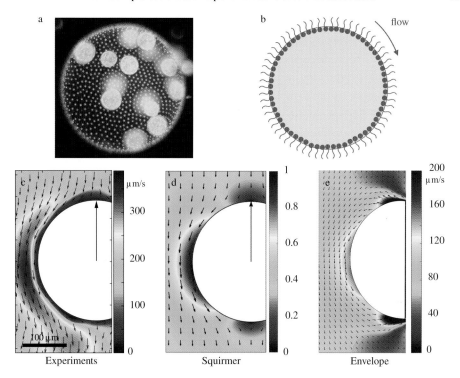

Figure 4.3 Experiments on flow field induced by the green alga *Volvox carteri* and comparison with squirmer and envelope models. (a) Picture of *Volvox carteri* (courtesy of Raymond Goldstein). (b) Schematic representation of the many flagellated cells located on the surface of the organism. (c) Measured time-averaged flow around the cell held in place by a pipette (Brumley et al., 2015). (d) Prediction of the squirmer model modified to account for the net force induced by the pipette holding the cell and including two nonzero modes (courtesy of Debasish Das). (e) Prediction of the rotating envelope model (analysis by Pedley et al. (2016) and figure courtesy of Douglas Brumley).

squirmer model from Sections 4.1–4.2. Because the cell is not free to swim but is instead held in place externally, it induces a net force on the surrounding fluid and therefore the flow solution contains a nonzero $1/r$-decaying mode. Keeping both the force and the stresslet modes and setting the others to zero leads to a squirmer flow given by

$$u_r(r,\theta) = B_1 \left[\left(\frac{a}{r} \right)^3 - \frac{a}{r} \right] \cos\theta + \frac{B_2}{2} \left[\left(\frac{a}{r} \right)^4 - \left(\frac{a}{r} \right)^2 \right] \left(3\cos^2\theta - 1 \right),$$

(4.53)

$$u_\theta(r,\theta) = \frac{B_1}{2} \left[\left(\frac{a}{r} \right)^3 + \frac{a}{r} \right] \sin\theta + B_2 \left[\left(\frac{a}{r} \right)^4 \right] \sin\theta\cos\theta.$$

(4.54)

The best fit between the experimental measurements and the model is obtained for $B_2/B_1 = 0.1$, with results shown in Fig. 4.3(d). This simple two-mode squirmer model is able to reproduce the main features of the experiments, including the small front-back asymmetry in the flow, the maximum flow near the equator and the overall structure of the flow field.

An additional first-principle prediction may be made using the envelope model outlined in Section 4.4 when modified to account for the additional azimuthal deformations of the effective envelope of the flagella tips. With deformation parameters taken directly from experiments (wavelengths, frequencies, amplitudes), Pedley et al. (2016) derived analytically the predictions of the second-order rotating envelope model. The resulting flow, when time-averaged, is reproduced in Fig. 4.3(e). Despite somewhat under-predicting the flow speeds (due to an over-estimation of back-flow occurring during backstroke motion of individual flagella in the envelope framework) and predicting strong flows at the poles (a mathematical consequence of forcing the travelling waves to be axisymmetric), the model is able to reproduce successfully the overall structure of the flow field and its variation near the equator.

Further Reading

A different solution method for the squirmer problem from the one presented in this chapter involves the use of Lamb's general solution for Stokes flows (Lamb, 1932). That methodology, based on spherical harmonics, allows us to recover the results above and to extend the squirmer model to a non-axisymmetric distribution of surface velocities, leading to three-dimensional swimming motion (Pak and Lauga, 2014). Squirmers in different geometries have also been proposed, either in simpler infinite setups (Blake, 1971c) or in the shape of a finite-volume torus (Purcell, 1977; Leshansky and Kenneth, 2008).

The optimisation of the envelope model, addressed in the large-amplitude limit by Michelin and Lauga (2010a), shows that surface deformation akin to metachronal waves leads to swimming with minimum energy expenditure. In general, swimming using metachronal waves may be tackled with the small-amplitude modelling framework from this chapter using an infinite number of interacting modes (Blake, 1971b; Pedley et al., 2016).

Squirmers have also proven to be adequate mathematical models to quantify the movement of so-called phoretic swimmers, whose locomotion results from self-generated chemical gradient and surface-driven phoretic flows (Golestanian et al., 2007; Lisicki et al., 2018). The chemical gradients, typically directly patterned on colloidal particles (Howse et al., 2007; Ebbens and Howse, 2011), can also emerge spontaneously from transport instability (Michelin et al., 2013) and have analogues in macroscopic systems of droplets (Thutupalli et al., 2011; Izri et al., 2014).

Exercises

1. Consider a spherical squirmer of radius a that swims by imposing instantaneously a velocity $\mathbf{u}'$ along its spherical boundary. The field $\mathbf{u}'$ is axisymmetric and everywhere tangential to the sphere, so that in spherical coordinates we may write $\mathbf{u}' = f(\theta)\mathbf{e}_\theta$ in the swimming frame. The swimming velocity is denoted $\mathbf{U} = U\mathbf{e}_z$ along the symmetry axis of the swimmer. Derive the only possible solution for $f(\theta)$ that leads to an irrotational flow, $\mathbf{u} = \nabla\phi$. Check that your result satisfies the reciprocal integral relationship derived in Chapter 2, linking $\mathbf{u}'$ to $\mathbf{U}$ for a sphere.

2. For the squirmer model with flow in Eq. (4.19), show that the instantaneous value of the dynamic pressure is given by

$$p = \mu \sum_{n \geqslant 2} \frac{2n-1}{n+1} (nA_n - 2B_n) \frac{a^n}{r^{n+1}} P_n(\cos\theta).$$

Calculate the surface traction (surface stress dotted by the outward unit normal) and deduce that the instantaneous rate of work of the squirmer, $\dot{W}$, is given by

$$\dot{W} = 2\pi\mu a \left[\frac{8}{3}A_1^2 + \frac{8}{3}B_1^2 + \frac{16}{3}A_1 B_1 \right.$$
$$\left. + \sum_{n \geqslant 2} \left(\frac{4n^2 + 6n + 8}{(2n+1)(n+1)}A_n^2 + \frac{8}{n(n+1)}B_n^2 + \frac{24}{(2n+1)(n+1)}A_n B_n \right) \right].$$

3. The swimming efficiency of a squirmer is defined as the ratio between the rate of work done to drag the spherical body at the swimming speed and the rate of work of the swimmer (i.e. $\dot{W}$ from Exercise 2). Consider a spherical squirmer with only tangential modes ($A_n = 0$). Show that the swimming efficiency is always less than $1/2$ and that this upper bound is reached by the potential squirmer derived in Exercise 1.

4. A two-dimensional version of the squirmer model can also be derived. Consider a disc of radius a swimming in a two-dimensional Stokes flow by imposing velocity boundary conditions in the moving frame as

$$u_r(a, \theta) = \sum_{n \geqslant 1} A_n \cos(n\theta), \quad u_\theta(a, \theta) = \sum_{n \geqslant 1} B_n \sin(n\theta).$$

Swimming occurs with speed U along the $\theta = 0$ axis and the flow is mirror-image symmetric about that direction. Show that $U = \frac{1}{2}(B_1 - A_1)$ and determine the flow field around the squirmer.

5. The squirmer model can be applied to study the motion of phoretic particles, which propel in viscous fluids using self-generated chemical gradients. Consider a spherical particle of radius a releasing from its surface (denoted $\mathbf{r}_s$) a chemical solute of concentration c at a fixed, spatially dependent rate $A(\mathbf{r}_s)$. The solute is assumed to obey the diffusion equation. Tangential solute gradients along the surface of the particle lead to phoretic motion of the fluid with velocity $\mathbf{u}'$, given in the moving frame by $\mathbf{u}'(\mathbf{r}_s) = M(\mathbf{r}_s)(\mathbf{1} - \mathbf{nn}) \cdot \nabla c$ (Anderson, 1989). Both the activity, A, and mobility, M, are assumed to be distributed axisymmetrically on the particle surface and can be written as sums of Legendre polynomials. Assuming that there is no solute at infinity, solve for the distribution of solute, c, around the particle. Deduce the value of $\mathbf{u}'$ and the swimming speed of the particle. In the particular case of a Janus particle with constant activities A_1, A_2 and mobilities M_1, M_2 on two hemispheres, solve for the swimming speed.

PART TWO

CELLULAR LOCOMOTION

5

Flagella and the Physics of Viscous Propulsion

The models of Chapters 3 and 4 provide a fluid mechanical introduction to the concept of cellular locomotion. While these models showed how a deforming surface may swim through the action of viscous stresses, they both make strong simplifying assumptions. The waving sheet in Chapter 3 is two-dimensional while the oscillating surface of Chapter 4 is spherical and axisymmetric. As seen in Chapter 1, biological locomotion for both major classes of organisms, prokaryotes and eukaryotes, is not only three-dimensional, but it is also dominated by the presence of slender filaments, flagella, which are rotating and waving in viscous fluids. In this chapter, we return to the biological movements seen in Chapter 1 and consider the relationship between the cellular shapes and their motion. Using elementary concepts from Stokes flows, we propose an intuitive physical and mathematical interpretation for the generation of propulsive forces by moving flagella and for the natural occurrence of helices and waves in cellular propulsion.

5.1 Kinematics of Flagellar Propulsion

In order to understand the physics of propulsion, we first need to look in detail at the deformation obtained in the biological world. We focus here on two canonical situations involving flagella, namely the helical propulsion of bacteria and the waving propulsion of spermatozoa.

5.1.1 Prokaryotes

As seen in Chapter 1, swimming bacteria generate propulsion by rotating helical flagellar filaments. This setup is illustrated in Fig. 5.1(a) in the case of a left-handed filament, which rotates in a counter-clockwise direction (CCW, as indicated by the arrow) when viewed from the right end towards the left. This would be, for

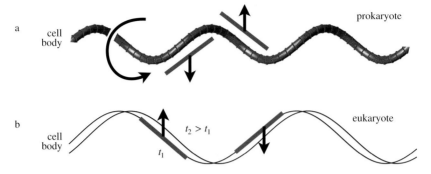

Figure 5.1 Zooming in on the motion of flagella, highlighting the relationship between geometry and velocity relative to the surrounding fluid. (a) Instantaneous rotation of a prokaryotic helical flagellar filament. (b) Planar waving motion of a eukaryotic flagellum shown at two times, t_1 and $t_2 > t_1$, in the swimming frame. Despite belonging to two different biological domains, both organisms exploit the same physics to self-propel in fluids.

example, the relevant geometry for the flagellar filament of *E. coli* with a cell body located to the left in the figure.

Zooming in on the helical shape, we highlight in Fig. 5.1(a) two specific points along the filament. The portion of filament located in the foreground of the figure has a locally positive slope (using the natural axis defined by the page) and, due to the direction of rotation of the helix, a negative instantaneous velocity. Similarly, examining a point along the helix located in the background of the figure, we see that it is in a configuration with a negative local slope and a positive velocity.

5.1.2 Eukaryotes

Eukaryotic cells such as spermatozoa generate propulsion by deforming flexible flagella as travelling waves. This situation is illustrated in Fig. 5.1(b) where we sketch, in the swimming frame, a sinusoidal travelling wave at two times, t_1 and $t_2 > t_1$. In the situation plotted, the cell body is located to the left and the wave travels to the right (i.e. from the proximal end to the distal end of the flagellum).

By considering the motion of material elements along the flagellum, we see that the portion of the wave where the slope is locally negative moves upwards whereas, at the points where the slope of the wave is positive, the material points move downwards.

Comparing Figs. 5.1(a) and (b), it is therefore evident that despite their different biological structures, both prokaryotes and eukaryotes actuate their flagella in a similar fashion. Specifically, we observe the same relationship between the local shape of a flagellum and its motion relative to the surrounding fluid.

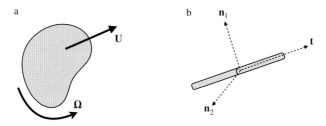

Figure 5.2 (a) Rigid-body motion of an arbitrary shape with translational velocity **U** and angular velocity **Ω**. (b) Particular case of a rigid rod, for which the
resistance matrix is easily written in the frame of reference based on the tangent
direction, **t**, and the two mutually orthogonal normals, $\mathbf{n}_1$ and $\mathbf{n}_2$.

5.2 Forces and Torques in Stokes Flows

In order to understand why the setup in Fig. 5.1 is able to create forward propulsion
for the cells, we have to consider how moving bodies experience forces in Stokes
flows.

5.2.1 Rigid Bodies

Consider an arbitrary rigid body moving in a viscous fluid with linear velocity **U**
and angular velocity **Ω** (Fig. 5.2(a)). The Stokes flow around the body leads to
stresses that, when integrated upon the surface of the body, induce a net hydrodynamic force, **F**, and torque, **T** (Happel and Brenner, 1965; Kim and Karrila, 1991;
Leal, 2007). As was touched upon in Chapter 2, since the equations of motion are
linear, both **F** and **T** depend linearly on **U** and **Ω** as

$$\begin{pmatrix} \mathbf{F} \\ \mathbf{T} \end{pmatrix}_{\text{body}} = -\mathbf{R} \cdot \begin{pmatrix} \mathbf{U} \\ \mathbf{\Omega} \end{pmatrix}_{\text{body}}, \tag{5.1}$$

which is the same as Eq. (2.24). The six-by-six resistance matrix **R** is positive definite, a consequence of the energy equation, Eq. (2.20), since the product $\mathbf{F} \cdot \mathbf{U} +$
$\mathbf{T} \cdot \mathbf{\Omega}$ is minus the rate of work of the rigid body on the fluid, and thus minus the
total (positive) viscous dissipation in the fluid.

Suppose now that the same rigid body undergoes a different motion with velocity
U* and angular velocity **Ω*** associated with hydrodynamic force **F*** and torque **T***.
Since the resistance matrix **R** depends only on the body geometry, we have by the
reciprocal theorem of Stokes flow, Eq. (2.27), that

$$\begin{pmatrix} \mathbf{F} \\ \mathbf{T} \end{pmatrix} \cdot \begin{pmatrix} \mathbf{U}^* \\ \mathbf{\Omega}^* \end{pmatrix} = \begin{pmatrix} \mathbf{F}^* \\ \mathbf{T}^* \end{pmatrix} \cdot \begin{pmatrix} \mathbf{U} \\ \mathbf{\Omega} \end{pmatrix}, \tag{5.2}$$

and therefore the requirement for the tensor $\mathbf{R}$ to be symmetric. Writing $\mathbf{R}$ in terms of three-by-three submatrices, we have

$$\mathbf{R} = \begin{pmatrix} \mathbf{A} & \mathbf{B} \\ \mathbf{B}^T & \mathbf{D} \end{pmatrix}, \tag{5.3}$$

and thus symmetry implies that $\mathbf{A} = \mathbf{A}^T$ and $\mathbf{D} = \mathbf{D}^T$, where ^{T} denotes the transpose of a matrix. Comparing the dimensions of $\mathbf{U}$ and $\mathbf{F}$, we deduce that $\mathbf{A}$ scales as the viscosity times a length. Denoting a characteristic length scale of the body by L, we can then write $\mathbf{A} = \mu L \hat{\mathbf{A}}$, where $\hat{\mathbf{A}}$ is dimensionless. Similarly, we may write $\mathbf{B} = \mu L^2 \hat{\mathbf{B}}$ and $\mathbf{D} = \mu L^3 \hat{\mathbf{D}}$ so that the relationship between dynamics and kinematics is written finally as

$$\begin{pmatrix} \mathbf{F} \\ \mathbf{T} \end{pmatrix}_{\text{body}} = -\mu \begin{pmatrix} L\hat{\mathbf{A}} & L^2\hat{\mathbf{B}} \\ L^2\hat{\mathbf{B}}^T & L^3\hat{\mathbf{D}} \end{pmatrix} \cdot \begin{pmatrix} \mathbf{U} \\ \mathbf{\Omega} \end{pmatrix}_{\text{body}}. \tag{5.4}$$

Importantly, in Eq. (5.4) each of the hat matrices $(\hat{\mathbf{A}}, \hat{\mathbf{B}}, \hat{\mathbf{D}})$ is dimensionless and depends solely on the shape of the body. Furthermore, since $\mathbf{R}$ is positive definite, both $\hat{\mathbf{A}}$ and $\hat{\mathbf{D}}$ are symmetric and positive definite. The exact value of the matrix $\hat{\mathbf{B}}$ depends on the point with respect to which torques are measured. It may be shown that for any shape there exists a special point termed the centre of mobility (or centre of diffusion) at which the tensor $\hat{\mathbf{B}}$ becomes itself symmetric (Brenner, 1967).

For bodies that are sufficiently symmetric, both $\hat{\mathbf{A}}$ and $\hat{\mathbf{D}}$ are isotropic tensors. This is the case, for example, for a sphere, a cube or a tetrahedron (Happel and Brenner, 1965). For a sphere, if L is chosen to be its radius, the classical Stokes formulae are $\hat{\mathbf{A}} = 6\pi\mathbf{1}$ and $\hat{\mathbf{D}} = 8\pi\mathbf{1}$. Nonzero values of the coupling tensor $\hat{\mathbf{B}}$ indicate a shape for which a rotation induces a nonzero force, and this arises in general for chiral bodies lacking mirror-image symmetry (Brenner, 1967). The prototypical example of such a shape is a helix, on which more details are provided in Chapter 8, where we address the swimming of bacteria powered by helical flagellar filaments.

Focusing on the sub-tensor $\hat{\mathbf{A}}$ linking forces to velocities, we can now consider the case of non-isotropic bodies. Since we wish to address the motion of slender flagella, a relevant shape to consider is that of a rod, as illustrated in Fig. 5.2(b). In that case $\hat{\mathbf{A}}$ is best written using a body-based coordinate system. Denoting by $\mathbf{t}$ the tangent vector along the rod and by $\mathbf{n}_1, \mathbf{n}_2$ two mutually orthogonal perpendicular normal vectors, we have by symmetry that

$$\hat{\mathbf{A}} = \alpha\mathbf{tt} + \beta\mathbf{n}_1\mathbf{n}_1 + \beta\mathbf{n}_2\mathbf{n}_2 = \alpha\mathbf{tt} + \beta(\mathbf{1} - \mathbf{tt}), \tag{5.5}$$

and the two positive coefficients α and β are known to satisfy $\beta > \alpha$, with a ratio of approximately 2 in the limit of slender rods (Kim and Karrila, 1991). The result in Eq. (5.5) means that the force required to move a rod at a given speed in a viscous fluid is about twice as large when the rod moves perpendicular to its elongated direction as when it moves along it. The fact that β and α have different values is usually referred to as the property of drag anisotropy for Stokes flows. Among other things, it is responsible for the sedimentation of rigid rods at an angle with respect to the direction of gravity when they are not aligned horizontally or vertically (see Exercises).

5.2.2 Slender Filaments

The result in Eq. (5.5), valid for rigid slender rods, turns out to be also applicable to flexible shape-changing filaments. The first section of Chapter 6 is devoted to deriving the asymptotic limit in which this is correct, but as long as both the length and the characteristic radius of curvature of the filament are much larger than its cross-sectional size (i.e. in the limit of slender, weakly bent filaments), the tensorial dependence from Eq. (5.5) can be applied in a pointwise fashion along the filament using for $\mathbf{t}$ its local, spatially dependent tangent vector. Denoting by $\mathbf{u}(s,t)$ the local velocity of a point on the flagellum relative to the background fluid, in general a function of the arclength along the filament s and of time t, the instantaneous hydrodynamic force acting on the filament per unit length, $\mathbf{f}(s,t)$, is given under this framework by

$$\mathbf{f}(s,t) = -\mu\left[\alpha\mathbf{t}(s,t)\mathbf{t}(s,t) + \beta\left[\mathbf{1} - \mathbf{t}(s,t)\mathbf{t}(s,t)\right]\right] \cdot \mathbf{u}(s,t), \qquad (5.6)$$

where $\mathbf{t}(s,t)$ is the local tangent, also a function of space and time. As shown in Chapter 6, the positive coefficients α and β depend only on the aspect ratio, ϵ, of the filament, and in the limit where $\epsilon \ll 1$ are given by $\beta \approx 2\alpha \approx 4\pi/\ln(2/\epsilon)$. Defining $c_{\parallel} \equiv \mu\alpha$ and $c_{\perp} \equiv \mu\beta$ as the positive parallel and perpendicular drag coefficients, we may then write Eq. (5.6) in the form often known as resistive-force theory (Cox, 1970; Lauga and Powers, 2009)

$$\mathbf{f}(s,t) = -\left[c_{\parallel}\mathbf{t}(s,t)\mathbf{t}(s,t) + c_{\perp}\left[\mathbf{1} - \mathbf{t}(s,t)\mathbf{t}(s,t)\right]\right] \cdot \mathbf{u}(s,t), \qquad (5.7)$$

where it is important to remember that $c_{\perp} > c_{\parallel}$ (see Section 6.3). Note that the result in Eq. (5.7) assumes that there is no flow at infinity, and we will discuss in Chapters 6 and 10 the modifications required to include the presence of a background flow.

5.3 Physics of Drag-Based Propulsion

Using the Stokes force-velocity relationship in Eq. (5.7) we can now explain why the flagellar movement in Fig. 5.1 leads to forward propulsion of the cells. As a

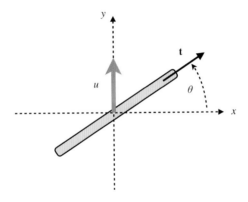

Figure 5.3 Local portion of a moving flagellum modelled as a rigid rod oriented at an angle θ from the horizontal (x) and moving with speed u in the vertical direction (y). Resolving the components of the hydrodynamic force acting on the rod allows us to interpret the directions of the biological propulsive forces arising in Fig. 5.1.

model of the local motion of a small portion of a flagellum, consider the setup illustrated in Fig. 5.3 where a small rod making an angle θ with the horizontal (x direction) moves in the y direction with speed u.

Resolving Eq. (5.7) we can compute all components of the force experienced locally by the flagellum. The tangent vector and the velocity relative to the fluid are given by

$$\mathbf{t} = \begin{pmatrix} \cos\theta \\ \sin\theta \end{pmatrix}, \quad \mathbf{u} = \begin{pmatrix} 0 \\ u \end{pmatrix}. \tag{5.8}$$

We thus have $\mathbf{u} \cdot \mathbf{t} = u \sin\theta$ and

$$(\mathbf{u} \cdot \mathbf{t})\mathbf{t} = \begin{pmatrix} u \sin\theta \cos\theta \\ u \sin^2\theta \end{pmatrix}. \tag{5.9}$$

Using Eq. (5.7), we may then calculate the force per unit length acting on the flagellum as

$$\mathbf{f} = -\begin{pmatrix} (c_\| - c_\perp)u \sin\theta \cos\theta \\ c_\perp u + (c_\| - c_\perp)u \sin^2\theta \end{pmatrix}. \tag{5.10}$$

The important result predicted in Eq. (5.10) concerns the propulsive force, i.e. the component of the force induced in the direction perpendicular to the motion. This force component is given explicitly by

$$\mathbf{f} \cdot \mathbf{e}_x = (c_\perp - c_\|)u \sin\theta \cos\theta. \tag{5.11}$$

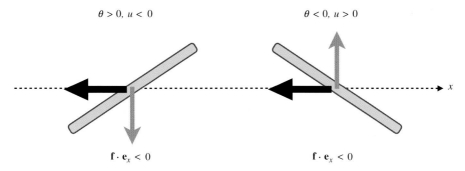

Figure 5.4 Relationship between the geometry of a slender filament and its local velocity (grey arrow) as arising biologically in Fig. 5.1. Propulsive viscous forces are induced on the filament in the direction perpendicular to the motion (black arrows). Both cases induce propulsion in the same direction, providing an explanation for the locomotion of flagellated cells.

Since $c_\perp \neq c_\parallel$, a propulsive force is generated in a manner that depends critically on the relationship between the local orientation (θ) and the motion relative to the background fluid (u). Note that exploiting viscous stresses to generate propulsion is sometimes described as drag-based thrust (see Chapter 2), a denomination reminiscent of the friction-based thrust exploited by slithering snakes (Hu et al., 2009).

We can now interpret physically the diagram of flagellar motion shown in Fig. 5.1 and predict the direction of propulsive forces. We sketch in Fig. 5.4 the two kinematic situations apparent in Fig. 5.1, and we use a black arrow to denote the direction of propulsion induced by the fluid forces as predicted by Eq. (5.11). In the situation illustrated on the left of Fig. 5.4, the local slope along the filament is positive ($\theta > 0$) and the velocity is downward ($u < 0$), and as a result the propulsive force is directed to the left ($\mathbf{f} \cdot \mathbf{e}_x < 0$). Similarly, on the right of Fig. 5.4, we see that a negative slope ($\theta < 0$) combined with positive velocity ($u > 0$) also leads to a propulsive force pointing to the left.

This simple mathematical result has profound implications for the locomotion of cells. Comparing the local force generation illustrated in Fig. 5.4 to the biological actuation in Fig. 5.1, we see that everywhere along both the helical filaments of bacteria (Fig. 5.1(a)) and the flexible flagella of eukaryotes (Fig. 5.1(b)), the coordination between shape and movement leads to the creation of propulsive forces that are always directed to the left and are thus pushing the cell body forward. Cells exploit therefore the anisotropy of viscous drag offered by slender filaments and actuate their shape in such a way that the individual propulsive forces that are generated along their flagella add constructively and all point in the same direction.

As a final note, we remark that the y component of Eq. (5.10) shows that

$$\mathbf{f} \cdot \mathbf{u} = -(c_\perp u^2 \cos^2 \theta + c_\parallel u^2 \sin^2 \theta), \tag{5.12}$$

which is clearly always negative. Given Eq. (2.20), this was in fact expected since the product $-\mathbf{f} \cdot \mathbf{u}$ is equal to the rate of work on the surrounding fluid (per unit length of the flagellum), which itself is equal to the positive rate of viscous dissipation in the fluid.

5.4 Helices and Travelling Waves Are Optimal

The simple calculation in the previous section shows intuitively how cells generate propulsion by exploiting the anisotropy of drag at low Reynolds numbers. Using the same framework, we now propose a physical interpretation for the use of rigid helices and flexible travelling waves as a means of coordinating this propulsion over the whole length of a flagellum.

5.4.1 Rigid Three-Dimensional Flagella: Helices

If the local tangent to the filament has the angle $\theta = 0$ (parallel configuration) or $\theta = \pi/2$ (perpendicular) then no propulsive force is being extracted from the fluid flow. The best possible shape for a viscous propeller is the one that, for a given velocity, maximises Eq. (5.11). Since $\sin \theta \cos \theta = \frac{1}{2} \sin 2\theta$, the propulsive force in Eq. (5.11) is maximal for a unique angle $\theta = 45°$.

What filament shape is characterised by a constant angle between its local tangent and another prescribed direction while moving at a constant velocity? The answer, which is unique, is that of a rotating helix, or 'corkscrew' motion (Purcell, 1977). The best viscous propeller therefore takes the shape of a helix rotating about its axis and the direction of its propulsive force depends only on the chirality of the helix (left- vs. right-handed). In Chapter 8, we will consider in detail the locomotion of bacteria using helical filaments and revisit this issue of chirality.

5.4.2 Flexible Planar Flagella: Waves

If a flexible flagellum can only undergo planar deformation, it cannot have a smooth shape while maintaining a constant angle between its tangent and the propulsion direction (see further details in Chapter 7). In that case, we might ask what type of flexible deformation is optimal from a propulsive standpoint. Motivated by energetic considerations for cellular movement, we consider the mode of deformation of the flagellum that maximises its total propulsive force subject to a fixed rate of work on the fluid (further details on the energy budget of eukaryotic flagella are provided in Chapter 7). Using the method of calculus of variations, we now use a simple model to show that travelling waves arise naturally as the mathematical solution to this optimisation problem.

Consider an infinite flagellum whose shape is described in Cartesian coordinates by material points $(x, y(x, t))$ in the swimming frame. For simplicity, we assume that the flagellum undergoes only gentle long-wavelength deformations and thus work under the assumption that $|\partial y/\partial x| \ll 1$ everywhere. The flagellum is further assumed to be spatially periodic with period λ in the x direction and its motion is periodic in time with period T.

Since $|\partial y/\partial x| \ll 1$, the tangent angle θ is approximately

$$\theta \approx \frac{\partial y}{\partial x}, \tag{5.13}$$

and thus the propulsive force induced along the x direction, Eq. (5.11), depends quadratically on the flagellum amplitude as

$$\mathbf{f} \cdot \mathbf{e}_x = (c_\perp - c_\parallel) \frac{\partial y}{\partial x} \frac{\partial y}{\partial t}. \tag{5.14}$$

Integrating in space over one wavelength of the periodic shape and calculating the average in time leads to a total propulsive force of

$$F_x = \frac{1}{T} \int_0^T \int_0^\lambda \mathbf{f} \cdot \mathbf{e}_x \, dx \, dt = \frac{(c_\perp - c_\parallel)}{T} \int_0^T \int_0^\lambda \frac{\partial y}{\partial t} \frac{\partial y}{\partial x} \, dx \, dt. \tag{5.15}$$

The constraint used to maximise Eq. (5.15) is the total rate of work on the fluid. The pointwise rate of work is given by $\dot{w} = -\mathbf{f} \cdot \mathbf{u}$, which is given by Eq. (5.12) and becomes, at leading order in the flagellum amplitude,

$$\dot{w} = c_\perp \left(\frac{\partial y}{\partial t} \right)^2. \tag{5.16}$$

The total rate of work over one wavelength and averaged in time is therefore given by the integral

$$\dot{W} = \frac{c_\perp}{T} \int_0^T \int_0^\lambda \left(\frac{\partial y}{\partial t} \right)^2 \, dx \, dt. \tag{5.17}$$

The optimal propulsion problem can then be posed mathematically as the following: Find the time-varying flagellum shape $y(x, t)$ maximising F_x (Eq. (5.15)) subject to a fixed value of $\dot{W}$, which we call $\dot{W}_0$ (Eq. (5.17)). Following the classical framework of calculus of variations (Riley et al., 1999) we define a Lagrangian

$$\mathcal{L}[y] \equiv F_x + \Gamma(\dot{W} - \dot{W}_0), \tag{5.18}$$

where Γ is a constant Lagrange multiplier used to enforce that $\dot{W} = \dot{W}_0$ for all shapes. The Euler–Lagrange equation satisfied by $y(x, t)$ may be obtained by considering a small periodic change in the shape of the flagellum, $y \rightarrow y + \delta y$, and calculating the resulting change in the Lagrangian, $\mathcal{L} \rightarrow \mathcal{L} + \delta \mathcal{L}$. A simple algebraic manipulation leads to a first variation given by

$$\delta \mathcal{L} = \int_0^T \int_0^\lambda \left[\frac{(c_\perp - c_\parallel)}{T} \left(\frac{\partial y}{\partial t} \frac{\partial \delta y}{\partial x} + \frac{\partial \delta y}{\partial t} \frac{\partial y}{\partial x} \right) + \frac{2 \Gamma c_\perp}{T} \frac{\partial y}{\partial t} \frac{\partial \delta y}{\partial t} \right] dx \, dt. \quad (5.19)$$

We may then use integration by parts for each of the terms in that integral, together with the requirement that all fields be periodic in space and time, to obtain the perturbation to the Lagrangian as

$$\delta \mathcal{L} = \frac{2}{T} \int_0^T \int_0^\lambda \delta y \left[(c_\parallel - c_\perp) \frac{\partial^2 y}{\partial t \partial x} - \Gamma c_\perp \frac{\partial^2 y}{\partial t^2} \right] dx \, dt. \quad (5.20)$$

The first-order optimality condition is that, at the optimal point, the first variation of the Lagrangian is zero, $\delta \mathcal{L} = 0$, for all perturbations δy (Riley et al., 1999), and thus the flagellum deformation must satisfy

$$(c_\parallel - c_\perp) \frac{\partial^2 y}{\partial t \partial x} = \Gamma c_\perp \frac{\partial^2 y}{\partial t^2}. \quad (5.21)$$

This equation can be integrated once in time to

$$(c_\parallel - c_\perp) \frac{\partial y}{\partial x} = \Gamma c_\perp \frac{\partial y}{\partial t}, \quad (5.22)$$

where the integration constant has been set to zero since the resulting equation needs to average to zero over time. The general solutions to Eq. (5.22), which is a one-dimensional conservation equation, for the optimal flagellum kinematics can then be written as

$$y(x, t) = \mathcal{Y} \left[x - \frac{(c_\perp - c_\parallel)}{\Gamma c_\perp} t \right], \quad (5.23)$$

for some arbitrary shape function $\mathcal{Y}$. The planar flagellar kinematics creating the largest propulsive force at a fixed total energetic cost therefore take the form of travelling waves.

Another optimisation analysis in Chapter 7 allows us to derive the best waveform and relate it to the best helical shape. Clearly, however, planar deformations are always less efficient than a rotating helix. While a helix is able to maintain a constant angle between its local tangent and the direction of its propulsion, a waving filament displays variations of this angle including some portions that are locally horizontal and produce no thrust.

Further Reading

Additional details on the resistance matrices of rigid bodies and their symmetries can be found in the textbooks of Happel and Brenner (1965) and Kim and Karrila (1991). Taylor's movie on low Reynolds number hydrodynamics offers a pedagogical view on the use of drag anisotropy for viscous propulsion (Taylor, 1967).

The appearance of flagellar waves as the solution to the optimal locomotion problem was treated formally by Pironneau and Katz (1974), while optimal flapping in Stokes flows was addressed by Was and Lauga (2013).

Exercises

1. A straight rigid rod of uniform mass per unit length sediments in a viscous fluid at low Reynolds number under the action of gravity. The rod is oriented at an angle θ to the vertical. For general values of the drag coefficients $c_\perp$ and $c_\parallel$, compute the angle α between the direction of sedimentation of the rod and the vertical. Interpret physically the limits $c_\perp = c_\parallel$ and $c_\perp \gg c_\parallel$. A geometrical illustration of this result is offered in Taylor's movie on low Reynolds number flows (Taylor, 1967).

2. A slender swimmer is located in a hypothetical isotropic-drag fluid in which the hydrodynamic force per unit length is proportional to the local centreline velocity as $\mathbf{f} = -\alpha\mathbf{u}$, where α is a constant with dimensions of viscosity. The swimmer's centreline is inextensible (total length L) and has uniform mass per unit length. Show that force-free swimming implies that no instantaneous motion of the swimmer's centre of mass is possible.

3. Mastigonemes are short hairs located along some eukaryotic flagella that enable a flagellum to have an effective tangential drag coefficient larger than its perpendicular one, and thus allow cells to reverse their swimming direction. Consider a straight flagellum characterised by drag coefficients $c_\parallel < c_\perp$. Straight and rigid slender hairs are uniformly distributed along the flagellum with a number n per unit flagellum length. Each hair has length ℓ on either side of the flagellum, perpendicular to which it is clamped and is characterised by its own drag coefficients $\xi_\parallel < \xi_\perp$. Ignoring any hair-hair and hair-flagellum hydrodynamic interactions, compute the two effective drag coefficients of the flagellum with its mastigonemes. Show that there is a critical value for n leading to a ratio of tangential to perpendicular drag coefficients larger than 1.

4. A straight rigid rod of length L is made to oscillate in position and orientation, due to time-varying external forces and torques, around the configuration where it is aligned with the y direction and its centre is located at the origin. The centre

of the rod moves along x as $x(t) = \epsilon a \cos(\omega t)$, where a has dimension of length and ϵ is a dimensionless parameter. The orientation of the rod oscillates in the (x, y) plane with angle $\theta = \epsilon \cos(\omega t + \phi)$ to the y direction, where ϕ is a constant. Using symmetry arguments, show that the time-averaged force induced by the rod on the fluid in the x direction is zero. Assuming that $\epsilon \ll 1$, calculate the time-averaged force on the fluid in the y direction, at leading order in ϵ. Why is there no force when $\phi = 0$ or π? Compute the time-averaged rate of work of the rod on the fluid at leading order in ϵ.

5. The instantaneous propulsive force induced by a rigid rod in the direction perpendicular to its velocity, $f_\perp$, was obtained as

$$f_\perp = (c_\perp - c_\parallel) u \sin \theta \cos \theta,$$

where u is the rod velocity, θ the angle between the rod and the velocity, and $c_\perp$ and $c_\parallel$ are the drag coefficients. This formula has maximum value at $\theta = \pi/4$ but the angles between many types of flagella and their direction of propulsion tend to be slightly smaller than that. Propose a physical argument rationalising the existence of orientation angles below what appears to be the mathematical optimum.

6. We illustrated in Figs. 5.1 and 5.4 the geometrical relationship between the local slope of flagella and the direction in which they move. Specifically, flagella with locally negative slope move up and those with positive slope move down. Demonstrate this result algebraically by considering a deforming flagellum with centreline located at $y(x, t) = f(x - ct), c > 0$.

6

Hydrodynamics of Slender Filaments

In Chapter 5 we showed that the anisotropy of the viscous drag acting on a moving filament was the key to predicting correctly the direction of flagellar propulsive forces. In this chapter we provide the mathematical basis for this result by evaluating asymptotically the hydrodynamic forces experienced by slender filaments. After revisiting the classical solution for Stokes flow due to the translation of a rigid sphere and interpreting it as being due to the superposition of two hydrodynamic singularities, we capture the flow along weakly bent, slender filaments using a centreline distribution of similar singularities. We show that in the slender limit the flow near any cross section of the filament is dominated by a local anisotropic force density, which arises from hydrodynamic singularities in the vicinity of the cross section, whose magnitudes depend logarithmically on the aspect ratio of the filament (resistive-force theory). The introduction of additional singularities far from the local cross section of the filament allows us to derive an improved nonlocal integral relationship relating the velocity of the filament to the distribution of hydrodynamic forces (slender-body theory). We close by comparing the predictions of the nonlocal theory with experiments on rotating helices as models for interacting flagellar filaments.

6.1 Revisiting Stokes Flow Past a Sphere

While a number of different derivations exist to capture the hydrodynamics of slender filaments, we follow in this chapter the arguments of Hancock (1953) and Lighthill (1975), who tackled the problem with a mix of physical ingredients and mathematical sophistication. Their approaches allow us to recover the local resistive-force theory of Cox (1970) in the very slender limit and to derive the nonlocal slender-body theory of Lighthill (1976). The alternative, more mathematical approach of Johnson (1980) is summarised in Section 6.4.2.

The basis of Hancock and Lighthill's physical argument for the flow past a filament is that the instantaneous hydrodynamic response of a filament must not be so different from that of a sphere. In fact, a large collection of connected beads has been proposed as a computational model of slender filaments (Lowe, 2003).

The solution for the flow induced by a moving sphere at zero Reynolds number is a classical result of fluid dynamics, first solved by Stokes (Lamb, 1932; Happel and Brenner, 1965; Kim and Karrila, 1991). If the sphere has radius a, is located instantaneously at the origin and translates with velocity $\mathbf{U}$ with no rotation, the flow velocity in the surrounding fluid at location $\mathbf{r}$ (magnitude r) away from the centre of the sphere is given by

$$\mathbf{u}(\mathbf{r}) = \frac{3}{4}a \left(\frac{1}{r} + \frac{\mathbf{rr}}{r^3} \right) \cdot \mathbf{U} + \frac{1}{4}a^3 \left(\frac{1}{r^3} - \frac{3\mathbf{rr}}{r^5} \right) \cdot \mathbf{U}. \tag{6.1}$$

This solution can be interpreted physically as being due to the instantaneous superposition of two flow singularities, namely a stokeslet (first term, decaying spatially as $1/r$) and the potential source dipole (second term, decaying spatially as $1/r^3$) (Chwang and Wu, 1975).

The stokeslet is the name given to the Green's function for Stokes flows, induced by a point force $\mathbf{F}$ applied locally as a Dirac delta function (Kim and Karrila, 1991). If the singularity is located at the origin, the fluid velocity and pressure are solutions to

$$\mathbf{0} = -\nabla p_F + \mu \nabla^2 \mathbf{u}_F + \delta(\mathbf{r})\mathbf{F}, \quad \nabla \cdot \mathbf{u}_F = 0, \tag{6.2}$$

and given by

$$\mathbf{u}_F(\mathbf{r}; \mathbf{F}) = \frac{1}{8\pi\mu} \left(\frac{1}{r} + \frac{\mathbf{rr}}{r^3} \right) \cdot \mathbf{F}, \quad p_F(\mathbf{r}; \mathbf{F}) = \frac{\mathbf{F} \cdot \mathbf{r}}{4\pi r^3}; \tag{6.3}$$

see also Exercises and Section 9.1. Together with the classical Stokes drag law for the total hydrodynamic force experienced by the fluid, $\mathbf{F} = 6\mu\pi a\mathbf{U}$, Eq. (6.3) allows us to recover directly the first term on the right-hand side of Eq. (6.1). Note that the flow in Eq. (6.3) is anisotropic and, for a given value of r, the component of the flow in the direction of $\mathbf{F}$ is twice as large when $\mathbf{r}$ is aligned with $\mathbf{F}$ compared to when it is perpendicular to it.

The second flow in Eq. (6.1) is a potential source dipole. If such a dipole of strength $\mathbf{M}$ is located at the origin, the corresponding flow singularity is given by

$$\mathbf{u}_{SD}(\mathbf{r}; \mathbf{M}) = \frac{1}{4\pi} \left(\frac{1}{r^3} - \frac{3\mathbf{rr}}{r^5} \right) \cdot \mathbf{M}. \tag{6.4}$$

This flow can be interpreted physically as the limit of a point source and a point sink merging together. Consider a three-dimensional point source flow whose strength m measures the flow rate injected by the source into the fluid. If the source is located at the origin, the corresponding flow, with velocity denoted $\mathbf{u}_m$, is potential (i.e. irrotational) and is given classically by

$$\mathbf{u}_m(\mathbf{r}; m) = \frac{m}{4\pi} \frac{\mathbf{r}}{r^3},$$
(6.5)

which may be verified using spherical symmetry and the condition that the volume flux through any surface containing $\mathbf{0}$ is m. A source dipole is obtained by placing a sink of strength $-m$ at a position $\mathbf{d}$ relative to a source m, and taking the limit $d \equiv |\mathbf{d}| \to 0$ while the product md remains finite. That flow is obtained by assuming a linear superposition and by using a Taylor expansion so that

$$\mathbf{u}_m(\mathbf{r}; m) + \mathbf{u}_m(\mathbf{r} - \mathbf{d}; -m) \approx \mathbf{d} \cdot \nabla \mathbf{u}_m = \frac{1}{4\pi} \left(\frac{1}{r^3} - \frac{3\mathbf{r}\mathbf{r}}{r^5} \right) \cdot (m\mathbf{d}),$$
(6.6)

which is the source dipole of Eq. (6.4) with strength $\mathbf{M} = m\mathbf{d}$. Note that this flow is irrotational and thus, in contrast to the stokeslet, it is associated with no dynamic pressure in the Stokes flow limit.

6.2 Line Superposition of Hydrodynamic Singularities

6.2.1 Physical Motivation

Since the flow due to a moving sphere is given by a combination of a stokeslet and a source dipole, Hancock and Lighthill argued physically that the same should hold for the flow induced by the motion of filaments. Specifically, with the centreline of the filament playing a role analogous to the centre of the sphere, the flow past a slender filament should be well approximated by a line distribution of these two flow singularities of appropriate strengths. The distribution of stokeslets would then have a clear physical interpretation, namely that it would represent the density of hydrodynamic forces acting on the filament. In contrast, the strengths of the source dipoles are unknown and would have to be computed. For a sphere, we note from Eq. (6.1) that we have $\mathbf{M}/(4\pi) = a^3\mathbf{U}/4$ and thus $\mathbf{M} = a^2\mathbf{F}/(6\mu)$. In the case of a filament we expect to derive a similar formula where a is replaced by the cross-sectional radius of the filament. In addition, the direction of the local force density relative to the orientation of the filament should play a role since, unlike a sphere, a filament does not have an isotropic shape.

6.2.2 Local vs. Nonlocal Contributions

In order to compute the distribution of hydrodynamic singularities, we make the further assumption that the filament is only weakly bent. Mathematically, if L denotes the total linear length of the filament and if the maximum curvature of its centreline is denoted by κ, we assume that $\kappa L \lesssim O(1)$. Straight filaments have $\kappa = 0$ and so satisfy the constraint while a bent filament would have, at most, a radius of curvature on the order of L under this assumption, and no kinks.

This weakly bent approximation allows the geometry near any particular point on the filament to be separated into a 'local' region, on which the filament is almost straight, and a 'nonlocal' region outside it. Since we assume that the filament is slender, the radius of its cross section, a, is much smaller than its length, i.e. $a \ll L$. It is therefore possible mathematically to find an intermediate length scale, b, separating these two regions. For example, the choice $b \sim (aL)^{1/2}$ gives a length scale satisfying $b \ll L$ and thus on which the filament is straight (since $\kappa b \ll 1$) while at the same time still being long and slender since $a \ll b$.

Note that for a periodic shape, the relevant length scale might not be its total length L but instead a wavelength λ, and in this case the intermediate length scale needs to satisfy $a \ll b \ll \lambda$. For example, a eukaryotic spermatozoon has a typical cross-sectional radius $a = 50$ nm with a wavelength $\lambda = 20\,\mu$m (see Chapters 1 and 8), a difference of over two orders of magnitude. For a bacterial flagellar filament, we would instead have $a = 20$ nm while $\lambda = 2\,\mu$m (see Chapters 1 and 9), so also a difference of two orders of magnitude. Thus, in both cases we have a robust separation of scales and the intermediate length b can be found.

6.2.3 Local Perpendicular Motion

Since Stokes flows are linear in their boundary conditions, the velocity vector at any point along the filament $\mathbf{u}$ may be decomposed into two components, $\mathbf{u} = \mathbf{u}_\perp + \mathbf{u}_\parallel$, with one component perpendicular to the local tangent vector ($\mathbf{u}_\perp$, Fig. 6.1(a)) and the other one parallel to it ($\mathbf{u}_\parallel$, Fig. 6.1(b)). Here we consider the perpendicular case first.

We assume that the cross section of the filament has constant radius a. Since in the local region the filament is straight, it takes the form of a cylinder and we use Cartesian coordinates to describe this cylindrical geometry (Fig. 6.1(a)). The axis of the cylinder is taken along the z direction and therefore the cross section is described by $x^2 + y^2 = a^2$. We choose x and y such that the translation of the filament occurs in the x direction and assume that the local portion of the filament ranges from $z = -b$ to $z = +c$ with $b, c \gg a$.

In order to compute the flow, we consider a specific point of interest on the filament's cross section at position $(x, y, 0)$. We then calculate the total fluid velocity induced at this location by a superposition of hydrodynamic singularities that are distributed uniformly on the axis and are therefore positioned at points $(0, 0, Z)$ with $-b \leqslant Z \leqslant c$.

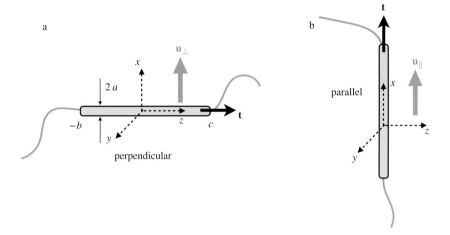

Figure 6.1 Setup used to compute the hydrodynamic singularities (both stokeslets and source dipoles) required to model the local component of the flow past slender filaments. The local cylindrical geometry of the filament is assumed to have radius a and length $b + c$. (a) Translation of filament perpendicular to local tangent. (b) Translation parallel to it.

Stokeslets

By symmetry, we expect the local forcing on the fluid to be aligned with the direction of motion. We thus consider a uniform distribution of stokeslets with strengths $(f_\perp, 0, 0)$ per unit length along the z direction. Since the fluid velocity for a stokeslet is given in Eq. (6.3), a collection of such stokeslets on a small line element dZ at location $(0, 0, Z)$ along the filament leads to a flow, $\mathbf{u}_\perp^f$, at the position $\mathbf{r} = (x, y, z)$ with components given in Cartesian coordinates by

$$\mathbf{u}_\perp^f(\mathbf{r}) = \frac{f_\perp dZ}{8\pi\mu} \left(\frac{x^2 + r^2}{r^3}, \frac{xy}{r^3}, \frac{x(z - Z)}{r^3} \right), \tag{6.7}$$

with $r^2 = |\mathbf{r} - Z\mathbf{e}_z|^2 = x^2 + y^2 + (z - Z)^2$.

We are interested in computing the resulting velocity at location $(x, y, 0)$ along the surface of the cylinder. The total flow from a superposition of stokeslets distributed between $Z = -b$ and $Z = c$ is given by the integral

$$\mathbf{u}_\perp^f(x, y, 0) = \frac{f_\perp}{8\pi\mu} \int_{-b}^{c} \left(\frac{x^2 + r^2}{r^3}, \frac{xy}{r^3}, \frac{-xZ}{r^3} \right) dZ, \tag{6.8}$$

where r^2 simplifies to $r^2 = a^2 + Z^2$.

Examining Eq. (6.8) we see that there are three integrals to evaluate, namely

$$I_1 = \int \frac{dZ}{r^3}, \quad I_2 = \int \frac{Z\,dZ}{r^3}, \quad I_3 = \int \frac{dZ}{r}. \tag{6.9}$$

For I_1, using a change of variable and using the fact that $b \gg a$, we obtain

$$I_1 = \int_{-b}^{c} \frac{dZ}{(a^2 + Z^2)^{3/2}} = \frac{1}{a^2} \int_{-b/a}^{c/a} \frac{du}{(1 + u^2)^{3/2}} = \frac{1}{a^2} \left[\frac{u}{(1 + u^2)^{1/2}} \right]_{-b/a}^{c/a} \approx \frac{2}{a^2},$$
(6.10)

plus higher-order terms in a/c and a/b. For I_2 we have

$$I_2 = \int_{-b}^{c} \frac{Z\, dZ}{(a^2 + Z^2)^{3/2}} = \frac{1}{a} \int_{-b/a}^{c/a} \frac{u\, du}{(1 + u^2)^{3/2}} = \frac{1}{a} \left[\frac{-1}{(1 + u^2)^{1/2}} \right]_{-b/a}^{c/a},$$
(6.11)

and therefore $I_2 \approx 0$ since a/c and a/b are both small. Finally I_3 may be integrated exactly as

$$I_3 = \int_{-b}^{c} \frac{dZ}{(a^2 + Z^2)^{1/2}} = \frac{1}{a} \int_{-b/a}^{c/a} \frac{du}{(1 + u^2)^{1/2}} = \frac{1}{a} \left[\sinh^{-1}(u) \right]_{-b/a}^{c/a}.$$
(6.12)

We then use the identity

$$\sinh^{-1}(u) = \ln(u + \sqrt{1 + u^2}),$$
(6.13)

for which, in the limit $|u| \gg 1$,

$$\sinh^{-1}(u) = \ln(2|u|) + O\left(\frac{1}{u^2}\right), \quad u > 0,$$
(6.14a)

$$\sinh^{-1}(u) = -\ln(2|u|) + O\left(\frac{1}{u^2}\right), \quad u < 0,$$
(6.14b)

so that in the slender limit we obtain

$$I_3 \approx \ln\left(\frac{2c}{a}\right) + \ln\left(\frac{2b}{a}\right) = \ln\left(\frac{4bc}{a^2}\right),$$
(6.15)

plus terms at higher order in a/b and a/c.

With these three integrals evaluated in the slender limit, the local velocity of a point $(x, y, 0)$ on the cross section, Eq. (6.8), may be computed and we obtain

$$u_{\perp}^{f}(x, y, 0) = \frac{f_{\perp}}{8\pi\mu}\left(\frac{2x^2}{a^2} + \ln\left(\frac{4bc}{a^2}\right), \frac{2xy}{a^2}, 0\right),$$
(6.16)

plus higher-order terms in a/c and b/c. Clearly this velocity is not uniform along the cross section but depends on the specific values of x and y. A distribution of perpendicular stokeslets is therefore not sufficient to capture the transverse motion of a rigid filament.

Source Dipoles

In order to correct the result in Eq. (6.16) and obtain velocities that are uniform on the surface of the filament, we take inspiration from the case of a rigid sphere and add a distribution of source dipoles. As for a sphere, we assume that the dipoles are oriented parallel to the stokeslets, i.e. along the x direction, and have uniform strength $(g, 0, 0)$ per unit length. The value of g is then solved for in order to ensure that every point on the cross section of the filament moves with identical rigid-body velocity.

The fluid velocity associated with a source dipole is given by Eq. (6.4), so that a small portion of the filament of size dZ induces a flow at $\mathbf{r}$

$$\mathbf{u}_\perp^g (\mathbf{r}) = \frac{g\, dZ}{4\pi} \left(\frac{r^2 - 3x^2}{r^5}, -\frac{3xy}{r^5}, \frac{-3x(z - Z)}{r^5} \right), \tag{6.17}$$

with the same r as above. At the specific location $(x, y, 0)$ on the cross section, the total flow given by the linear superposition of dipoles is thus written

$$\mathbf{u}_\perp^g (x, y, 0) = \frac{g}{4\pi} \int_{-b}^{c} \left(\frac{r^2 - 3x^2}{r^5}, -\frac{3xy}{r^5}, \frac{3xZ}{r^5} \right) dZ. \tag{6.18}$$

Given Eq. (6.18), we see that two new integrals need to be computed, namely

$$I_4 = \int \frac{dZ}{r^5}, \quad I_5 = \int \frac{Z\, dZ}{r^5}, \tag{6.19}$$

both of which can be evaluated analytically. For I_4 we obtain

$$I_4 = \int_{-b}^{c} \frac{dZ}{(a^2 + Z^2)^{5/2}} = \frac{1}{a^4} \int_{-b/a}^{c/a} \frac{du}{(1 + u^2)^{5/2}} = \frac{1}{a^4} \left[\frac{\frac{2}{3}u^3 + u}{(1 + u^2)^{3/2}} \right]_{-b/a}^{c/a} \approx \frac{4}{3a^4}, \tag{6.20}$$

plus higher-order terms. Similarly for I_5, we obtain

$$I_5 = \int_{-b}^{c} \frac{Z\, dZ}{(a^2 + Z^2)^{5/2}} = \frac{1}{a^3} \int_{-b/a}^{c/a} \frac{u\, du}{(1 + u^2)^{5/2}} = \frac{1}{3a^3} \left[\frac{-1}{(1 + u^2)^{3/2}} \right]_{-b/a}^{c/a}, \tag{6.21}$$

and thus $I_5 \approx 0$.

Using these results, together with the value for I_1 above, we obtain that the leading-order flow on the cross section is given by

$$\mathbf{u}_\perp^g (x, y, 0) = \frac{g}{4\pi} \left(\frac{2}{a^2} - \frac{4x^2}{a^4}, -\frac{4xy}{a^4}, 0 \right), \tag{6.22}$$

plus higher-order terms in a/c and a/b.

Total Flow

Adding the stokeslet flow in Eq. (6.16) and the dipolar contribution from Eq. (6.22) and rearranging the terms, we find that the total flow is

$$\mathbf{u}_\perp(x, y, 0) = \frac{1}{8\pi} \left(\frac{f_\perp}{\mu} \ln \left(\frac{4bc}{a^2} \right) + \frac{4g}{a^2}, 0, 0 \right) + \left(\frac{f_\perp}{4\pi\mu} - \frac{g}{\pi a^2} \right) \left(\frac{x^2}{a^2}, \frac{xy}{a^2}, 0 \right).$$

(6.23)

It is possible to choose a value of the dipole magnitude, g, to ensure that the final velocity is uniform on the cross section and thus independent of x and y. Specifically we have to pick

$$\frac{g}{\pi a^2} = \frac{f_\perp}{4\pi\mu} \rightarrow g = \frac{f_\perp a^2}{4\mu},$$

(6.24)

a result reminiscent of the value for translation of a rigid sphere, $\mathbf{M} = a^2\mathbf{F}/(6\mu)$. With this choice for g, the total flow at the cross section is given by the uniform value

$$\mathbf{u}_\perp(x, y, 0) = \frac{f_\perp}{8\pi\mu} \left(\ln \left(\frac{4bc}{a^2} \right) + 1, 0, 0 \right).$$

(6.25)

An appropriate linear superposition of stokeslets and dipoles is thus able to capture the local motion of a rigid filament mathematically.

6.2.4 Local Parallel Motion

The same analysis may be used to address the configuration in which the filament moves in the direction parallel to its local tangent. In that case, the stokeslets are directed along the tangent vector, resulting in an axisymmetric flow that is uniform around the cross section and thus source dipoles are not needed.

The geometry in this case is illustrated in Fig. 6.1(b). The locally straight cylindrical filament now has its axis along the x direction, ranging from $x = -b$ to $x = c$, and its cross section is characterised by the equation $y^2 + z^2 = a^2$. We consider a specific point of interest on the cross section with coordinates $(0, y, z)$ and compute the flow induced there by a uniform density of stokeslets of strengths $(f_\parallel, 0, 0)$ (per unit length along x) located at $(X, 0, 0)$ with $-b \leqslant X \leqslant c$.

Using Cartesian coordinates, stokeslets at $X\mathbf{e}_x$ along a small line element dX of the filament lead to the flow at location $\mathbf{r} = (x, y, z)$ given by

$$\mathbf{u}_\parallel^f(\mathbf{r}) = \frac{f_\parallel \, dX}{8\pi\mu} \left(\frac{(x - X)^2 + r^2}{r^3}, \frac{(x - X)y}{r^3}, \frac{(x - X)z}{r^3} \right),$$

(6.26)

with $r^2 = |\mathbf{r} - X\mathbf{e}_x|^2 = (x - X)^2 + y^2 + z^2$. The total velocity at the cross section where $x = 0$ is thus given by the integral

$$\mathbf{u}_\parallel^f(0, y, z) = \frac{f_\parallel}{8\pi\mu} \int_{-b}^{c} \left(\frac{X^2 + r^2}{r^3}, -\frac{Xy}{r^3}, -\frac{Xz}{r^3} \right) dX, \qquad (6.27)$$

where $r^2 = X^2 + a^2$ since $x = 0$ and $y^2 + z^2 = a^2$ on the cross section.

Using arguments similar to the results in the perpendicular case, we obtain that the integrals for the second and third components of Eq. (6.27) are approximately zero. In order to calculate the first integral we use the fact that $X^2 = r^2 - a^2$ and thus have to evaluate

$$\mathbf{e}_x \cdot \mathbf{u}_\parallel^f = \frac{f_\parallel}{8\pi\mu} \int_{-b}^{c} \left(\frac{2r^2 - a^2}{r^3} \right) dX = \frac{f_\parallel}{8\pi\mu} \int_{-b}^{c} \left(\frac{2}{r} - \frac{a^2}{r^3} \right) dX. \qquad (6.28)$$

Using the already-computed integrals I_1 and I_3 we thus obtain the total velocity due to the stokeslets at the point $(0, y, z)$ on the cross section:

$$\mathbf{u}_\parallel^f(0, y, z) = \frac{f_\parallel}{4\pi\mu} \left(\ln\left(\frac{4bc}{a^2} \right) - 1, 0, 0 \right). \qquad (6.29)$$

As was expected by symmetry, the velocity is uniform across the section of the filament and no dipoles are required in the parallel case, so that the total flow is $\mathbf{u}_\parallel = \mathbf{u}_\parallel^f$.

Comparing the parallel configuration with Eq. (6.25) we note two differences, namely the factor of 2 in the overall magnitude and the change in sign from $+1$ to -1 in the constant next to the logarithm. Furthermore, while we assume in our calculations that a is constant, it is straightforward to see that the results for $\mathbf{u}_\perp$ and $\mathbf{u}_\parallel$ remain valid for a varying cross section, $a(s)$, where s is the arclength along the filament, provided that the typical length scale on which a varies is at least of order L (i.e. $da/ds \lesssim O(a/L)$).

6.2.5 Application to Prolate Spheroids

Before addressing the general curved-filament configuration, we examine the case of a straight geometry. In that case, ignoring end effects, the integration of uniform flow singularities in the locally straight region can be extended over the entire length of the filament. If the flow is evaluated at a particular arclength s, the boundaries of the integral spanning the filament start thus at $b = s$ and finish at $c = L - s$.

Examining Eqs. (6.25) and (6.29) we see that the analysis predicts a drag coefficient, i.e. a force per unit velocity, that depends logarithmically on the ratio $bc/(a^2)$. In the particular case of long prolate spheroids, this ratio turns out to be a constant along the shape. Indeed, consider a prolate spheroid of major axis $\bar{b}$ and minor axis $\bar{a}$. The local radius of the cross section at arclength $0 \leqslant s \leqslant L$ satisfies

$$\frac{a^2(s)}{\bar{a}^2} = \frac{s(L-s)}{\bar{b}^2}, \tag{6.30}$$

and therefore the ratio $bc/(a^2)$ is constant and equal to

$$\frac{b(s)c(s)}{a^2(s)} = \frac{s(L-s)}{a^2(s)} = \frac{\bar{b}^2}{\bar{a}^2}. \tag{6.31}$$

The force–velocity relationship in Eqs. (6.25) and (6.29) therefore leads to the prediction that the translation of prolate spheroids is characterised by two drag coefficients given by

$$c_\perp = \frac{4\pi\mu}{\ln(2\bar{b}/\bar{a}) + 1/2}, \tag{6.32}$$

$$c_\| = \frac{2\pi\mu}{\ln(2\bar{b}/\bar{a}) - 1/2}. \tag{6.33}$$

This asymptotic result, derived in the limit $\bar{b}/\bar{a} \gg 1$, is in fact precisely the leading-order approximation to the exact solution for prolate spheroids (Kim and Karrila, 1991); note that the next-order term in the expansion of the exact solution is of order $\bar{a}^2/(\bar{b}^2)$ and is consistent with our calculations above. A distribution of stokeslets and source dipoles therefore provides us with an excellent model to capture the hydrodynamics of straight filaments.

6.3 Local Hydrodynamics: Resistive-Force Theory (RFT)

The results in Eqs. (6.25) and (6.29) give the relationship between the velocity of the filament and the hydrodynamic force acting on it as a result of the integration of singularities along a locally straight portion. This can in turn be used to derive a local theory for the hydrodynamics of filaments. The first work along these lines was carried out by Cox (1970) and Batchelor (1970b) and was originally referred to as 'slender-body theory'. We now reserve that appellation for the nonlocal theory with algebraically small errors (see Section 6.4) and refer to the local theory with logarithmically small errors in a/L as 'resistive-force theory'.

In the framework of resistive-force theory the dominant hydrodynamic forces arise from the local analysis in Section 6.2. To quantify the limit in which this is a valid approximation, we have to estimate the magnitude of the nonlocal contribution. Let us consider the geometry in Fig. 6.1 and assume for simplicity that $b = c$. Since $b \gg a$, we can use a scaling argument to show that the nonlocal hydrodynamic contributions are dominated by the point forces, while the contributions from nonlocal dipoles may be safely neglected. Indeed, the condition $\kappa L \lesssim O(1)$ ensures that all points on the nonlocal portion of the filament are located at least a distance $\sim b$ away from the point of evaluation of the flow. Outside this region of size b, individual stokeslets of strength f decaying as $1/r$ lead to a flow velocity at most of order $f/(\mu b)$ in the local region. In contrast, source dipoles of strength g lead to velocities of order $g/(b^3)$. Since it was shown above that the dipole strength was related to the stokeslet strength with scaling $g \sim a^2 f/\mu$, we get dipolar flows on the order of $fa^2/(\mu b^3)$, which have magnitude $(a/b)^2 \ll 1$ smaller than the stokeslets flows and may thus be neglected.

In order to estimate the magnitude of the flow induced by the nonlocal stokeslets, u_{NL}, we use the fact that any point in the nonlocal portion of the filament is at least at a distance b away on either side of the arclength s where the velocity is being evaluated. By considering the integrals on each side of the arclength, we obtain that the total flow induced by the nonlocal stokeslets has a typical magnitude

$$u_{NL}(s) \sim \int_{r>b} \frac{f}{\mu r} \, ds' = O\left(\frac{f}{\mu} \ln\left(\frac{L-s}{b}\right)\right) + O\left(\frac{f}{\mu} \ln\left(\frac{s}{b}\right)\right). \qquad (6.34)$$

For the motion of the filament perpendicular to its tangent direction, we obtain that the total velocity of the cross section at position s is given by the sum of the local velocity, Eq. (6.25), and the nonlocal one, in Eq. (6.34), leading to

$$u_{\perp}(s) = \frac{f_{\perp}(s)}{4\pi\mu}\left(\ln\left(\frac{2b}{a}\right) + \frac{1}{2}\right) + O\left(\frac{f}{\mu} \ln\left(\frac{L-s}{b}\right)\right) + O\left(\frac{f}{\mu} \ln\left(\frac{s}{b}\right)\right). \qquad (6.35)$$

To evaluate this relationship in the slender limit, we rearrange this formula to have the filament aspect ratio, a/L, appear explicitly. This leads to the equation

$$u_{\perp}(s) = \frac{f_{\perp}(s)}{4\pi\mu} \ln\left(\frac{L}{a}\right) + O\left(\frac{f}{\mu} \ln\left(\frac{L-s}{b}\right)\right) + O\left(\frac{f}{\mu} \ln\left(\frac{s}{b}\right)\right)$$
$$+ O\left(\frac{f}{\mu}\right) + O\left(\frac{f}{\mu} \ln\left(\frac{L}{b}\right)\right). \qquad (6.36)$$

In the limit where the filament is asymptotically slender, $L/a \to \infty$, the magnitude of the first term increases logarithmically while all the other terms remain of order one, and therefore at leading order in $\ln(L/a)$ the perpendicular force on the fluid is given by

$$f_{\perp}(s) \approx c_{\perp} u_{\perp}(s), \quad c_{\perp} = \frac{4\pi\mu}{\ln(L/a)}. \qquad (6.37)$$

For translation of the filament in the direction parallel to its tangent, a similar calculation using Eqs. (6.29) and (6.34) leads to

$$u_{\parallel}(s) = \frac{f_{\parallel}(s)}{2\pi\mu} \ln\left(\frac{L}{a}\right) + O\left(\frac{f}{\mu} \ln\left(\frac{L-s}{b}\right)\right) + O\left(\frac{f}{\mu} \ln\left(\frac{s}{b}\right)\right)$$
$$+ O\left(\frac{f}{\mu}\right) + O\left(\frac{f}{\mu} \ln\left(\frac{L}{b}\right)\right), \tag{6.38}$$

which in the slender limit gives a leading-order relationship between the parallel force on the fluid and the velocity of the filament as

$$f_{\parallel}(s) \approx c_{\parallel}u_{\parallel}(s), \quad c_{\parallel} = \frac{2\pi\mu}{\ln(L/a)}. \tag{6.39}$$

The results in Eqs. (6.37) and (6.39) provide the leading-order magnitude of each component of the force density acting on the fluid. The drag force acting on the filament is instantaneously equal and opposite to the force on the fluid. As a consequence, the relationship between filament velocity, $\mathbf{u}$, and drag per unit length, $\mathbf{f}$, can be written formally as

$$\mathbf{f}(s,t) = -\left[c_{\parallel}\mathbf{tt} + c_{\perp}(\mathbf{1} - \mathbf{tt})\right] \cdot \mathbf{u}(s,t), \tag{6.40}$$

where $\mathbf{t} \equiv \mathbf{t}(s,t)$ is the instantaneous tangent to the slender filament at position s, and where the tensors $\mathbf{tt}$ and $(\mathbf{1} - \mathbf{tt})$ allow pointwise projections of the velocity in the directions parallel and perpendicular to the filament, respectively. Note that this is the result we anticipated in Eq. (5.7). Importantly, the leading-order relationship predicted by resistive-force theory is local, shape-dependent (through the variation of $\mathbf{t}$ along the shape) and anisotropic with $c_{\perp}/c_{\parallel} = 2$. This factor of 2 originates from the anisotropy of the Green's function itself in Eq. (6.3). While Eq. (6.40) assumes that there is no background flow, it is straightforward to incorporate it and in that case $\mathbf{u}$ measures the velocity of the body centreline relative to the velocity of the background fluid (see Chapter 10).

It is often said that resistive-force theory is logarithmically correct. This may be seen by inspecting Eqs. (6.36) and (6.38). Since the next term in the velocity is of order one, the relative errors are $O(1/\ln(L/a))$. Alternatively, and more rigorously, the results of resistive-force theory may be derived formally as the solution to a matched-asymptotics mathematical problem for the flow around filaments with circular (Cox, 1970) and non-circular cross sections (Batchelor, 1970b). Under that framework the inner solution, valid on the length scale a, is that due to a translating cylinder. This is characterised by a logarithmic increase in the flow magnitude far from it (a generic feature of two-dimensional solutions of Stokes flows with a net force). In the outer solution, relevant to the length scale L, the filament is well approximated by a thin line whose flow can be captured by a distribution of hydrodynamic stokeslets, also associated with a logarithmic increase when approaching the line. Matching between the inner and outer problems allows us to compute hydrodynamic forces in powers of $[1/\ln(L/a)]^n$. At leading order in $1/\ln(L/a)$, the result is resistive-force theory, i.e. Eqs. (6.37) and (6.39), while the next-order term occurs at order $[1/\ln(L/a)]^2$ and involves nonlocal hydrodynamic interactions between different points along the filament.

As a final remark, we note that some work has gone into attempting to improve on the accuracy of resistive-force theory without losing its analytical simplicity, an idea that Lighthill systematically referred to as 'suboptimal' (Lighthill, 1976). In his work analysing the hydrodynamics of slender filaments, Lighthill proposed that the periodicity of the shape of waving flagella could be included in the model by requiring the force distribution to be spatially periodic with wavenumber k, i.e. $f \propto \cos(kz)$, where z is the direction along the periodic flagellum of wavelength $\lambda = 2\pi/k$ (Lighthill, 1975). Applying this assumption to the derivations in Section 6.2 changes the value of the integral I_3 in Eq. (6.9), which now becomes a convergent integral

$$I_3 = \int_{-b}^{c} \frac{\cos(kZ)}{r} \, dZ \approx 2 \int_{0}^{\infty} \frac{\cos(kat)}{\sqrt{1+t^2}} \, dt = 2K_0(ka), \qquad (6.41)$$

where K_0 is a modified Bessel function of the second kind. In the long-wavelength limit $ka \ll 1$, we use the asymptotic result $K_0(x) \sim \ln(2/x) - \gamma$, where γ is Euler's constant (≈ 0.577), and thus obtain

$$I_3 \approx 2\ln\left(\frac{2}{kae^\gamma}\right) \approx 2\ln\left(\frac{\Delta}{a}\right), \qquad (6.42)$$

with $\Delta = \lambda/(\pi e^\gamma) \approx 0.18\lambda$. For a periodic flagellum, Lighthill therefore concluded that the total length, L, appearing in the resistance coefficients, Eqs. (6.37) and (6.39), should be replaced by the length scale, Δ, itself only a small fraction of the wavelength.

6.4 Nonlocal Hydrodynamics: Slender-Body Theory

The integration of flow singularities in the straight portion of a filament allows us to derive, in the slender limit, the local and logarithmically accurate resistive-force theory. By accounting for the flows arising from the nonlocal portion of the filament, an improved theory may be obtained. This 'slender-body theory' turns out to be algebraically accurate, hence an important improvement over resistive-force theory, but it is nonlocal and involves integrals that usually can only be evaluated numerically.

Different versions of slender-body theory exist. We first outline in Section 6.4.1 the one due to Lighthill (1976), which was derived by exploiting physical arguments and the analogy with the flow past a sphere. As later shown by Childress

(1981), the flow resulting from Lighthill's integral formulation is accurate to order $\epsilon^{1/2}$, with $\epsilon = a/L$ being the aspect ratio of the filament. To get increased precision in the modelling approach, other formulations have been proposed, including the ones from Keller and Rubinow (1976) and from Johnson (1980), which were derived using matched asymptotic expansions. The latter is the most accurate, with relative errors for the force distribution of order $\epsilon^2 \ln \epsilon$, and we quote its result in Section 6.4.2.

6.4.1 Lighthill's Slender-Body Theory

Consider a point on the cross section of the slender filament. Following the approach outlined above, we may write the velocity of the filament there as the sum of the velocity induced by the local singularities, $\mathbf{u}_L$, and the velocity due to the hydrodynamic singularities in the nonlocal portion of the filament, $\mathbf{u}_{NL}$.

Rewriting the Local Contribution

Since the local analysis has shown that a combination of stokeslets and source dipoles leads to uniform flows over the cross section, we may identify the velocity of the centreline with this uniform velocity. We may then manipulate the results in Eqs. (6.25) and (6.29) to rewrite this local centreline velocity as a non-divergent line integral of stokeslets only. Consider first the parallel flow component from Eq. (6.29) and take $c = b$. Using elementary properties of the logarithm, the velocity may be rewritten as

$$u_{L,\|}(s) = \frac{f_\|}{4\pi\mu}\left[2\ln\left(\frac{2b}{a}\right) - 1\right] \equiv \frac{f_\|}{8\pi\mu}\left[4\ln\left(\frac{b}{\delta}\right)\right], \tag{6.43}$$

where $\delta = \frac{1}{2}a\sqrt{e} \approx 0.82a$. We then recognise this to be formally equal to

$$u_{L,\|}(s) = \frac{1}{8\pi\mu}\int_{\delta \leqslant |s-s'| \leqslant b}\frac{2f_\|}{r}\,ds', \tag{6.44}$$

where $r = |s - s'|$. This is an integral of stokeslets parallel to the local configuration of the filament in the two regions where $\delta \leqslant r \leqslant b$ (one region on either side of the cross section; see illustration in Fig. 6.2).

Similarly, for the perpendicular case, the flow component in Eq. (6.25) with the choice $c = b$ can be manipulated to be written as

$$u_{L,\perp}(s) = \frac{f_\perp}{8\pi\mu}\left[2\ln\left(\frac{2b}{a}\right) + 1\right] = \frac{f_\perp}{4\pi\mu} + \frac{f_\perp}{8\pi\mu}\left[2\ln\left(\frac{b}{\delta}\right)\right]. \tag{6.45}$$

This can therefore be rewritten as

$$u_{L,\perp}(s) = \frac{f_\perp}{4\pi\mu} + \frac{1}{8\pi\mu}\int_{\delta \leqslant |s-s'| \leqslant b}\frac{f_\perp}{r}\,ds'. \tag{6.46}$$

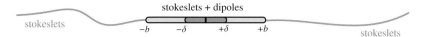

stokeslets + dipoles

stokeslets

$-b$ $-\delta$ $+\delta$ $+b$

stokeslets

Figure 6.2 Mathematical construction to compute the flow induced by slender filaments. The flow induced at the cross section of the filament (small vertical line) is due to a superposition of flow singularities: stokeslets and source dipoles within a distance b from the evaluation point (local part) and solely stokeslets further away (nonlocal part). The integration of both stokeslets and dipoles in the local region is equivalent to the integration of only stokeslets located from δ to b away from the evaluation point.

In this case, the second term is an integral of stokeslets perpendicular to the local orientation of the filament at distances $\delta \leqslant r \leqslant b$ away from the cross section.

Formally, the components of the local flow in Eqs. (6.44) and (6.46) may be combined to write the velocity in vector form. If $\mathbf{r}_0(s)$ denotes the location of the centreline of filament at arclength s and if we use the vector $\mathbf{r}(s, s')$ to connect the points at s' and s, i.e. $\mathbf{r}(s, s') = \mathbf{r}_0(s) - \mathbf{r}_0(s')$ with $r = |\mathbf{r}|$, then

$$\mathbf{u}_{\mathrm{L}}(s) = \frac{1}{4\pi\mu}(\mathbf{1} - \mathbf{tt}) \cdot \mathbf{f}(s) + \frac{1}{8\pi\mu} \int_{\delta \leqslant |s-s'| \leqslant b} \left(\frac{\mathbf{1}}{r} + \frac{\mathbf{rr}}{r^3} \right) \cdot \mathbf{f}(s')\, ds', \quad (6.47)$$

where we have related the perpendicular force component, $\mathbf{f}_\perp$, to the total force density by projecting it as $\mathbf{f}_\perp = (\mathbf{1} - \mathbf{tt}) \cdot \mathbf{f}$, where $\mathbf{t}(s)$ is the local unit tangent vector. The result in Eq. (6.47) shows that the local velocity is the sum of a local term plus a non-singular integral of stokeslets along the centreline. In essence, the presence of source dipoles has removed the singularity from the distribution of pure stokeslets.

Nonlocal Contribution

In the nonlocal region (i.e. further away than a distance b from the cross section at s), we saw in Section 6.3 that the flow contribution from the source dipoles is of order $(a/b)^2$ smaller than that due to the stokeslets and can be neglected. As a consequence, the total nonlocal flow is well approximated by the stokeslet integral

$$\mathbf{u}_{\mathrm{NL}} = \frac{1}{8\pi\mu} \int_{b \leqslant |s-s'|} \left(\frac{\mathbf{1}}{r} + \frac{\mathbf{rr}}{r^3} \right) \cdot \mathbf{f}(s')\, ds'. \quad (6.48)$$

Total Filament Velocity

Strikingly, Eqs. (6.47) and (6.48) involve the same hydrodynamic kernel evaluated on two complementary portions of the filament. They may then be combined to give the total flow, $\mathbf{u} = \mathbf{u}_{\mathrm{L}} + \mathbf{u}_{\mathrm{NL}}$ as

$$\mathbf{u}(s) = \frac{1}{4\pi\mu}(\mathbf{1} - \mathbf{tt}) \cdot \mathbf{f}(s) + \frac{1}{8\pi\mu} \int_{\delta \leqslant r} \left(\frac{\mathbf{1}}{r} + \frac{\mathbf{rr}}{r^3} \right) \cdot \mathbf{f}(s') \, \mathrm{d}s', \qquad (6.49)$$

which is notably independent of the particular value of b chosen.

Discussion

Lighthill's slender-body theory is the final result in Eq. (6.49). It provides a linear, Fredholm-type integral relationship (of the second kind) between the distribution of hydrodynamic force acting on the fluid per unit length, $\mathbf{f}$, and the local velocity of points along the filament, $\mathbf{u}$. Note that the first term in Eq. (6.49) only involves the local perpendicular component of the force because this is the only component that requires force dipoles.

A visual summary of the modelling approach leading to Eq. (6.49) is offered in Fig. 6.2. In the local region of size b on either side of the point of interest (shown by the vertical line), the flow is due to both stokeslets and source dipoles, the total combination of which is equivalent to a local forcing term plus a line distribution of stokeslets on two regions of sizes $b - \delta$. The addition of the nonlocal contributions from stokeslets at a distance b away from the point of interest leads to the velocity on the centreline of the filament as a sum of a local term plus a non-singular centreline distribution of point forces.

For problems in which the distribution of forces is known, e.g. in the case where $\mathbf{f}$ balances a gravitational force, the formula may be evaluated directly to obtain $\mathbf{u}$. In many biophysical problems involving flagellar locomotion, however, the problem of interest is the inverse, whereby the velocity of a flagellum is known from experimental measurements and Eq. (6.49) needs to be inverted to solve for the value of $\mathbf{f}$. In most cases, both the direct and inverse uses of Eq. (6.49) require the use of numerical integration.

6.4.2 Johnson's Slender-Body Theory

Johnson considered a filament of cross-sectional radius $a(s)$ and length L and derived the relationship between the velocity of one point along the filament, $\mathbf{u}(s)$, and the distribution of point forces of magnitude $\mathbf{f}(s')$ using a matched asymptotic expansion. Beyond the stokeslet and source dipole singularities used by Lighthill, Johnson also included point sources and higher-order force singularities (dipoles and quadrupoles). He further assumed that the ends of the body were approximately elliptic, leading to the integral relationship for the velocity at arclength $-L/2 \leqslant s \leqslant L/2$,

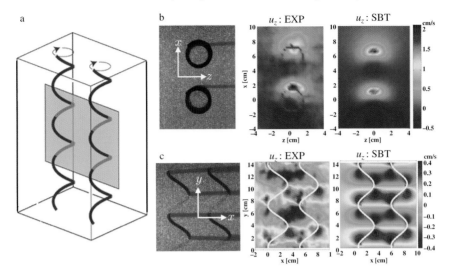

Figure 6.3 Comparison between the predictions of Lighthill's slender-body theory and experimental flow measurements on a macroscopic model of rotating flagellar filaments (Kim et al., 2004). (a) Two rigid helices rotate in a viscous fluid and the flow is measured using particle-image velocimetry. (b) Experimental measurements and theoretical predictions for the flow in the plane perpendicular to the axis of the helices. (c) Flow in the plane of the axis. Adapted by permission from Kim et al. (2004).

$$8\pi\mu\mathbf{u}(s) = \ln\left(\frac{L^2 - 4s^2}{a^2(s)}\right)(\mathbf{1} + \mathbf{tt}) \cdot \mathbf{f}(s) + [\mathbf{1} - 3\mathbf{tt}] \cdot \mathbf{f}(s)$$

$$+ \int_{-L/2}^{L/2}\left[\left(\frac{1}{r} + \frac{\mathbf{rr}}{r^3}\right) \cdot \mathbf{f}(s') - \frac{1 + \mathbf{tt}}{|s' - s|} \cdot \mathbf{f}(s)\right]ds', \qquad (6.50)$$

where we recall that $\mathbf{t}$ is the unit tangent at arclength s along the filament and where, as above, the vector $\mathbf{r}(s, s')$ connects the points at s' and s, i.e. $\mathbf{r}(s, s') = \mathbf{r}_0(s) - \mathbf{r}_0(s')$ with $r = |\mathbf{r}|$. Note that the integrand in the second term appears formally to be singular but is in fact not because the two $1/r$-diverging terms cancel each other out in the limit $s' \to s$. Note also that the logarithmic term in Eq. (6.50) does not depend on s for a body with an elliptic cross section (prolate spheroid).

Beyond its increased accuracy, another advantage of Johnson's formulation is that the integral operator in Eq. (6.50) may be transformed into one that has known eigenfunctions in the form of Legendre polynomials (Götz, 2000). Johnson's slender-body theory is thus naturally amenable to a numerical solution using a Galerkin method on a basis of Legendre polynomials.

6.4.3 Comparison with Experiments: Rotated Helices

In order to demonstrate the accuracy of slender-body theory in reproducing the flow induced by slender filaments, we turn to experiments on macroscopic models of bacterial flagellar filaments (Kim et al., 2004). The setup is illustrated in Fig. 6.3(a). Two rigid metallic helices parallel to each other are rotated in a high-viscosity fluid (silicon oil) and particle-image velocimetry is used to measure the flow field in the fluid. A numerical implementation of Lighthill's slender-body theory, following Higdon (1979a), is used to compute the strength of the flow singularities along the centreline of each helix. The fluid flow is then obtained as a linear superposition of the flow fields created by each singularity (Kim and Powers, 2004). The comparison between the experimental measurements and the simulations is displayed in Fig. 6.3 for flow in the plane perpendicular to the axis of the helices (Fig. 6.3(b), u_z component only) and flow in the plane that includes both axes (Fig. 6.3(c), u_x component only). In each case the same colour code is used to quantify the magnitude of the velocity components, measured in cm/s. Despite the inevitable experimental noise, the predictions of slender-body theory for the spatial distribution and magnitude of the flow are in good agreement with the measurements.

Further Reading

In contrast to the method used in this chapter where the nature of the two centreline singularities was postulated, slender-body theory can be derived directly from the boundary integral formulation of Stokes flows (see Exercises) using matched asymptotic expansions in the slender limit (Koens and Lauga, 2018).

A detailed comparison between the predictions of resistive-force theory and those of slender-body theory was carried by Johnson and Brokaw (1979) and Rodenborn et al. (2013). While the equations of slender-body theory in general have to be inverted numerically (Tornberg and Shelley, 2004), the force distribution is constant in the special case where the filament shape is helical and infinite, and some further analytical simplifications may be exploited (Lighthill, 1976, 1996a,b).

Analogous versions of resistive-force theory were derived to tackle the electrophoresis of elongated charged particles (Solomentsev and Anderson, 1994) and the locomotion of slender phoretic swimmers (Schnitzer and Yariv, 2015). Both resistive-force theory and slender-body theory have also been extended to tackle the hydrodynamics of slender ribbons characterised by three well-separated length scales (Koens and Lauga, 2016, 2017).

Exercises

1. The Stokes flow, $\mathbf{u}_F$, due to a point force, $\mathbf{F}$, located at the origin satisfies

$$0 = -\nabla p_F + \mu \nabla^2 \mathbf{u}_F + \delta(\mathbf{r})\mathbf{F}, \quad \nabla \cdot \mathbf{u}_F = 0.$$

Show that the velocity and pressure fields are given by the stokeslet solution,

$$\mathbf{u}_F(\mathbf{r}; \mathbf{F}) = \frac{1}{8\pi\mu} \left(\frac{\mathbf{1}}{r} + \frac{\mathbf{rr}}{r^3} \right) \cdot \mathbf{F}, \quad p_F(\mathbf{r}; \mathbf{F}) = \frac{\mathbf{F} \cdot \mathbf{r}}{4\pi r^3},$$

and that the stress field is

$$\sigma_F(\mathbf{r}; \mathbf{F}) = -\frac{3}{4\pi} \frac{\mathbf{rrr}}{r^5} \cdot \mathbf{F}.$$

2. Show that a source dipole singularity,

$$\mathbf{u}_{SD}(\mathbf{r}; \mathbf{M}) = \frac{1}{4\pi} \left(\frac{\mathbf{1}}{r^3} - \frac{3\mathbf{rr}}{r^5} \right) \cdot \mathbf{M},$$

is proportional to the Laplacian of a stokeslet from Question 1.

3. Two incompressible Stokes flows, $\mathbf{u}$ and $\hat{\mathbf{u}}$, satisfy the Stokes equations subject to body force densities, $\mathbf{f}$ and $\hat{\mathbf{f}}$,

$$0 = \nabla \cdot \sigma + \mathbf{f}, \quad \nabla \cdot \mathbf{u} = 0,$$
$$0 = \nabla \cdot \hat{\sigma} + \hat{\mathbf{f}}, \quad \nabla \cdot \hat{\mathbf{u}} = 0.$$

For a flow in a fluid volume V bounded by a closed surface S with unit normal $\mathbf{n}$ pointing into the fluid, derive the reciprocal theorem

$$\iint_S \mathbf{u} \cdot \hat{\sigma} \cdot \mathbf{n} \, dS - \iint_S \hat{\mathbf{u}} \cdot \sigma \cdot \mathbf{n} \, dS = \iiint_V \mathbf{u} \cdot \hat{\mathbf{f}} \, dV - \iiint_V \hat{\mathbf{u}} \cdot \mathbf{f} \, dV.$$

Deduce the result in Eq. (2.27) in the absence of body forces in the fluid.

Apply the reciprocal theorem to the stokeslet solution from Exercise 1 above (with velocity and stress fields written formally as $\mathbf{u}_F(\mathbf{r}; \mathbf{F}) \equiv \mathbf{G}(\mathbf{r}) \cdot \mathbf{F}$ and $\sigma_F(\mathbf{r}; \mathbf{F}) \equiv \Sigma(\mathbf{r}) \cdot \mathbf{F}$) and a second arbitrary, unforced flow solution. Hence, demonstrate that the velocity in the bulk of an arbitrary Stokes flow can be written instantaneously as the sum of surface integrals of velocities and tractions:

$$\mathbf{u}(\mathbf{r}) = -\iint_S \mathbf{u}(\mathbf{x}) \cdot \Sigma(\mathbf{r} - \mathbf{x}) \cdot \mathbf{n}(\mathbf{x}) \, dS_{\mathbf{x}} - \iint_S \mathbf{G}(\mathbf{r} - \mathbf{x}) \cdot \sigma(\mathbf{x}) \cdot \mathbf{n}(\mathbf{x}) \, dS_{\mathbf{x}},$$

where $dS_{\mathbf{x}}$ is the surface element along $\mathbf{x}$. This is called the boundary integral representation of Stokes flows.

4. A regularised stokeslet is a solution to the Stokes equations with a forcing that, instead of being a singular Dirac delta as in Exercise 1, is distributed over a smooth function $\phi(r)$ with spherical symmetry:

$$\mathbf{0} = -\nabla p + \mu \nabla^2 \mathbf{u} + \phi(r)\mathbf{F}, \quad \nabla \cdot \mathbf{u} = 0.$$

Calculate the velocity and pressure field of a regularised stokeslet for the particular choice of regularisation function

$$\phi(r) = \frac{15\epsilon^4}{8\pi(r^2 + \epsilon^2)^{7/2}}.$$

5. The centreline of a slender filament of cross-sectional radius a takes the shape of a circular ring of radius $b \gg a$. The filament moves at constant speed U along the axis of the ring in a viscous fluid at low Reynolds number. Use Lighthill's nonlocal slender-body theory, Eq. (6.49), to compute the total hydrodynamic drag force on the filament. Show that the same result is predicted by Johnson's slender-body theory, Eq. (6.50).

6. Resistive-force and slender-body theory quantify the distribution of forces arising from local translations of the centreline of a slender filament. However, neither includes the hydrodynamic torques generated by local rotation of slender filaments around their centreline. This can be done by adding a term of the form $-\zeta_r(\omega \cdot \mathbf{t})\mathbf{t}$ to the hydrodynamic torque density acting on the slender filament. To compute the value of ζ_r, solve for the flow around a cylinder of radius a rotating with angular speed ω in a two-dimensional Stokes flow. Show that the hydrodynamic torque per unit length is given by $-4\pi a^2 \mu \omega$ and therefore $\zeta_r = 4\pi a^2 \mu$.

7

Waving of Eukaryotic Flagella

Building on biological observations, the previous chapters reviewed the physics of cellular propulsion and introduced a framework for computing the hydrodynamic forces on slender flagella. In the next two chapters, we apply these tools to address the two fundamental modes of locomotion of flagellated cells. Motivated by the propulsion of spermatozoa, the current chapter focuses on the planar waving of eukaryotic flagella. We first show how to apply resistive-force theory to compute the swimming speeds of simple flagellar waves and detail how wave geometry influences locomotion. Next we introduce a measure of swimming efficiency and show how to use it in order to derive the shape of the optimal waving motion. Finally we model the eukaryotic flagellum as an active filament where actuation from molecular motors is accounted for in a continuum manner and the waving motion is obtained as a mechanical balance between molecular forcing, fluid dynamics and passive elasticity. We close with a demonstration of how that model allows us to quantify cellular energy consumption.

7.1 Swimming of a Periodically Waving Flagellum

7.1.1 Setup

In this section we consider an idealised model for the swimming of a spermatozoon equipped with a single flagellum, as illustrated in Fig. 7.1. We assume that the centreline of the flagellum remains in the (x, y) plane, is infinite and moves in the shape of a travelling wave of frequency ω. The wave has wavelength $\lambda = 2\pi/k$ as measured along the x axis and $\Lambda \geqslant \lambda$ as measured along the arclength of the swimmer. The travelling wave propagates with speed $V \equiv \omega/k$ in the x direction. The shape of the wave is assumed to be $y \leftrightarrow -y$ symmetric, so that swimming occurs solely along x. We denote the unknown swimming velocity by $-U\mathbf{e}_x$ so

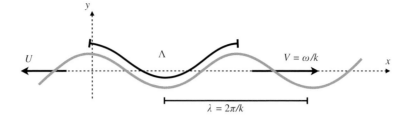

Figure 7.1 Setup for the swimming of a waving planar flagellum. An infinite planar flexible flagellum deforms as a travelling wave of frequency ω, wavenumber
$k = 2\pi/\lambda$ and wave speed $V = \omega/k$. The wavelength measured along the centreline of the flagellum is denoted Λ $(\geqslant \lambda)$. Swimming occurs with magnitude U in
the direction opposite to that of wave propagation.

that a positive value of U indicates swimming in the direction opposite to that of
wave propagation.

In order to compute the value of U analytically, we employ resistive-force
theory. The hydrodynamic force density acting on the flagellum, $\mathbf{f}$, is given by
Eq. (6.40). We need to evaluate the velocity of the flagellum relative to the background fluid, $\mathbf{u}$, i.e. we need to characterise the waving kinematics. The value of U
is then determined by the requirement for the cell to be force-free at all times. In
the absence of a cell body, for an infinitely periodic flagellum this is captured by
the statement that

$$\int_0^\Lambda \mathbf{f} \cdot \mathbf{e}_x \, ds = 0, \tag{7.1}$$

and the other force components are zero by symmetry. In the presence of a cell
body, the net hydrodynamic force acting on the flagellum would instead balance
drag on the cell body (see Section 7.1.4).

7.1.2 Wave Kinematics

The first step in the calculation consists of determining the kinematics of the flagellum. Following Lighthill (1976), the simplest way to carry out this calculation uses
two changes of frame. In the lab frame, the velocity at infinity is zero, while the
wave appears to propagate at speed $V - U$ due to the swimming of the flagellum
with velocity $-U\mathbf{e}_x$ (Fig. 7.2(a)). Jumping into the swimming frame, the fluid at
infinity now moves with velocity $U\mathbf{e}_x$ while the wave propagates with velocity $V\mathbf{e}_x$
(Fig. 7.2(b)). We next move into the frame travelling with the wave and in which
the fluid at infinity has velocity $(U - V)\mathbf{e}_x$ (Fig. 7.2(c)). Notably, in this frame

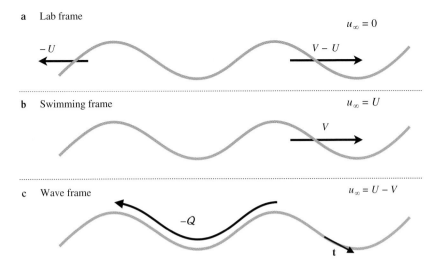

a   Lab frame

$u_\infty = 0$

$-U$ $V - U$

b   Swimming frame

$u_\infty = U$

V

c   Wave frame

$u_\infty = U - V$

$-Q$

t

Figure 7.2 Kinematics of a planar swimmer deforming as a travelling wave, obtained using three different frames of reference. (a) Lab frame in which the swimmer moves with velocity $-U\mathbf{e}_x$ and the wave propagates at speed $(V - U)$ in a quiescent fluid. (b) Swimming frame in which the wave propagates with speed V and the fluid at infinity moves with velocity $U\mathbf{e}_x$. (c) Wave frame in which the fluid at infinity moves with velocity $(U - V)\mathbf{e}_x$ while material points on the flagellum move tangentially with velocity $-Q\mathbf{t}$, where $\mathbf{t}$ is the unit tangent.

the wave appears 'frozen' and therefore material points can only move tangentially along the flagellum. At any point along the filament, the velocity in the wave frame is thus given by $-Q\mathbf{t}$, where Q is constant for an inextensible flagellum. Jumping back into the lab frame, we thus obtain that the velocity of any point on the flagellum is given by

$$\mathbf{u} = (V - U)\mathbf{e}_x - Q\mathbf{t}. \tag{7.2}$$

To determine the value of Q, we use the periodicity of the wave. The period of the wave is given by Λ/Q when measured in the wave frame and λ/V in the swimming frame. These two must be equal and thus we have $Q = \Lambda V/\lambda$, so that the velocity of any point on the flagellum, Eq. (7.2), is ultimately given by

$$\mathbf{u} = (V - U)\mathbf{e}_x - V\frac{\Lambda}{\lambda}\mathbf{t}. \tag{7.3}$$

Note that, for a given waveform and wave speed, all terms in Eq. (7.3) are known except for the value of U.

7.1.3 Force-Free Swimming

The hydrodynamic force density, $\mathbf{f}$, predicted from resistive-force theory, Eq. (6.40), is given by

$$-\mathbf{f} = c_\perp \mathbf{u} + (c_\| - c_\perp)(\mathbf{u} \cdot \mathbf{t})\mathbf{t}, \qquad (7.4)$$

which can be evaluated using the kinematics derived in Eq. (7.3). Using

$$(\mathbf{u} \cdot \mathbf{t})\mathbf{t} = (V - U)(\mathbf{e}_x \cdot \mathbf{t})\mathbf{t} - Q\mathbf{t}, \qquad (7.5a)$$

$$\mathbf{u} \cdot \mathbf{e}_x = (V - U) - Q(\mathbf{t} \cdot \mathbf{e}_x), \qquad (7.5b)$$

the x component of the force density acting on the fluid becomes

$$-\mathbf{f} \cdot \mathbf{e}_x = c_\perp(V - U) - c_\| Q(\mathbf{t} \cdot \mathbf{e}_x) + (c_\| - c_\perp)(V - U)(\mathbf{e}_x \cdot \mathbf{t})^2. \qquad (7.6)$$

We now have to integrate Eq. (7.6) over one wavelength in order to enforce the free-swimming requirement, Eq. (7.1), which leads to

$$c_\perp(V - U)\Lambda - c_\| Q \int_0^\Lambda (\mathbf{t} \cdot \mathbf{e}_x)\, ds + (c_\| - c_\perp)(V - U) \int_0^\Lambda (\mathbf{e}_x \cdot \mathbf{t})^2\, ds = 0. \qquad (7.7)$$

In order to evaluate the two integrals in Eq. (7.7), we consider the geometry of a wave whose shape is described by $\mathbf{r}(s) = (x(s), y(s))$. Since by definition $\mathbf{t} = d\mathbf{r}/ds$, the first integral is given by

$$\int_0^\Lambda (\mathbf{t} \cdot \mathbf{e}_x)\, ds = \int_0^\Lambda \frac{dx}{ds}\, ds = x(\Lambda) - x(0) = \lambda. \qquad (7.8)$$

The second integral can not be evaluated analytically but may be written as

$$\int_0^\Lambda (\mathbf{e}_x \cdot \mathbf{t})^2\, ds \equiv \beta\Lambda, \qquad (7.9)$$

where the coefficient β is a function of the wave geometry and, because $\mathbf{t}$ is a unit vector, satisfies $\beta < 1$.

Using Eqs. (7.8)–(7.9), the force balance in Eq. (7.7) can now be written as

$$c_\perp(V - U)\Lambda - c_\| Q\lambda + (c_\| - c_\perp)(V - U)\beta\Lambda = 0. \qquad (7.10)$$

We then use the fact that $Q\lambda = V\Lambda$ to simplify the above as

$$[c_\perp + (c_\| - c_\perp)\beta](V - U) = c_\| V, \qquad (7.11)$$

and finally

$$\frac{U}{V} = \frac{(c_\perp - c_\|)(1 - \beta)}{c_\perp + (c_\| - c_\perp)\beta}, \qquad (7.12)$$

which constitutes an explicit solution for U in terms of β and the two drag coefficients.

7.1.4 Discussion

Introducing the ratio of drag coefficients, $\rho = c_{\|}/c_{\perp}$, the result for the swimming speed, Eq. (7.12), may be rewritten as

$$\frac{U}{V} = \frac{(1-\rho)(1-\beta)}{1+(\rho-1)\beta}.$$

(7.13)

For a travelling wave, we could have predicted *a priori* that the swimming speed would depend on the parameters:

$$U = F(c_{\perp}, c_{\|}, \omega, \lambda, \text{shape}).$$

(7.14)

Using dimensional analysis, this may be recast as

$$\frac{U}{V} = F(\rho, \text{shape}),$$

(7.15)

which agrees with Eq. (7.13) and shows that the only dependence on the shape is through the value of β.

If $\rho = 1$ (i.e. if the drag coefficients are equal, $c_{\perp} = c_{\|}$) then our theory predicts that $U = 0$. Anisotropic drag is therefore crucial for swimming, confirming the result discussed intuitively in Section 5.3.

Since $\rho < 1$ for a smooth flagellum and $\beta < 1$ by geometry, the denominator in Eq. (7.13), equal to $1 - \beta + \rho\beta$, is always positive. Consequently, $U/V > 0$ and swimming always occurs in the direction opposite to the direction of wave propagation, as is observed experimentally. Non-smooth flagella equipped with protrusions termed mastigonemes may instead have $\rho > 1$, reversing the direction of swimming, which occurs now in the same direction as that of the wave (see Chapter 5, Exercises).

For a cell with a smooth flagellum, the ratio U/V is always less than 1, typically on the order of 10–30%. In the lab frame, one then observes swimming in one direction while the wave appears to propagate at speed $V - U > 0$ in the opposite direction.

While the result in Eq. (7.13) assumes that the flagellum itself is force-free, it may be modified to account for the presence of a cell body. In this case, the total force acting on a flagellum of linear length L has to be balanced by the drag force on the body of the cell. Using the results from Section 5.2.1, the force acting on the cell body of typical size a can be written as $ca\mu U \mathbf{e}_x$, where c is a dimensionless constant whose value depends solely on the cell shape. Introducing for simplicity the notation $ca\mu U \equiv c_{\perp} L U \delta$, balancing the drag from the cell body with that acting on the flagellum in order to have zero total force leads to the new value for the swimming speed,

$$\frac{U}{V} = \frac{(1-\rho)(1-\beta)}{1+(\rho-1)\beta+\delta};$$ (7.16)

see Exercises. The presence of the head is seen to always slow down the organism $(dU/d\delta < 0)$, a result consistent with the relatively small sizes of the cell bodies of spermatozoa, compared for example with swimming bacteria (Brennen and Winet, 1977).

The only dependence of the swimming speed on the shape of the travelling wave is through the parameter β. Its value is calculated easily using Cartesian coordinates and assuming that the waveform can be characterised by a single-valued periodic function $y = f(x)$. In that case we have $ds = \sqrt{1+f'^2}\,dx$ along a small line element of the flagellum, so that the wavelength Λ is given by

$$\Lambda = \int_0^\lambda \sqrt{1+f'^2}\,dx,$$ (7.17)

and the β-dependence of Eq. (7.16) may be expressed as

$$\beta = \frac{1}{\Lambda}\int_0^\lambda \frac{1}{\sqrt{1+f'^2}}\,dx, \quad 1-\beta = \frac{1}{\Lambda}\int_0^\lambda \frac{f'^2}{\sqrt{1+f'^2}}\,dx.$$ (7.18)

The formulae of Eqs. (7.17)–(7.18), together with the result of Eq. (7.13), allow us to recover the seminal Cartesian calculations of Gray and Hancock (1955), who showed that the theoretical prediction using sinusoidal waves, $f(x) = A\sin(kx)$, gives remarkable quantitative agreement with the measured locomotion speeds of sea-urchin spermatozoa.

In the long-wavelength limit, our prediction for the swimming speed can be compared directly to the one derived for the small-amplitude waving sheet in Chapter 3. The long-wavelength assumption can be expressed as $|f'| \ll 1$, and in that limit we have $\lambda/\Lambda = 1+O(f'^2)$ so that

$$\beta \approx 1, \quad 1-\beta \approx \frac{1}{\lambda}\int_0^\lambda f'^2\,dx.$$ (7.19)

Comparing with the sinusoidal wave from Chapter 3, we consider a waveform given by $f(x) = A\sin(kx)$ with $Ak \ll 1$ and then get

$$1-\beta \approx \frac{1}{\lambda}\int_0^\lambda (Ak)^2\cos^2(kx)\,dx = \frac{1}{2}(Ak)^2.$$ (7.20)

In that case, and using the asymptotic result $\rho \approx 1/2$ for slender flagella, Eq. (7.13) predicts the approximate result for long wavelengths

$$\frac{U}{V} = \frac{1}{2}(Ak)^2. \tag{7.21}$$

This three-dimensional result agrees perfectly with the swimming speed of a small-amplitude transversely waving two-dimensional sheet obtained in Section 3.3.

7.2 Hydrodynamically Optimal Travelling Wave

The geometry of the wave affects the locomotion speed, Eq. (7.13), through the value of β. From a purely hydrodynamical standpoint, we might enquire about the shape of the most efficient waveform. Biologically, not all expended internal energy is lost to the surrounding fluid in the form of viscous dissipation; some of the energy would for example be dissipated internally or stored in the elastic deformation of the flagellum. Nevertheless, it is enlightening to derive the shape of the hydrodynamically optimal wave, as carried out by Lighthill (1975), and compare it with observed waveforms. For simplicity we focus here on the limit where the cell body is small enough to be neglected hydrodynamically. For the case where the cell body is included, see Exercises.

7.2.1 Rate of Work

For a slender filament whose hydrodynamics is governed by resistive-force theory, the energy equation for the rate of work over one wavelength, $\dot{W}$, Eq. (2.18) becomes

$$\dot{W} = \int_0^\Lambda -\mathbf{f} \cdot \mathbf{u}\, ds, \tag{7.22}$$

where $\mathbf{f}$ is the density of hydrodynamic forces acting on the flagellum. Using Eq. (6.40), this can be written as

$$\dot{W} = \int_0^\Lambda c_\perp (\mathbf{u} \cdot \mathbf{n})^2 + c_\|(\mathbf{u} \cdot \mathbf{t})^2\, ds = \int_0^\Lambda c_\perp \mathbf{u} \cdot \mathbf{u} + (c_\| - c_\perp)(\mathbf{u} \cdot \mathbf{t})^2\, ds. \tag{7.23}$$

Given Eq. (7.2) we have

$$\mathbf{u} \cdot \mathbf{u} = (V - U)^2 + Q^2 + 2(U - V)Q(\mathbf{t} \cdot \mathbf{e}_x) \tag{7.24}$$

and

$$(\mathbf{u} \cdot \mathbf{t})^2 = (V - U)^2(\mathbf{t} \cdot \mathbf{e}_x)^2 + Q^2 + 2(U - V)Q(\mathbf{t} \cdot \mathbf{e}_x), \tag{7.25}$$

so that the rate of work per wavelength of the flagellum is

$$\frac{\dot{W}}{\Lambda} = c_\| \frac{V^2}{\alpha^2} + 2c_\| V(U - V) + c_\perp (U - V)^2 + (c_\| - c_\perp)(U - V)^2 \beta, \tag{7.26}$$

where we have defined $\alpha \equiv V/Q < 1$. Using Eq. (7.13) we finally obtain the dimensionless rate of work as

$$\frac{\dot{W}}{c_\perp \Lambda V^2} = \frac{\rho}{\alpha^2} - \frac{\rho^2}{1 + \beta(\rho - 1)}. \qquad (7.27)$$

7.2.2 Swimming Efficiency

The hydrodynamically optimal flagellar wave is the one maximising the swimming speed for a given amount of energy dissipated in the flow (see also Section 5.4.2 for an analytical approach at small amplitude). It is then customary to define a low Reynolds number swimming efficiency, $\mathcal{E}$, as the ratio of energy requirements for forced vs. force-free motion (Childress, 1981, 2012). Specifically, the efficiency is taken to be the ratio between (i) the rate of work of an external force required to drag the swimmer at a velocity equal to the swimming speed, given thus by $c_\parallel \Lambda U^2$ per wavelength of a straight flagellum, and (ii) the rate of work of one wavelength of the flagellum, $\dot{W}$, to propel itself at the same speed, i.e.

$$\mathcal{E} = \frac{c_\parallel \Lambda U^2}{\dot{W}}, \qquad (7.28)$$

and we look to find the flagellar shape that maximises $\mathcal{E}$.

7.2.3 Optimal Wave

Using the results in Eq. (7.13) for the swimming speed and Eq. (7.27) for the rate of work, the efficiency is given by

$$\mathcal{E} = \frac{\rho(1 - \rho)^2(1 - \beta)^2}{[1 + \beta(\rho - 1)]^2} \frac{1}{\dfrac{\rho}{\alpha^2} - \dfrac{\rho^2}{1 + \beta(\rho - 1)}}. \qquad (7.29)$$

We may then find an upper bound for $\mathcal{E}$ by exploiting the Cauchy–Schwarz inequality applied to the waveform, namely $\alpha^2 \leqslant \beta$. The upper bound is therefore obtained by replacing α^2 in Eq. (7.29) by β, leading to

$$\mathcal{E} \leqslant \frac{\rho(1 - \rho)^2(1 - \beta)^2}{[1 + \beta(\rho - 1)]^2} \frac{1}{\dfrac{\rho}{\beta} - \dfrac{\rho^2}{1 + \beta(\rho - 1)}} = \frac{\beta(1 - \rho)^2(1 - \beta)}{1 + \beta(\rho - 1)}. \qquad (7.30)$$

The right-hand side of Eq. (7.30) is zero for both $\beta = 0$ and $\beta = 1$, and thus an optimal value of β exists. By computing its first derivative it is elementary to show that this maximum is obtained at $\beta = \beta_{\max}$ where

$$\beta_{\max} = \frac{1}{1 + \sqrt{\rho}}, \tag{7.31}$$

for which the upper bound for the swimming efficiency, $\mathcal{E}_{\max}$, is

$$\mathcal{E}_{\max} = (1 - \sqrt{\rho})^2. \tag{7.32}$$

Since a good estimate for ρ is $1/2$, the upper bound for the swimming efficiency is approximately $\tilde{\mathcal{E}}_{\max} = 3/2 - \sqrt{2} \approx 8.6\%$, with $\beta_{\max} = 2 - \sqrt{2} \approx 0.586$.

Can this upper bound ever be reached? The shapes for which this maximum efficiency is attained are such that the Cauchy–Schwarz inequality becomes an equality, $\alpha^2 = \beta$, and therefore such that the integrand in β from Eq. (7.9) is constant. This corresponds to waveforms whose angle θ between the local slope and the swimming direction, $\theta = \arccos(\mathbf{t} \cdot \mathbf{e}_x)$, remains constant all along the flagellum and is given by

$$\theta = \arccos(\sqrt{\beta_{\max}}) \approx 40^\circ. \tag{7.33}$$

In three dimensions, such a shape is a smooth helix (see also Section 5.4.1 and Chapter 8). In two dimensions, the corresponding shape is a non-smooth sawtooth wave, which may be regularised by taking into account either the elastic energy stored in the bending deformation of the flagellum (Spagnolie and Lauga, 2010) or the irreversible nature of internal work done by molecular motors (Lauga and Eloy, 2013). The optimal angle given by this analytical optimisation leads to a relationship between the amplitude of the wave, h, and its wavelength given approximately by $h/\lambda \approx 0.21$, which is close to experimental observations for spermatozoa with small heads (Brennen and Winet, 1977; Lauga and Eloy, 2013).

7.3 Swimming of a Finite Flagellum

7.3.1 Swimming Kinematics

If the planar flagellum has finite size, swimming no longer occurs along a straight line but has in general three degrees of freedom, namely a swimming velocity with components $\mathbf{U} = (U_\parallel, U_\perp)$ in the (x, y) plane of beating and an angular velocity, $\Omega \mathbf{e}_z$, perpendicular to it. As a consequence, the kinematics in Eq. (7.2) are modified to

$$\mathbf{u} = (V - U_\parallel)\mathbf{e}_x - Q\mathbf{t} + U_\perp \mathbf{e}_y + (\Omega \mathbf{e}_z) \times \mathbf{r}, \tag{7.34}$$

where $\mathbf{r}$ denotes the instantaneous location of points along the flagellum relative to an origin moving with velocity $\mathbf{U}$. The swimming kinematics are obtained by enforcing the force- and torque-free conditions. For a cell with a negligible body these are written as

$$\left(\int_0^L \mathbf{f}\, ds \right) \cdot \mathbf{e}_i = 0, \quad i = x, y, \tag{7.35a}$$

$$\left(\int_0^L \mathbf{r} \times \mathbf{f}\, ds \right) \cdot \mathbf{e}_z = 0, \tag{7.35b}$$

while in the presence of a cell body these forces and torques are balanced by forces and torques on the cell.

The resulting equations can be written formally as an instantaneous linear system that can be inverted to obtain the three degrees of freedom $(U_\parallel, U_\perp, \Omega)$. This is in general too tedious to carry out by hand but can be done numerically (Berman et al., 2013). If we further assume that the flagellum beats with a small amplitude ϵ, the solution may be obtained analytically with both $U_\perp$ and Ω of order ϵ while the mean value of $U_\parallel$ is of order ϵ^2 (see Exercises).

Depending on the shape of the waveform, different swimming motions are predicted theoretically and observed experimentally. If the waveform is $y \leftrightarrow -y$ symmetric, swimming occurs on average along a straight line but is accompanied in general by a time-periodic wobbling motion in the plane (i.e. $\langle \Omega \rangle = 0$ but $\Omega(t) \neq 0$). Swimming with $\langle \Omega \rangle \neq 0$ leads to circular trajectories, and may result either from asymmetries in the cell body or from curvature of the flagellar waveform (Woolley, 2003; Friedrich et al., 2010). Some spermatozoa also display non-planar flagellar beating, which results in three-dimensional swimming kinematics (Brennen and Winet, 1977; Woolley and Vernon, 2001).

7.3.2 *Comparison with Experiments: Bull Spermatozoon*

The success of the theoretical predictions is highlighted in Friedrich et al. (2010). In that study, the waving motion of the flagellum and cell body of a bull spermatozoon were tracked in detail near a surface. Due to the asymmetric flagellar waveform of bull sperm, the cell swims along circular trajectories superimposed with small-amplitude oscillations. The measured flagellar movements were then used, along with resistive-force theory, to predict the swimming kinematics of the cell. The comparison between these predictions (red dashed line) and the measured kinematics (black solid line) are shown in Fig. 7.3 as a function of time. Specifically, we plot the instantaneous value of the component of the swimming velocity aligned with the cell body ($U_\parallel$, Fig. 7.3(a)), the velocity component perpendicular to it ($U_\perp$, Fig. 7.3(b)) and the angular velocity (Ω, Fig. 7.3(c)). A remarkable agreement is obtained for both $U_\perp$ and Ω; while the $U_\parallel$ component contains

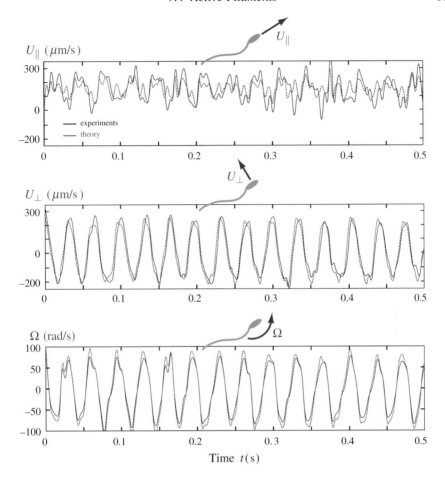

Figure 7.3 Comparison between the instantaneous swimming kinematics of a bull spermatozoon (black solid line) and predictions from resistive-force theory based on waveform measurements (red dashed line). Top: component of the swimming velocity along the cell body ($U_\parallel$). Middle: swimming component perpendicular to the cell body ($U_\perp$). Bottom: rotation rate of the cell body (Ω). Adapted from Friedrich et al. (2010) with permission.

non-trivial high-frequency motion at two and three times the flagellar frequency, it is still well predicted by the theoretical approach.

7.4 Active Filaments

The approach in this chapter so far has been to assume that the waveform is known and to deduce from it the swimming kinematics. Real eukaryotic flagella, however, are active filaments whose observed waveform is governed by a balance between internal actuation from molecular motors and forces from both fluid dynamics and elasticity. In this last part of the chapter, we present a continuum model of such an active filament and use the model to obtain simple estimates for the internal energetic requirements (ATP consumption) of eukaryotic flagella.

7.4.1 Elasticity

In Chapter 6, we computed the hydrodynamic forces acting on slender filaments. In this section, we now focus on the derivation of the elastic forces resisting the deformation of flexible flagella. We adopt the classical Kirchhoff framework for the elasticity of slender rods (Landau and Lifshitz, 1986). The calculation of the elastic force density acting on a deformable filament starts with deriving the energetic cost of bending it away from its preferred orientation. We show that the energy is given by the mean squared curvature of the flagellum, which in turn may be used to compute the elastic force by exploiting variational arguments.

Strains and Stresses

For simplicity we restrict all deformations to two dimensions; for the extension to the three-dimensional case see e.g. Audoly and Pomeau (2010) and Powers (2010). Consider a slender filament of cross-sectional radius a whose centreline is bent in a plane with local radius of curvature R (see illustration in Fig. 7.4). The limit of slender rods, in which the classical theory is derived, assumes that everywhere along the filament we have $a \ll R$. The dimensionless elastic strain at any point inside the elastic filament, ε, is defined as the dimensionless ratio

$$\varepsilon = \frac{\ell - \ell_0}{\ell_0}, \tag{7.36}$$

where ℓ is the length of a bent line element parallel to the centreline whose rest length is ℓ_0.

By definition, an inextensible filament needs to have $\varepsilon = 0$ anywhere on its centreline. Then, as illustrated in Fig. 7.4, the elastic material located in the lower half of the filament (the half nearer to the centre of curvature) is compressed while the material in the upper half (further away from the centre of curvature) is being stretched. We use y to measure the distance away from the centreline of the filament while x is locally tangent to it (see notation in Fig. 7.4). Since R is the radius of curvature of the centreline, a line element of size $R\delta\theta$, where $\delta\theta$ is a small angular portion measured from the local centre of curvature of the filament, can be taken as the rest length ℓ_0. In the limit $a \ll R$, the length at a location y away from the centreline is approximately $\ell \approx (R + y)\delta\theta$ and the elastic strain is thus given by

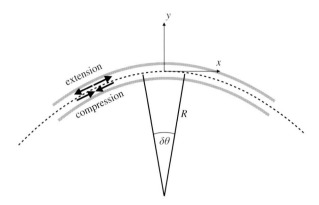

Figure 7.4 Local, planar bending of an elastic filament. The radius of curvature of the centreline is denoted R and y measures the local distance to the centreline. The half of the filament in the $y > 0$ region undergoes extension while the half in the $y < 0$ region is compressed. A small angular portion of size $\delta\theta$ measured from the local centre of curvature of the filament is used to compute the elastic strains in the filament.

$$\varepsilon = \frac{\ell - \ell_0}{\ell_0} \approx \frac{(R + y)\delta\theta - R\delta\theta}{R\delta\theta} \approx \frac{y}{R}. \tag{7.37}$$

When $y > 0$ we have $\varepsilon > 0$, indicating extension, while in the bottom half of the filament, $y < 0$, we have $\varepsilon < 0$ and thus compression.

Within the framework of linear elasticity the stress in the filament is everywhere proportional to the strain, with a proportionality constant, E, termed the Young's modulus of the elastic material, i.e.

$$\sigma = E\varepsilon \approx E\frac{y}{R}. \tag{7.38}$$

Elastic Energy

In order to postulate the bending energy, E_B, we use the analogy of a simple one-dimensional spring whose elastic potential energy is one half of the spring constant times the square of the deformation. The bending energy of the elastic rod of length L can then be written as the integral of the energy density along the arclength of the centreline,

$$E_B = \int_0^L e_B \, \mathrm{d}s, \tag{7.39}$$

where e_B is the total integral over the cross section A of one half of the spring constant (here, E) times the square of the strain (here, ε), i.e.

$$e_B = \frac{1}{2} \int_A E\varepsilon^2 \, \mathrm{d}A. \tag{7.40}$$

Since $E\varepsilon$ is nothing other than the elastic stress, this represents in physical terms the work done on the material in order to induce the deformation. Using Eqs. (7.37)–(7.38) the bending energy density can be written as

$$e_B = \frac{1}{2} \int_A E \left(\frac{y}{R}\right)^2 \, \mathrm{d}A \equiv \frac{1}{2}\frac{EI}{R^2}, \quad I = \int_A y^2 \, \mathrm{d}A, \tag{7.41}$$

where we have introduced the moment of inertia, I, of the cross section (for a circular cross section of radius a we have classically $I = \pi a^4/4$). The bending modulus of the material, B, is then defined as $B \equiv EI$ and is assumed to be constant. We denote by $\kappa = 1/R$ the local curvature, which is equal formally to $\kappa = |\partial^2 \mathbf{r}/\partial s^2|$ for a filament whose centreline is located at $\mathbf{r}(s)$ (we use partial derivatives along s because we allow all fields to also vary in time). We then obtain finally the total bending elastic energy of the filament as

$$E_B = \frac{1}{2}B \int_0^L \frac{1}{R^2} \, \mathrm{d}s = \frac{1}{2}B \int_0^L \kappa^2 \, \mathrm{d}s, \tag{7.42}$$

which is proportional to the mean square curvature.

Tension

For a filament that cannot undergo any net compression or stretching, the bending energy needs to be supplemented by an additional energy term enforcing inextensibility. If the filament is inextensible, then at every point along its length if we denote by $\mathbf{u}$ the velocity of material points on the centreline, tangential velocity gradients $\partial \mathbf{u}/\partial s$ cannot have a component along the tangent vector $\mathbf{t} = \partial \mathbf{r}/\partial s$, as otherwise compression or stretching would ensue. This is written mathematically as $\mathbf{t} \cdot \partial \mathbf{u}/\partial s = 0$ and since $\mathbf{u} = \partial \mathbf{r}/\partial t$ this is equivalent to

$$\frac{\partial}{\partial t}(\mathbf{t} \cdot \mathbf{t}) = 0. \tag{7.43}$$

Inextensibility is therefore equivalent to requiring that the vector $\mathbf{t}$ remains a unit vector at every point along the filament. This can be enforced mathematically by adding a second energy term, E_T, which includes a Lagrange multiplier T with dimension of force (denoted the tension) that enforces pointwise inextensibility as

$$E_T = \frac{1}{2} \int_0^L T \left(\frac{\partial \mathbf{r}}{\partial s} \cdot \frac{\partial \mathbf{r}}{\partial s} - 1\right) \, \mathrm{d}s. \tag{7.44}$$

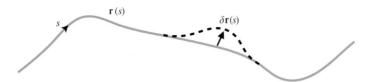

Figure 7.5 Elastic filament undergoing a small deformation $\delta\mathbf{r}(s)$, leading to a restoring elastic force such that the change in total elastic energy is given by the work done by the force.

The total elastic energy, the sum of Eq. (7.42) and Eq. (7.44), is thus finally given by

$$E = E_B + E_T = \frac{1}{2}B \int_0^L \kappa^2 \, ds + \frac{1}{2} \int_0^L T \left(\frac{\partial \mathbf{r}}{\partial s} \cdot \frac{\partial \mathbf{r}}{\partial s} - 1 \right) ds. \tag{7.45}$$

Elastic Force Density

From the energy in Eq. (7.45), the elastic force density may be obtained using a variational method. Consider a small perturbation to the shape of the filament $\mathbf{r}(s) \rightarrow \mathbf{r}(s) + \delta\mathbf{r}(s)$ (Fig. 7.5). This change in shape leads to an increase in the elastic energy of the form

$$\delta E = \int -(\mathbf{f}_e \cdot \delta\mathbf{r}) \, ds, \tag{7.46}$$

plus additional boundary terms. The integrand in Eq. (7.46) is a work density and thus the vector $\mathbf{f}_e$ may be interpreted as the elastic force density. If $\mathbf{f}_e \cdot \delta\mathbf{r} < 0$ then the force does elastic work on the local perturbation and the elastic energy increases, whereas if $\mathbf{f}_e \cdot \delta\mathbf{r} > 0$ then the perturbation is favourable and elastic energy is released.

We start by perturbing the bending contribution to the elastic energy, E_B. The change in energy, δE_B, due to a small change in shape is given by

$$\delta E_B = \int_0^L B \frac{\partial^2 \mathbf{r}}{\partial s^2} \cdot \frac{\partial^2 \delta\mathbf{r}}{\partial s^2} \, ds, \tag{7.47}$$

which, after two integrations by parts, can be written as

$$\delta E_B = \int_0^L B \left[\frac{\partial}{\partial s} \left(\frac{\partial^2 \mathbf{r}}{\partial s^2} \cdot \frac{\partial \delta\mathbf{r}}{\partial s} \right) - \frac{\partial}{\partial s} \left(\frac{\partial^3 \mathbf{r}}{\partial s^3} \cdot \delta\mathbf{r} \right) + \frac{\partial^4 \mathbf{r}}{\partial s^4} \cdot \delta\mathbf{r} \right] ds. \tag{7.48}$$

Similarly, for the tension contribution to the elastic energy, E_T, we obtain the perturbation

$$\delta E_T = \int_0^L T \frac{\partial \mathbf{r}}{\partial s} \cdot \frac{\partial \delta\mathbf{r}}{\partial s} \, ds + \frac{1}{2} \int_0^L \delta T (\mathbf{t} \cdot \mathbf{t} - 1) \, ds. \tag{7.49}$$

The second term in Eq. (7.49) is identically zero since $|\mathbf{t}| = |\partial \mathbf{r}/\partial s| = 1$, while the first term may be simplified using integration by parts, giving

$$\delta E_T = \int_0^L \left[\frac{\partial}{\partial s} \left(T \frac{\partial \mathbf{r}}{\partial s} \cdot \delta \mathbf{r} \right) - \frac{\partial}{\partial s} \left(T \frac{\partial \mathbf{r}}{\partial s} \right) \cdot \delta \mathbf{r} \right] ds. \tag{7.50}$$

Combining Eqs. (7.48) and (7.50), and identifying the force density as in Eq. (7.46), we obtain

$$\mathbf{f}_e = -B \frac{\partial^4 \mathbf{r}}{\partial s^4} + \frac{\partial}{\partial s} \left(T \frac{\partial \mathbf{r}}{\partial s} \right), \tag{7.51}$$

with additional boundary terms arising in the variations of the energy and corresponding to external forces acting at the end points of the filament (at $s = 0$ and $s = L$).

7.4.2 Active Elasto-hydrodynamics

In an active filament, an internal molecular forcing balances with dissipative processes and with elastic forces to generate the observed shapes. For eukaryotic flagella, the internal energy source is ATP, whose hydrolysis by a motor protein (dynein) converts chemical energy into mechanical work by promoting the relative sliding of polymeric filaments (microtubule doublets). Let us denote by $\mathbf{f}_a$ the density of active forces resulting from these internal processes. Ignoring for simplicity sources of internal dissipation and thus assuming that all dissipation comes from work done on the surrounding fluid, the shape of the waveform is found by balancing instantaneously $\mathbf{f}_a$ with the hydrodynamic forces ($\mathbf{f}$, Eq. (6.40)) and elasticity ($\mathbf{f}_e$, Eq. (7.51)). The equation of active elasto-hydrodynamics can thus be written as

$$-[c_\| \mathbf{tt} + c_\perp (\mathbf{1} - \mathbf{tt})] \cdot \frac{\partial \mathbf{r}}{\partial t} - B \frac{\partial^4 \mathbf{r}}{\partial s^4} + \frac{\partial}{\partial s} \left(T(s,t) \frac{\partial \mathbf{r}}{\partial s} \right) + \mathbf{f}_a = \mathbf{0}, \tag{7.52}$$

with a second instantaneous equation for the tension T to ensure inextensibility of the filament (see Exercises). Note that the active term, $\mathbf{f}_a$, has dimensions of force per unit length, and may thus preferably be viewed as a gradient of active torque density, whose time-dependent magnitude can be estimated from detailed tracking data obtained from swimming spermatozoa (Gaffney et al., 2011).

For a filament whose deformation is described in Cartesian coordinates as $\mathbf{r} = (x, y(x, t))$, the tension T is at least quadratic in the filament amplitude (see Exercises) and thus, in the small-amplitude limit, the leading-order elasto-hydrodynamic force balance in the y direction is

$$c_\perp \frac{\partial y}{\partial t} + B \frac{\partial^4 y}{\partial x^4} = f_a(x, t). \tag{7.53}$$

In the absence of activity, $f_a = 0$, and the resulting equation is the classical hyper-diffusion equation for the amplitude of an elastic filament immersed in a viscous fluid (Wiggins and Goldstein (1998); see also Exercises). The relative importance of elastic vs. viscous forces in determining the shape of the filament is captured by a dimensionless number called the sperm number, Sp. Assuming periodic motion (e.g. waving) of a filament of length L with frequency ω, we then have the scaling

$$\frac{c_\perp \frac{\partial y}{\partial t}}{B\frac{\partial^4 y}{\partial x^4}} \sim \frac{c_\perp y\omega}{By/(L^4)} \sim \frac{c_\perp L^4 \omega}{B}. \tag{7.54}$$

This dimensionless ratio defines the sperm number, which is given by

$$Sp = L\left(\frac{c_\perp \omega}{B}\right)^{1/4} \equiv \frac{L}{\ell_\omega}, \quad \ell_\omega = \left(\frac{B}{c_\perp \omega}\right)^{1/4}, \tag{7.55}$$

where ℓ_ω is the so-called elasto-viscous penetration length. Physically, a periodically actuated elastic filament is deformed on a length scale of magnitude $\sim \ell_\omega$ and thus filaments for which Sp $\ll 1$ are barely bent by hydrodynamic forces while in the opposite limit, Sp $\gg 1$, viscous stresses induce significant elastic deformation.

In the case of an active filament, the value of the sperm number is key to understanding the balance between activity and mechanical resistance to motion. For a waving flagellum, an estimate for the penetration length may be given as follows. Measurements of the bending rigidity of the flagella of sea-urchin spermatozoa reported $B \approx 0.9 \times 10^{-21}$ Nm^2 (Howard, 2001), with very similar numbers for *Chlamydomonas* flagella (Xu et al., 2016). Using the classical kinematic measurements from Gray and Hancock (1955) for sea-urchin spermatozoa, the flagellum of radius $a \approx 100$ nm has length $L \approx 30$ μm. The drag coefficient for such a slender filament in water is then given approximately by $c_\perp \approx 4\pi\mu/\ln(L/a) \approx 2 \times 10^{-3}$ Pa s. With oscillation at frequency $\omega = 2\pi f$ with $f \approx 35$ Hz, the elasto-viscous penetration length takes the approximate value $\ell_\omega \approx 6.5$ μm. Since Sp is defined as the ratio of L to ℓ_ω, we obtain that, while cilia are typically characterised by Sp $= O(1)$, most spermatozoa flagella are in the Sp $\gg 1$ limit.

7.4.3 Internal Forces and Energy Consumption

We may use the active-filament model to obtain order-of-magnitude estimates for the density of internal flagellar forces developed by the molecular motors and for the number of ATP molecules that have to be hydrolysed in order to obtain flagellar beating consistent with experimental observations.

Consider a simple bending wave of internal activity written in complex notation as $f_a = \mathrm{Re}\{\tilde{f}_a e^{i(kx-\omega t)}\}$. Denoting similarly the Fourier transform of the flagellum amplitude by $\tilde{y}$, the force balance in Eq. (7.53) leads to

$$(-i\omega c_\perp + Bk^4)\tilde{y} = \tilde{f}_a, \tag{7.56}$$

whose solution is

$$\tilde{y} = \frac{\tilde{f}_a}{\omega c_\perp [(\ell_\omega k)^4 - i]}. \tag{7.57}$$

Substituting parameters from the original study of Gray and Hancock (1955), the flagellum amplitude is on the order of $\tilde{y} \approx 4\ \mu\mathrm{m}$, $k = 2\pi/\lambda$ with wavelength $\lambda \approx 24\ \mu\mathrm{m}$, and with the value $\ell_\omega \approx 6.5\ \mu\mathrm{m}$ calculated above we estimate the magnitude of the active force density as $\tilde{f}_a \approx 11\ \mathrm{pN}/\mu\mathrm{m}$.

To obtain estimates of the energy budget, we calculate separately the rate of energy dissipation in the fluid and the rate at which bending energy is periodically stored in the elastic flagellum. The fluid contribution can be estimated first by evaluating the integral in Eq. (5.17), which is valid at leading order in the waving amplitude. The rate at which the flagellum is doing work on the viscous fluid is therefore approximately given per wavelength by

$$\dot{W}_{\mathrm{fluid},\lambda} \approx c_\perp (\tilde{y}\omega)^2 \lambda. \tag{7.58}$$

In contrast, the rate at which the elastic energy in Eq. (7.42) is periodically stored in the bending of the flagellum is given per wavelength by

$$\dot{W}_{\mathrm{elastic},\lambda} \approx \omega B \kappa^2 \lambda \approx \omega B \tilde{y}^2 k^4 \lambda, \tag{7.59}$$

where we have used the estimate $\kappa \approx \tilde{y}k^2$ for the typical curvature of the flagellum. The relative importance of viscous and elastic energy rates is therefore given by the ratio

$$\frac{\dot{W}_{\mathrm{fluid},\lambda}}{\dot{W}_{\mathrm{elastic},\lambda}} \approx \frac{c_\perp \omega}{Bk^4} = \frac{1}{(\ell_\omega k)^4}, \tag{7.60}$$

which provides us with another way to interpret the length scale ℓ_ω. Using the numbers from Gray and Hancock (1955) together with $\ell_\omega \approx 6.5\ \mu\mathrm{m}$, we obtain $\dot{W}_{\mathrm{elastic},\lambda}/\dot{W}_{\mathrm{fluid},\lambda} \approx 10$ and therefore the majority of the work done by the molecular motors is devoted to deforming the flagellum periodically. Using the same numbers, we can compute the total rate at which energy is being expended per period of the flagellum as the sum

$$\dot{W}_\lambda = \dot{W}_{\mathrm{fluid},\lambda} + \dot{W}_{\mathrm{elastic},\lambda} \approx 3.9 \times 10^{-13}\ \mathrm{W}. \tag{7.61}$$

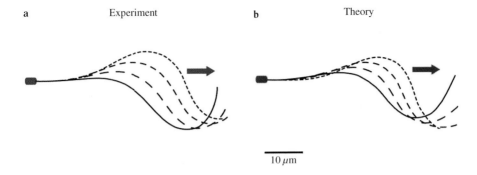

Experiment Theory

10 μm

Figure 7.6 Comparison between the predictions of the active elasto-hydrodynamics model and experiments. Left: beat patterns of flagella for a clamped bull spermatozoon over one quarter of the period. Right: theoretical predictions of the active elasto-hydrodynamics model for the same parameters. Reproduced from Riedel-Kruse et al. (2007).

With a beating frequency of 35 Hz, the required energy per beat and wavelength is therefore approximately

$$\Delta E_{\lambda,\text{beat}} = \frac{\dot{W}_{\lambda}}{\text{beat}} \approx 1.1 \times 10^{-15} \text{ J}. \tag{7.62}$$

The work done by the dynein molecular motors arises from the hydrolysis of ATP into ADP, a process that releases approximate 50 kJ per mole of ATP (Alberts et al., 2007). Since one mole is approximately 6×10^{23} molecules (Avogadro's number) the energy released by the hydrolysis of each ATP molecule, ΔE, is approximately

$$\Delta E_{\text{ATP}} \approx 8.2 \times 10^{-20} \text{ J}. \tag{7.63}$$

Comparing Eq. (7.62) with Eq. (7.63), we see that the energy required to bend the flagellum periodically in the viscous fluid is equivalent to an internal consumption of approximately $\Delta E_{\lambda,\text{beat}}/\Delta E_{\text{ATP}} \approx 1.4 \times 10^5$ molecules of ATP per wavelength and per beat. Direct ATP consumption by sea-urchin flagella was measured at a single-cell level by Chen et al. (2015) using a technique in which the ATP concentration could be monitored by fluorescence. In that case, measurements reported between 1×10^5 and 3×10^5 ATP molecules consumed per beat and wavelength, in good agreement with our theoretical estimate.

7.4.4 Comparison with Experiments: Bull Spermatozoon

An early consideration of active filaments was used to demonstrate that the flagellum of a spermatozoon could not be passively actuated from one of its ends but that, instead, actuation had to be spatially distributed along the flagellum length (Machin, 1958). The success of the active-filament model may be illustrated further by comparing its predictions to experimental measurements. This is shown in Fig. 7.6 in the case of a bull spermatozoon, adapted from a study by Riedel-Kruse et al. (2007). In Fig. 7.6(a), the measured flagellar shapes are shown in the case of a clamped cell during the first quarter of its beat. In Fig. 7.6(b), the theoretical predictions of the active-filament model are plotted using material properties inferred from past measurements and with two fitting parameters related to the internal friction and regulation of the molecular motor. Agreement between the theory and the experiment is within 5% in a least-square sense, emphasising the ability of such a continuum modelling approach to capture the biophysics of the internally driven flagella of eukaryotes accurately.

Further Reading

Spermatozoa cells were discovered in the late seventeenth century by van Leeuwenhoek (Gest, 2004), and a biological overview of their structure is presented by Fawcett (1975). Seminal papers on the modelling of spermatozoa locomotion include an early mathematical model of three-dimensional flagella (Taylor, 1952) (see also Exercises), a quantitive study comparing mathematical modelling and experiments (Gray and Hancock, 1955), the consideration of the role of elasticity and the distribution of internal forces (Machin, 1963) and an early model incorporating hydrodynamic interactions between flagellum and cell body (Higdon, 1979a). The fluid mechanics of spermatozoa was reviewed at the single-cell level by Gaffney et al. (2011) and in the context of mammalian reproduction by Fauci and Dillon (2006).

Mathematical optimisation of flagellar movement showed that travelling waves lead to a minimum of energy dissipation in the fluid (Pironneau and Katz, 1974). The hydrodynamically optimal singular waveform derived in Section 7.2 can be regularised either by introducing an energetic penalty for bending deformation of the flagellum (Spagnolie and Lauga, 2010), by using a different energy measure based on the irreversible nature of the work done by molecular motors (Lauga and Eloy, 2013) or by considering the full three-dimensional motion of finite-size cells (Tam and Hosoi, 2011). Hydrodynamic interactions between cell body and flagellum also lead to optimal head-to-tail ratios (Higdon, 1979a; Tam and Hosoi, 2011).

Bending of the flagellum is due to mechanical work done by dynein molecular motors inside the flagellum (axoneme), which leads to the relative sliding of microtubule doublet filaments (Brokaw, 1972, 1989; Gibbons and Gibbons, 1972). Models for the collective behaviour of molecular motors were reviewed by Jülicher et al. (1997) and applied to spermatozoa by Camalet and Jülicher (2000). For a critical review of the regulation mechanisms of motor activity, we refer to Sartori et al. (2016).

Exercises

1. Consider the high-amplitude passive elasto-hydrodynamics equation

$$-[c_\parallel \mathbf{t}\mathbf{t} + c_\perp(1 - \mathbf{t}\mathbf{t})] \cdot \frac{\partial \mathbf{r}}{\partial t} - B\frac{\partial^4 \mathbf{r}}{\partial s^4} + \frac{\partial}{\partial s}\left(T(s,t)\frac{\partial \mathbf{r}}{\partial s}\right) = \mathbf{0},$$

   with all symbols defined in Section 7.4. Given that the tension T is obtained by enforcing that the filament is inextensible (i.e. $\mathbf{t} \cdot \mathbf{t} = 1$ for all times where $\mathbf{t} = \partial \mathbf{r}/\partial s$), derive the ordinary differential equation satisfied by T in terms of $\mathbf{r}$ and its spatial derivatives. If the filament amplitude is of order $\epsilon \ll 1$, use that equation to show that the tension is of order ϵ^2 and thus can be neglected in the leading-order dynamic balance, resulting in the linear mechanical balance,

$$c_\perp \frac{\partial y}{\partial t} + B\frac{\partial^4 y}{\partial x^4} = 0,$$

   in the direction perpendicular to the filament.

2. An organism swims in a two-dimensional plane by waving its planar flagellum with small dimensionless amplitude ϵ. Let $\mathbf{e}_x$ denote the unit vector directed instantaneously along the swimmer and $\mathbf{e}_y$ the unit vector instantaneously orthogonal to it in the plane of swimming (with conventional choice). The organism swims with velocity $\epsilon^2 U_0 \mathbf{e}_x + \epsilon V_0 \cos(\omega t)\mathbf{e}_y$, and rotates about the direction $\mathbf{e}_z = \mathbf{e}_x \times \mathbf{e}_y$ with angular velocity $\Omega(t) = \epsilon\Omega_0 \cos(\omega t + \phi)$. Compute the time-averaged swimming speed of the organism at order ϵ^2 in the lab frame. Interpret the sign of the result for $\phi = \pm\pi/2$.

3. The flagellar waveform optimisation carried in Section 7.2 out for a swimmer with no cell body can be generalised to incorporate the drag on the body. Writing the magnitude of the drag force on the cell body as $c_\perp LU\delta$, where δ is a dimensionless number, show that the swimming speed U of a flagellated swimmer with a cell body is given by

$$\frac{U}{V} = \frac{(1 - \rho)(1 - \beta)}{1 + (\rho - 1)\beta + \delta},$$

   where V is the wave speed, ρ is the ratio of drag coefficients and β is the geometrical wave parameter defined in Eq. (7.9). The swimming efficiency, $\mathcal{E}$, is defined as the ratio between the rate of work of an external force dragging the entire swimmer at speed U and the rate of work of the whole flagellum to propel the cell at the same speed. Calculate the maximum efficiency $\mathcal{E}_{max}$ as a function of δ.

4. An active slender filament of length L undergoes prescribed planar deformation in a Newtonian fluid, which results in swimming at low Reynolds number. In a frame (x, y, z) moving with the swimmer, the material points on the filament are located at $(x_s, y_s, z_s) = (x, \epsilon g(x, t), 0)$, for $0 \leqslant x \leqslant L$, where $\epsilon \ll 1$ is a dimensionless parameter. The swimming frame (x, y, z) is defined relative to the filament by prescribing that the shape function g satisfies $g(0, t) = \partial g/\partial x(0, t) = 0$. The swimming motion is described by the instantaneous translational speed, $U\mathbf{e}_x + V\mathbf{e}_y$, and instantaneous angular velocity, $\Omega\mathbf{e}_z$, of the origin and orientation of the swimming frame, measured with respect to the frame in which the fluid is at rest at infinity. Determine the distribution of velocities, $\mathbf{u}(x, t)$, of points along the filament relative to the background fluid, for all points $0 \leqslant x \leqslant L$. Anticipating that the swimming kinematics are to be solved as a perturbation expansion, $\{U, V, \Omega\} = \epsilon\{U_1, V_1, \Omega_1\} + O(\epsilon^2)$, explain why it is only necessary to characterise the tangent vector along the filament to $O(1)$ in order to solve the problem at $O(\epsilon)$. Compute the total instantaneous hydrodynamic forces $F_x(t)$ and $F_y(t)$ and torque $T_z(t)$ on the filament at $O(\epsilon)$. Deduce that, for free-swimming motion, $U_1 = 0$. Compute the instantaneous values of V_1 and Ω_1 as functions of $\langle \partial g/\partial t \rangle$ and $\langle x(\partial g/\partial t) \rangle$ where $\langle \ldots \rangle \equiv \frac{1}{L}\int_0^L \ldots \, dx$. If the function $g(x, t)$ is periodic in time, show that both V_1 and Ω_1 time-average to zero.

5. An inextensible elastic slender filament of length L is actuated periodically and passively in a viscous fluid at low Reynolds number. Denoting the shape of the filament by $y(x, t)$, $0 \leqslant x \leqslant L$, in Cartesian coordinates, for small-amplitude motion the dynamic balance is given by

$$c_\perp \frac{\partial y}{\partial t} + B\frac{\partial^4 y}{\partial x^4} = 0.$$

Show that the total propulsive force induced by the fluid on the filament along the x direction is given by

$$\mathcal{F} = (c_\perp - c_\parallel) \int_0^L \frac{\partial y}{\partial x}\frac{\partial y}{\partial t} \, dx.$$

Using integration by parts, show that the value of $\mathcal{F}$ depends only on the dynamics at the boundary points $(x = 0, L)$. The filament is assumed to undergo periodic motion with frequency ω. Using the elasto-viscous penetration length, $\ell_\omega = (B/(c_\perp\omega))^{1/4}$, and ω^{-1} to non-dimensionalise lengths and times, show that the dimensionless filament dynamics satisfies

$$\frac{\partial y}{\partial t} = -\frac{\partial^4 y}{\partial x^4}.$$

The filament oscillates periodically at $x = 0$ with dimensionless boundary conditions

$$y(0, t) = y_0 \cos t, \qquad \left. \frac{\partial^2 y}{\partial x^2} \right|_{x=0,t} = 0.$$

The filament is long (i.e. $L \gg \ell_\omega$) and the dimensionless boundary condition on its other end is given by $y(x \to \infty, t) = 0$. Determine the solution for $y(x, t)$; show that it is the sum of two exponentially damped waves travelling in opposite directions and calculate their dimensionless wave speeds.

6. The year after his waving sheet paper, Taylor (1952) proposed a three-dimensional mathematical model of flagellar propulsion, which we now derive. Consider a cylindrical filament of radius ρ deforming as a periodic travelling wave in the (x, z) plane in the lab frame, where z is along the average filament axis and x is perpendicular to it. Denoting by δ the waving amplitude of the filament deformation in the x direction, the location of material points on the surface of the filament is given by

$$\mathbf{r}_s = (\delta + \rho \cos \theta) \mathbf{e}_x + \rho \sin \theta \, \mathbf{e}_y + z \, \mathbf{e}_z, \qquad \delta = b \sin(kz + kct),$$

where θ is the polar angle in the (x, y) plane with $0 \leqslant \theta < 2\pi$, k is the wavenumber and c is the wave speed. In the small-amplitude limit where $\epsilon = bk \ll 1$, solve the Stokes equations perturbatively and deduce the swimming speed of the filament at order ϵ^2. In the limit where $ak \ll 1$, show that this result agrees with the prediction from resistive-force theory, Eq. (7.21).

8

Rotation of Bacterial Flagellar Filaments

Motivated by the swimming of bacteria, we consider in this chapter the second fundamental mode of microorganism locomotion, namely that powered by the rotation of helical filaments. While many flagellated bacteria possess multiple flagellar filaments, we focus here on the canonical case of a cell body rotating a single filament, a situation that contains all the essential physics. Bacterial locomotion features two important differences from the swimming of spermatozoa addressed in Chapter 7. First, bacterial swimming does not involve time-varying shape changes but may be understood physically as due to the relative rotation of rigid bodies. Second, in order to balance hydrodynamic torques, a bacterium needs a cell body of finite size to swim. In order to explain these phenomena, we first derive the Stokes resistance matrix of a helix as predicted by resistive-force theory. We then use it to compute the translational and angular velocities of a bacterium moving along a straight line and compare our theoretical results with experimental measurements on *E. coli*. We next employ our theoretical estimates to address the energy expended by the molecular motor powering the rotation of filaments. Extending the theory to the full three-dimensional motion of finite-size bacteria, we obtain generically helical trajectories, which are compared to measurements for *Bacillus subtilis* (*B. subtilis*). We close by addressing the issue of optimisation, first by showing that there exists an optimal size of the cell body, and then by showing that we can define an intrinsic efficiency for the helical propeller, allowing us to rationalise the observed shapes of natural flagellar filaments.

8.1 Hydrodynamic Resistance of Helical Filaments

In order to propose a mathematical model for the locomotion of bacteria, we first need to quantify the hydrodynamics of helices. Consider the slender helical filament illustrated in Fig. 8.1 and characterised by a radius R (i.e. the radius of the cylinder on which its centreline is coiled), pitch λ (i.e. the wavelength measured

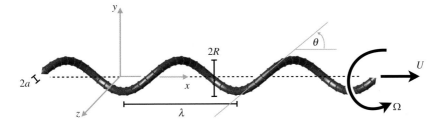

Figure 8.1 Rotation and translation of a slender helical filament of angular velocity Ω and linear speed U along the axis of the left-handed helix (x). The helix has radius R (radius of the cylinder on which its centreline is coiled) and cross-sectional diameter $2a$. The wavelength measured along the x axis is denoted by λ (pitch) while the angle between the tangent to the helix and the helical axis is denoted by θ.

along the x axis of the helix), cross-sectional diameter $2a$ and total linear length L. We denote by θ the constant angle between the local tangent along the helix and the helical axis (x), as illustrated in Fig. 5.3, i.e. such that $\cos\theta = \mathbf{t}\cdot\mathbf{e}_x$. We use the convention that $\theta > 0$ for a left-handed helix, while $\theta < 0$ for a right-handed one.

For simplicity we consider the case in which the helix is constrained to translate only along the x axis with linear velocity $U\mathbf{e}_x$ and rotate with angular velocity $\Omega\mathbf{e}_x$. The two important components of the resistance matrix are those allowing us to compute the hydrodynamic force, $F\mathbf{e}_x$, and torque, $T\mathbf{e}_x$, experienced by the helix along the same direction (we measure the torque about the x axis). From Eqs. (5.1) and (5.3) we know that the relationship is linear and can be written as

$$\begin{pmatrix} F \\ T \end{pmatrix}_{\text{helix}} = -\begin{pmatrix} A & B \\ B & D \end{pmatrix}\begin{pmatrix} U \\ \Omega \end{pmatrix}_{\text{helix}}. \tag{8.1}$$

In order to compute the values of the three coefficients A, B and D, we use the calculation from Chapters 5 and 6 on the dynamics of rods using resistive-force theory. Considering a point along the helix located at $(y = 0, z > 0)$ in Fig. 8.1, we see that the rotation of the helix about the x axis induces an instantaneous local velocity with y component $u_y = -\Omega R$. Using the force density given in Eq. (5.10), which takes the same value at any point along the helix, this leads to the total hydrodynamic force as

$$F = f_x L = (c_{\parallel} - c_{\perp})\sin\theta\cos\theta RL\Omega, \tag{8.2}$$

while the total hydrodynamic torque experienced by the helix is given by

$$T = -f_y RL = -(c_{\parallel}\sin^2\theta + c_{\perp}\cos^2\theta)R^2 L\Omega. \tag{8.3}$$

The force and torque due to translation of the helix at velocity $U\mathbf{e}_x$ may be deduced by exploiting symmetry and replacing in Eq. (5.10) the x direction by y, y by x and the angle θ by $\pi/2 - \theta$, leading to

$$F = -(c_\| \cos^2 \theta + c_\perp \sin^2 \theta)LU \qquad (8.4)$$

and

$$T = (c_\| - c_\perp) \sin \theta \cos \theta RU. \qquad (8.5)$$

Substituting the results of Eqs. (8.2)–(8.5) into Eq. (8.1), we get confirmation that the off-diagonal coefficients of the resistance matrix are identical, and obtain explicitly that

$$A = (c_\| \cos^2 \theta + c_\perp \sin^2 \theta)L, \qquad (8.6a)$$
$$B = (c_\perp - c_\|) \sin \theta \cos \theta RL, \qquad (8.6b)$$
$$D = (c_\| \sin^2 \theta + c_\perp \cos^2 \theta)R^2 L. \qquad (8.6c)$$

We always have $A > 0$, $D > 0$, which was expected since the resistance tensor in Eqs. (5.3) and (8.1) is positive definite. In contrast, the sign of B depends on the chirality of the helix. A left-handed helix has $\theta > 0$ and thus $B > 0$ so that a negative rotation around x ($\Omega < 0$) leads in this case to a positive propulsive hydrodynamic force, $F > 0$. As expected, in the limit where the helix becomes a rod ($\theta = 0$) or a circular ring ($\theta = \pi/2$), we get $B = 0$ and propulsion is lost.

We note that the relationship in Eq. (8.3) captures the hydrodynamic torque arising from the local velocity of every point on the helix induced by its axial rotation (i.e. $|u| = R|\Omega|$). A second contribution to the torque arises from the local rotation of each portion of the helix around its centreline (Chwang and Wu, 1971). The torque density arising in this case is constant, equal to $-4\pi\mu a^2 \tilde{\Omega}$, where $\tilde{\Omega} = \Omega \cos \theta$ is the component of the rotation rate along the tangent to the helix. This contributes to an additional hydrodynamic torque on the helix, T_m, given by

$$T_m = -4\pi\mu a^2 L\Omega \cos \theta. \qquad (8.7)$$

Considering the geometry of the normal flagellar filaments of *E. coli* as reported in Turner et al. (2000) and Darnton et al. (2007), we see that $a \approx 20$ nm, $R \approx 0.2\ \mu$m, so that $a/R \approx 0.1$, leading therefore to a ratio $T_m/(D\Omega) = O(10^{-2})$. This contribution to the hydrodynamic torque can therefore be safely neglected.

8.2 Swimming of a Flagellated Bacterium

8.2.1 Setup

The simplest setup to address the swimming of a bacterium is illustrated in Fig. 8.2. A cell body is equipped with a motor, which rotates a helical filament relative to

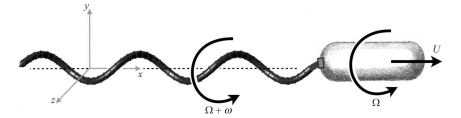

Figure 8.2 Propulsion induced by the rotation of a rigid helical filament. The helix, with notation from Fig. 8.1, rotates with angular velocity ω relative to the cell body rotating with angular velocity Ω. As a result of the relative rotation, the whole cell translates with speed U.

the body. Due to the off-diagonal propulsion component of the helical resistance matrix calculated above, this rotation induces a force on the cell body and leads to swimming. We assume for simplicity that both the cell and the filament rotate about and translate along a common straight line, and consider the more general three-dimensional situation in Section 8.3. Measuring all translational and angular velocities along the x axis, we denote by U the swimming speed and by Ω the angular velocity of the cell body. The helical filament rotates with angular velocity ω relative to the cell body, and thus with rate $\Omega + \omega$ when measured in the lab frame.

8.2.2 Free-Swimming Kinematics

We assume for simplicity that the cell body is sufficiently symmetric that its principal axes of hydrodynamic resistance are aligned with x, y, z and therefore that in this frame the resistance matrix is diagonal (this would be the case for a cell body axisymmetric around the x axis, for example). The hydrodynamic force and torque acting on the cell body may then be written as

$$\begin{pmatrix} F \\ T \end{pmatrix}_{body} = - \begin{pmatrix} A_0 & 0 \\ 0 & D_0 \end{pmatrix} \begin{pmatrix} U \\ \Omega \end{pmatrix}_{body}, \tag{8.8}$$

where A_0 and D_0 are two positive resistance coefficients, which are functions of the size and shape of the cell body and proportional to the fluid viscosity.

The swimming kinematics for the cell, i.e. the values of U and Ω, are obtained by requiring the overall balance of hydrodynamic force and torque. Enforcing this balance using Eqs. (8.1), with Ω replaced by $\Omega + \omega$, and (8.8) leads to the system

$$- \begin{pmatrix} A_0 & 0 \\ 0 & D_0 \end{pmatrix} \begin{pmatrix} U \\ \Omega \end{pmatrix} - \begin{pmatrix} A & B \\ B & D \end{pmatrix} \begin{pmatrix} U \\ \Omega + \omega \end{pmatrix} = \begin{pmatrix} 0 \\ 0 \end{pmatrix}, \tag{8.9}$$

which can be rearranged to get

$$
\begin{pmatrix} A_0 + A & B \\ B & D_0 + D \end{pmatrix} \begin{pmatrix} U \\ \Omega \end{pmatrix} = - \begin{pmatrix} B\omega \\ D\omega \end{pmatrix}, \tag{8.10}
$$

and inverted to obtain the values of the free-swimming translational and angular velocities as

$$
U = \frac{BD_0}{B^2 - (A_0 + A)(D_0 + D)} \omega, \tag{8.11a}
$$

$$
\Omega = \frac{D(A_0 + A) - B^2}{B^2 - (A_0 + A)(D_0 + D)} \omega. \tag{8.11b}
$$

From Eq. (8.11b) we also obtain the rotation rate of the helix in the lab frame as

$$
\Omega + \omega = \frac{D_0(A_0 + A)}{(A_0 + A)(D_0 + D) - B^2} \omega. \tag{8.12}
$$

8.2.3 Physical Interpretation

The simple results in Eqs. (8.11)–(8.12) have many fundamental implications. First, we note that if $c_\parallel/c_\perp = 1$, then we get from Eq. (8.6) that $B = 0$ and thus $U = 0$. Similar to the case of eukaryotic swimming, drag anisotropy is crucial to enable the locomotion of bacteria.

Second, we recall that in the case of the unidirectional motion of a flagellated spermatozoon the cell body always reduced the locomotion speed. In contrast, a cell body is required to enable bacterial swimming, as otherwise torques cannot be balanced. This may be seen by examining the results of Eq. (8.11a) in the case where the cell body disappears. In that case, both $A_0 = 0$ and $D_0 = 0$, leading to $U = 0$. Bacteria need therefore a body to swim, explaining the large relative sizes of bacterial cell bodies compared to those of spermatozoa (Brennen and Winet, 1977). Note that in the absence of a cell body, Eq. (8.12) leads to $\Omega + \omega = 0$, and the helix does not move in the lab frame; this result is consistent with Chapter 2, Exercise 4, showing that rigid bodies cannot swim.

What is the sign of the cell body rotation, Ω, in relation to the relative rotation of the flagellar filament, ω? Physical intuition on the balance of torques might suggest that the cell body should always counter-rotate, and thus that these angular velocities always have opposite signs. This can be seen by noting that the resistance matrix on the left-hand side of Eq. (8.10) is positive definite, hence the denominator of the fraction in Eq. (8.11b), which is minus its determinant, is negative. Since the numerator is the determinant of Eq. (8.1) plus a positive quantity, it is always positive. Therefore Ω and ω always have opposite signs. Furthermore, both numerator and denominator in the fraction of Eq. (8.12) are positive and therefore the angular velocity of the flagellum in the lab frame, $\Omega + \omega$, has a sign that is always opposite to that of the angular velocity of the cell body, Ω. This confirms the intuition that a counter-rotation is necessary to balance hydrodynamic torques.

Similarly, since both the denominator of Eq. (8.11a) and D_0 are positive, it is the sign of $B\omega$ that dictates the direction of locomotion. Therefore, the swimming direction depends on the chirality of the helix and on the rotation direction of the filament. The generic biological situation, for example for *E. coli* during normal swimming, is one where the filament is left-handed and the rotation of the filament is counter-clockwise (CCW) when measured from behind the cell looking along the axis of the helix. This means that $B > 0$ and $\omega < 0$ and therefore Eq. (8.21) predicts that $U > 0$. The rotation of the helical filament pushes the cell, which swims with cell body at the front and flagellum at the back. Changing the chirality of the helix, or the sign of the filament rotation, would reverse this direction so the cell would swim flagellum-first, as is observed for species of bacteria alternating between forward- and backward-swimming modes (Magariyama et al., 2005).

8.2.4 Orders of Magnitude

Since the orders of magnitude of Ω and U in Eq. (8.11) are not immediately obvious, we make a further simplification and follow Purcell (1997) to show that for a helix the ratio $B^2/(AD)$ is always small. Computing the ratio between B^2 and AD using Eq. (8.6) we obtain

$$\frac{B^2}{AD} = \frac{(c_\perp - c_\|)^2 \sin^2\theta\cos^2\theta}{(c_\| \cos^2\theta + c_\perp \sin^2\theta)(c_\| \sin^2\theta + c_\perp \cos^2\theta)}. \tag{8.13}$$

Since the fraction in Eq. (8.13) is symmetric under $\theta \leftrightarrow \pi/2 - \theta$, it can be shown that its maximum value is obtained at $\theta = \pi/4$, at which point both $\sin\theta$ and $\cos\theta$ are equal to $1/\sqrt{2}$ so

$$\frac{B^2}{AD} \leq \frac{(c_\perp - c_\|)^2}{(c_\| + c_\perp)^2}. \tag{8.14}$$

With the simple estimate $c_\|/c_\perp \approx 1/2$, Eq. (8.14) leads to the upper bound

$$\frac{B^2}{AD} \leq \frac{1}{9}. \tag{8.15}$$

With this result, we proceed to neglect B^2 in both the numerator and denominator of Eq. (8.11b), which leads to the simplifications

$$\Omega \approx -\frac{D}{(D_0 + D)}\omega, \tag{8.16}$$

and

$$\Omega + \omega \approx \frac{D_0}{(D_0 + D)}\omega. \tag{8.17}$$

The order of magnitude between the two rotational velocities depends clearly on the relative values of D and D_0. As pointed out by Purcell (1997), for standard flagellated bacteria we are typically in the limit where D/D_0 is a small number. This can be seen using measurements for *E. coli* (Darnton et al., 2007). The cell body has width $W \approx 0.88\,\mu m$ and body length $L_b \approx 2.5\,\mu m$, so the value of D_0 is well approximated by the result for a prolate spheroid of cross-sectional radius $R_b = W/2$ in the slender limit,

$$D_0 \approx \frac{8}{3}\pi\mu R_b^2 L_b, \tag{8.18}$$

with errors relative to the exact formulae on the order of the square of the aspect ratio, and thus of about 10% (Kim and Karrila, 1991). Assuming for the filament that $c_\parallel/c_\perp \approx 1/2$ and $\theta \approx \pi/4$, we have

$$D \approx \frac{3}{4}c_\perp R^2 L, \tag{8.19}$$

so that

$$\frac{D}{D_0} \approx \frac{9}{8\ln(L/a)}\frac{R^2 L^2}{R_b^2 L_b}. \tag{8.20}$$

With $R \approx 0.2\,\mu m$, $L \approx 10\,\mu m$ and $a \approx 20$ nm we then obtain the ratio $D/D_0 \approx 0.15$ and therefore, following Eq. (8.16), we have a difference of about one order of magnitude between the rotation of the flagellar filament and that of the cell body. With the flagellar rotation rate of 163 Hz reported by Darnton et al. (2007), our model predicts a cell body rotation at 24 Hz, which is exactly the value measured experimentally in that study.

The same two approximations used to derive an estimate for the value of Ω may also be used to simplify the result for the swimming speed. Assuming that both $B^2/(AD)$ and D/D_0 are small numbers, Eq. (8.11a) can be approximated as

$$U \approx -\frac{B}{(A_0 + A)}\omega, \tag{8.21}$$

and the order of magnitude for the swimming speed of the cell may be estimated using the same biological numbers as above. For a prolate spheroid we have approximately

$$A_0 \approx \frac{2\pi\mu L_b}{\ln(L_b/R_b) - 1/2}. \tag{8.22}$$

With $c_\parallel/c_\perp \approx 1/2$ and $\theta \approx \pi/4$ we also have

$$A \approx \frac{3}{4}c_\perp L, \quad B \approx \frac{1}{4}c_\perp RL, \tag{8.23}$$

so that Eq. (8.21) leads to

$$\frac{U}{R\omega} \approx \frac{1}{3 + 2\left(\frac{\ln(L/a)}{\ln(L_b/R_b)-1/2}\right)\frac{L_b}{L}} \approx \frac{1}{8.71}. \tag{8.24}$$

For a filament rotating with frequency 100 Hz, this leads to $U \approx 15\ \mu m/s$. For the experimental measurements of Darnton et al. (2007) with rotation at 163 Hz we predict $U \approx 23.5\ \mu m/s$, in excellent agreement with their measured value of $U \approx 25\ \mu m/s$.

8.2.5 Energy Expended by the Rotary Motor

The theoretical estimate for the hydrodynamics of the swimming bacterium allows us to estimate the power output of the bacterial rotary motor. The torque applied by the motor, N, is equal to the hydrodynamic torque acting on the rotating helical filament, given by the second term on the left-hand side of Eq. (8.9) as

$$N = BU + D(\Omega + \omega), \tag{8.25}$$

and therefore

$$N = \frac{B^2 - D(A_0 + A)}{B^2 - (A_0 + A)(D_0 + D)}D_0\omega. \tag{8.26}$$

Using the fact that both $B^2/(AD)$ and D/D_0 are small, Eq. (8.26) is simplified to $N \approx D\omega$. This is the same motor torque as that applied by a rotation of the helix at speed ω with no translation, showing that the additional contributions to the torque arising from translational and rotational swimming of the cell are negligible.

The power output of the motor, P_m, is thus obtained as

$$P_m = N\omega \approx \frac{3}{4}c_\perp R^2 L\omega^2. \tag{8.27}$$

With $c_\perp \approx 2 \times 10^{-3}$ Pa s, $R \approx 0.2\ \mu m$, and the rotation at 100 Hz of a filament of length $L \approx 10\ \mu m$, we obtain the estimate

$$P_m \approx 2 \times 10^{-16}\ \text{W}. \tag{8.28}$$

For a motor rotating at frequency 100 Hz, the work E_m required to power each revolution is

$$E_m = \frac{P_m}{100} \approx 2 \times 10^{-18}\text{J}. \tag{8.29}$$

Bacterial rotary motors are driven by either H^+ or Na^+ ion gradients, depending on the bacteria species. In the well-studied case of *E. coli*, it is the electrostatic work of H^+ ions crossing a membrane that powers the rotation of the motor. The charge of a proton is $e = 1.6 \times 10^{-19}$ C and with a transmembrane electric potential $\Delta V \approx 100$ mV, the maximum energy released by the passage of each proton is therefore

$$\Delta E = e \times \Delta V \approx 1.6 \times 10^{-20} \text{ J}. \tag{8.30}$$

Comparing Eq. (8.29) with Eq. (8.30) we see that for a motor at maximum efficiency, $N = E_m/\Delta E \approx 125$ protons have to cross the membrane per turn. This simple calculation shows that each ion powers a rotation of approximately $3°$, consistent with experimental estimates (Sowa and Berry, 2008).

8.3 Swimming Using Finite-Size Helical Filaments

8.3.1 Free Swimming

The model in Section 8.2 allows us to predict the swimming speed and angular velocity of a model bacterium translating and rotating along a straight line. In general, however, the directions of translational and rotational swimming are not aligned, and the one-dimensional approach needs to be generalised to three dimensions.

The resistance matrix of the helix, $\mathbf{R}_h$, relates linearly the force, $\mathbf{F}_h$, and torque, $\mathbf{T}_h$, acting on it to its instantaneous velocity, $\mathbf{U}_h$, and angular velocity, $\mathbf{\Omega}_h$, as

$$\begin{pmatrix} \mathbf{F}_h \\ \mathbf{T}_h \end{pmatrix} = -\mathbf{R}_h \begin{pmatrix} \mathbf{U}_h \\ \mathbf{\Omega}_h \end{pmatrix}, \tag{8.31}$$

where we now allow all vectors to be three-dimensional. For the cell body, we may write similarly the force and torque acting on it as linearly proportional to its translational and angular velocities, $\mathbf{U}$ and $\mathbf{\Omega}$, as

$$\begin{pmatrix} \mathbf{F}_b \\ \mathbf{T}_b \end{pmatrix} = -\mathbf{R}_b \begin{pmatrix} \mathbf{U} \\ \mathbf{\Omega} \end{pmatrix}. \tag{8.32}$$

By defining carefully the velocities $\mathbf{U}_h$ and $\mathbf{U}$ as those of a common point for the filament and the cell body (for example, the point of attachment of the filament on the body), the rotation at angular velocity ω of the filament relative to the body implies that the kinematics of filament and body are related as

$$\begin{pmatrix} \mathbf{U}_h \\ \mathbf{\Omega}_h \end{pmatrix} = \begin{pmatrix} \mathbf{U} \\ \mathbf{\Omega} \end{pmatrix} + \begin{pmatrix} 0 \\ \omega \end{pmatrix}. \tag{8.33}$$

Following the same approach as in Section 8.2, we determine the swimming kinematics by requiring the overall balance of both hydrodynamic force and torque.

Measuring torques at the same common point for both the helical filament and the cell body, we may write

$$-\mathbf{R}_b \begin{pmatrix} \mathbf{U} \\ \mathbf{\Omega} \end{pmatrix} - \mathbf{R}_h \begin{pmatrix} \mathbf{U} \\ \mathbf{\Omega} + \boldsymbol{\omega} \end{pmatrix} = \begin{pmatrix} \mathbf{0} \\ \mathbf{0} \end{pmatrix}, \tag{8.34}$$

leading to

$$\begin{pmatrix} \mathbf{U} \\ \mathbf{\Omega} \end{pmatrix} = -(\mathbf{R}_b + \mathbf{R}_h)^{-1} \mathbf{R}_h \begin{pmatrix} \mathbf{0} \\ \boldsymbol{\omega} \end{pmatrix}. \tag{8.35}$$

The solution in Eq. (8.35) is deceptively simple because all the details of the cell geometry are hidden in the values of the two six-by-six matrices, $\mathbf{R}_b$ and $\mathbf{R}_h$. For given shapes of the filament and the cell body, the swimming kinematics depend on the relative position of the helix and the body (since the velocity and torques are measured at a point common to both; see Exercises) and on their relative orientation. Furthermore, the value of $\boldsymbol{\omega}$ is known in the frame of the helical filament, in which the tensor $\mathbf{R}_h$ is constant, while the tensor $\mathbf{R}_b$ is fixed in the cell body frame. Since the filament rotates relative to the cell body, both $\mathbf{U}$ and $\mathbf{\Omega}$ are in general time-dependent.

8.3.2 Helical Trajectories

If the axes of the flagellar filament and of the cell body are fixed relative to one another, the swimming velocity, $\mathbf{U}$, and angular velocity, $\mathbf{\Omega}$, vary periodically around nonzero mean values. After each period, the whole cell has translated and rotated by some amount, and by periodicity we obtain three-dimensional trajectories resembling helices.

This can be made more precise by assuming for simplicity that the results in Eq. (8.35) lead to constant values of $\mathbf{U}$ and $\mathbf{\Omega}$. Except for the two particular cases $\mathbf{\Omega} = \mathbf{0}$ (swimming along a straight line with no rotation) and $\mathbf{U} = \mathbf{0}$ (rotation with no swimming), we can assume $\mathbf{U} \neq \mathbf{0}$ and $\mathbf{\Omega} \neq \mathbf{0}$. For notation convenience, we write in this case $\mathbf{\Omega} = \Omega \mathbf{e}_{\parallel}(t)$ and $\mathbf{U} = U_{\parallel} \mathbf{e}_{\parallel}(t) + U_{\perp} \mathbf{e}_{\perp}(t)$, where Ω denotes the magnitude of the angular velocity, $\mathbf{e}_{\parallel}$ is a unit vector in the direction of $\mathbf{\Omega}$, while $U_{\parallel}$ and $U_{\perp}$ are the projections of $\mathbf{U}$ parallel and perpendicular to $\mathbf{\Omega}$, respectively. The unit vectors $\mathbf{e}_{\parallel}(t)$ and $\mathbf{e}_{\perp}(t)$ would, *a priori*, be expected to change in time because of the rotation of the cell. For a unit vector $\mathbf{e}$ in the frame of the cell body, rigid-body rotation implies

$$\frac{d\mathbf{e}}{dt} = \mathbf{\Omega} \times \mathbf{e}. \tag{8.36}$$

Since $\mathbf{e}_\parallel$ and $\mathbf{\Omega}$ are parallel, we have

$$\frac{\mathrm{d}\mathbf{e}_\parallel}{\mathrm{d}t} = \mathbf{0}, \qquad (8.37)$$

and therefore the vector $\mathbf{e}_\parallel$ is constant.

Next, let $\mathbf{X}(t)$ be the position of the swimmer and define $X_\parallel(t) = \mathbf{X}(t) \cdot \mathbf{e}_\parallel$. By definition, the rate of change of $\mathbf{X}$ is the swimming velocity, i.e.

$$\frac{\mathrm{d}\mathbf{X}}{\mathrm{d}t} = \mathbf{U}. \qquad (8.38)$$

Projecting that equation along the constant $\mathbf{e}_\parallel$ we obtain linear motion as

$$X_\parallel(t) = U_\parallel t. \qquad (8.39)$$

Identifying the lab frame vector $\mathbf{e}_z \equiv \mathbf{e}_\parallel$, Eq. (8.38) then leads to circular motion in the directions (x, y) perpendicular to z with radius $R_\perp$ given by

$$R_\perp = \frac{U_\perp}{\Omega}, \qquad (8.40)$$

and frequency Ω. The combination of a linear motion along z with circular motion in (x, y) leads to helical trajectories with radius $R_\perp$ and pitch $2\pi U_\parallel/\Omega$.

8.3.3 *Comparison with Experiments: B. subtilis Locomotion*

As shown in the previous section, swimming bacteria undergo helical motion in general. The radius and pitch of the helical trajectories depend sensitively on the geometry of the cell and on the relative position and orientation of flagellum and cell body. Due to natural variations, among a group of cells a large variety of helices may therefore be observed. This is illustrated in Fig. 8.3 where we reproduce experimental measurements of the trajectories of a population of 72,087 *B. subtilis* bacteria reported by Hyon et al. (2012). We first show a sample of tracked cell motion as observed experimentally in Fig. 8.3(a), clearly illustrating a variety of helix radii and pitches. We next display the statistical distribution of helix pitch (Fig. 8.3(b)) and radius (Fig. 8.3(c)), further emphasising that although all cells undergo approximately helical motion, large variations are observed among a given population.

8.4 Optimal Helical Swimming

8.4.1 *Optimal Size of Cell Body*

We saw above that if a bacterium does not have a cell body, it is unable to swim. If the cell body is too large, then the swimming velocity of the cell also decreases

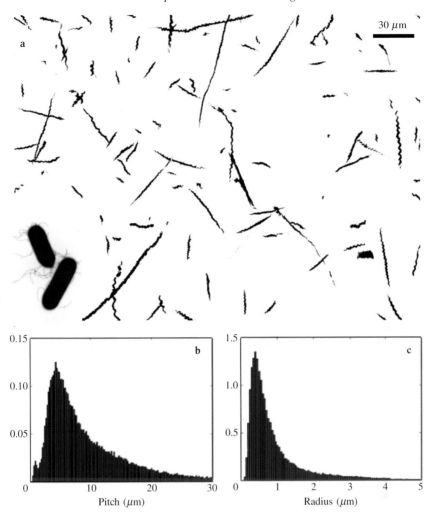

Figure 8.3 Helical trajectories of swimming bacteria. (a) Tracking of swimming *B. subtilis* cells in water showing helical trajectories with a variety of pitches and radii (inset: image of two cells, courtesy of Berkmen and Grossman, MIT). (b) Probability mass function for the distribution of pitch of helix for measurements on a population of 72,087 cells (average: 8.9 μm; standard deviation: 6 μm). (c) Same for the distribution of radius of helix (average: 0.91 μm; standard deviation: 0.81 μm). Adapted from Hyon et al. (2012), reproduced with permission.

to zero, which can be seen by examining Eq. (8.11a) in the limit of large cell body. If the size of the cell body is scaled by a factor δ, the dependence of the various components of the resistance matrix on size shown in Eq. (5.4) may be used to obtain in the limit of large δ that the numerator of Eq. (8.11a) scales as δ^3 whereas its denominator scales as δ^4, and thus the swimming speed decreases to zero as

$U \sim 1/\delta$. A body that is too small prevents the cell from balancing hydrodynamic torques but one that is too large induces too much drag.

If the velocity of a swimming bacterium goes to zero for both small and large cell bodies, an optimal size must exist. This question was addressed by Chwang and Wu (1971), who computed the optimal size of the cell body for a given helical filament. Examining their work in the case relevant to the flagellar filaments of *E. coli*, we see that their work predicts an optimal ratio $R_b/a \approx 20$. Since the thickness of a filament is $a = 20$ nm, this leads to an optimal body of width $2R_b = 0.8\ \mu m$, a result very close to the measured width of *E. coli* cells of 0.88 μm (Darnton et al., 2007).

8.4.2 Intrinsic Propeller Efficiency

The argument pertaining to the size of the cell body can be adapted to show that helices of optimal shape must also exist. Indeed, consider the helical filament from Fig. 8.1 with the value of R fixed and vary the wavelength λ. In the limit $\lambda/R \ll 1$ the helix becomes a tight circular ring, which is symmetric and cannot propel the cell; similarly, in the limit $\lambda/R \gg 1$, the helix becomes a straight filament that is not propelling the cell either. An optimal helical shape must therefore exist.

In order to derive the optimal helix, we use dimensional analysis to characterise the intrinsic properties of each helical shape. Consider a helical filament whose hydrodynamic resistance coefficients are given by the matrix in Eq. (8.1). An intrinsic dimensionless hydrodynamic efficiency of the helix, ε, should be a function of all three components of the resistance matrix, i.e.

$$\varepsilon = f(A, B, D). \tag{8.41}$$

Since ε is dimensionless, it must be a function of a dimensionless combination of A, B and D. As the components of the matrix are only functions of length and viscosity, there is only one combination of them that is dimensionless and we thus expect the helical efficiency to be given by

$$\varepsilon = f\left(\frac{B^2}{AD}\right). \tag{8.42}$$

Furthermore, since a helix with a larger value of B (propulsive force) and smaller values of both A and D (viscous friction) is expected to be more efficient, the function f in Eq. (8.42) should be monotonic.

One method to derive the value of ε was presented by Purcell (1997) in a posthumous paper. Purcell proposed to define the efficiency as the ratio between the useful power required to move the cell body at speed U, i.e. $A_0 U^2$, and the power output of the rotary motor, $N\omega$, so that

$$\varepsilon = \frac{A_0 U^2}{N\omega}.$$ (8.43)

Using Eqs. (8.11) and (8.26) in order to evaluate the values of U and N, we find that the efficiency is given by

$$\varepsilon = \frac{A_0 D_0 B^2}{\left[B^2 - D(A_0 + A)\right] \left[B^2 - (A_0 + A)(D_0 + D)\right]}.$$ (8.44)

We may further simplify Eq. (8.44) by using the two approximations in the relative values of the resistance coefficients derived in Section 8.2.4, namely $B^2 \ll AD$ and $D \ll D_0$, to obtain the simplified expression

$$\varepsilon = \frac{A_0 B^2}{D(A_0 + A)^2}.$$ (8.45)

Among all possible cell bodies, we saw in the previous section that the swimming speed was maximal for a specific value of the body size. The same result holds for the swimming efficiency, which is zero in the limits of both small ($A_0 \ll A$) and large ($A_0 \gg A$) cell body relative to the helical filament. For a given helical shape, the maximum possible efficiency may be obtained by using a scaling of the size of the cell body by a dimensionless factor δ so that $A_0 \rightarrow \delta A_0$. The value of δ leading to the maximum of Eq. (8.45) is easily seen to be $\delta = A/A_0$, corresponding to a value of the maximum efficiency, $\bar{\varepsilon}$, being

$$\bar{\varepsilon} = \frac{1}{4} \frac{B^2}{AD}.$$ (8.46)

The result in Eq. (8.46), the intrinsic propelling efficiency of each helical shape, is independent of the length of the helix and is consistent with the expected dimensionless scaling predicted by Eq. (8.42). Note that due to the upper bound derived in Eq. (8.14), we expect the efficiency to be a small number, and certainly $\bar{\varepsilon} \lesssim 3\%$.

8.4.3 Efficiency of Polymorphism

The flagellar filaments of bacteria are polymers whose monomer (flagellin) exists in two distinct conformation states. As a consequence of the molecular architecture of the filaments, when monomers with different conformation states are mixed, the flagellar filaments take one of 11 so-called polymorphic helical shapes. These shapes are illustrated in Fig. 8.4(a) and drawn so that all filaments have a total length of 5 μm. Out of the 11 possible shapes, 9 have been observed experimentally; 1 is straight (when all monomers are in the same conformation state), while

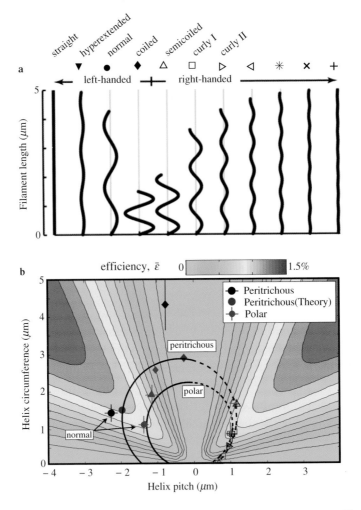

Figure 8.4 Efficiency of the polymorphic shapes of bacterial flagellar filaments. (a) The 11 polymorphic shapes in the case of peritrichous bacteria, of which 1 is straight, 3 are left-handed and 7 are right-handed, drawn so that all filaments have total length of 5 μm (adapted from Darnton and Berg (2007), with permission from Elsevier). (b) Contour plot of intrinsic propelling efficiency, $\bar{\varepsilon}$, as a function of the pitch of the helix and its circumference (colours), overlaid with symbols showing the geometrical characteristics of the polymorphic shapes of peritrichous bacteria using the same symbols as in (a) (measurements shown in black and theory shown in red) and polar flagellar (measurements shown in blue) (adapted from Spagnolie and Lauga (2011) with permission, Copyright 2011 by the American Physical Society).

3 of the helical shapes are left-handed and 7 are right-handed (Calladine, 1975; Kamiya et al., 1980; Hasegawa et al., 1998). Note that most of what we know about

polymorphism concerns 'peritrichous' bacteria with multiple flagella, and *E. coli* in particular. Bacteria from the 'polar' family equipped with a single flagellar filament assembled from a different flagellin protein follow a similar polymorphic sequence of helical shapes (Fujii et al., 2008).

The different polymorphic shapes of flagellar filaments come into play during locomotion of the cells. As exemplified by the well-studied behaviour of *E. coli*, wild-type bacteria explore their environment by alternating periods of straight swimming ('runs') with quick reorientation events ('tumbles'). During a swimming run, flagellar filaments are always in the left-handed 'normal' shape (third from the left in Fig. 8.4(a)). In contrast, during a tumble some of the motors switch their direction of rotation, inducing a polymorphic change for the corresponding filaments, typically to the right-handed curly I (sixth shape from the left in Fig. 8.4(a)) and semi-coiled configuration (fifth from the left). Occasionally, other polymorphic shapes are also involved in reorientation events (Turner et al., 2000).

The use of various polymorphic shapes by the swimming cells, and in particular the prevalence of the normal form during forward swimming, may be rationalised using the intrinsic propelling efficiency derived in Eq. (8.46). Since each polymorphic shape is a different helix, we may calculate the value of $\bar{\varepsilon}$ for each shape and compare them. In addition, we may compute the intrinsic propelling efficiency of any theoretical helix, and contrast it with the efficiency of the observed ones. The results are shown in Fig. 8.4(b) where we plot iso-values of the propelling efficiency, $\bar{\varepsilon}$, as a function of the pitch of the helix and its circumference (i.e. the perimeter of the cylinder on which the helix is coiled). Overlaid on this colour map are the measured geometrical characteristics of the polymorphic shapes of peritrichous bacteria (black), using the same symbols as in Fig. 8.4(a) (Spagnolie and Lauga, 2011). Similar results for polar flagellar filaments are shown in blue, while the red symbols show the theoretical peritrichous shapes, which are predicted to lay on circles in the pitch–circumference plane (Calladine, 1975, 1978).

Remarkably, the 'normal' polymorphic shapes in all cases (peritrichous, polar and theory) are the most efficient among all helical shapes available to the cell, approximately 25% more efficient than the next two shapes, semi-coiled and curly I, both of which are used for reorientation. In addition, we see that the experimental normal polymorphic shapes are very close to the overall optimal helix, obtained graphically as the intersection between the circles and the warm portion of the colour map. This hydrodynamic optimality of the normal polymorph, a conclusion that is in fact robust to a change in the effective flagellum diameter or length (Spagnolie and Lauga, 2011), suggests that, although the energetic costs of locomotion are known to be small for bacteria (Purcell, 1977), fluid mechanical forces may have played a significant role in the evolution of bacterial flagella.

Further Reading

Bacteria, discovered by van Leeuwenhoek in the seventeenth century (Gest, 2004), can differ greatly in their morphology and the arrangement of their flagella (Leifson, 1960). Peritrichous bacteria, such as *E. coli*, have flagella projecting in all directions, which bundle behind the cell during locomotion (Macnab, 1977; Berg, 2004). In contrast, polar bacteria have motors located only at the poles of the cell body. Of those, monotrichous bacteria have only one motor (e.g. *Pseudomonas aeruginosa*), lophotrichous bacteria have a tuft of flagellar filaments originating from a single pole (e.g. *Photobacterium fischeri*), while amphitrichous bacteria have flagellar filaments on both poles (e.g. *Ectothiorhodospira halochloris*). An additional class of bacteria, called spirochetes (e.g. *Borrelia burgdorferi*, causing Lyme disease), have flagella hidden in the small space between the outer membrane and the cytoplasmic membrane of the cell body and swim using whole-body helical undulations resulting from flagellar rotation (Vig and Wolgemuth, 2012).

The precise actuation mechanism whereby the locomotion of bacteria results from the passive rotation of helical flagellar filaments by rotary motors, instead of active helical deformation, was elucidated in the 1970s (Berg and Anderson, 1973; Silverman and Simon, 1974). Many molecular details of the bacterial rotary motor are now well understood (Berg, 2003). In particular, the torque–frequency relationship of the motor has been characterised for multiple species (Sowa and Berry, 2008). This torque is approximately constant at low motor frequency and then above a critical frequency it decreases linearly until a maximum frequency corresponding to the no-torque limit (both frequencies on the order of a few hundred hertz). Joining the rotary motor to the flagellar filament, and transmitting its rotation, is the short flexible hook. The hook is under compression when the flagellar filament pushes on the cell body, which can lead to buckling (Son et al., 2013) and asymmetry between the forward and backward motion of monotrichous bacteria (Xie et al., 2011). The bending of the hook is also crucial to allow locomotion of peritrichous bacteria (Riley et al., 2018).

The visualisation of flagellar filaments in action was achieved by Turner et al. (2000) using a novel fluorescence staining method. This method allowed them to observe polymorphic changes in real time, and was used alongside optical tweezers to measure force–extension relationships for different flagellar polymorphs (Darnton and Berg, 2007). Polymorphic changes, which have long been known to be induced by a variety of environmental factors, including external flows (Hotani, 1982), can be understood using the tools of solid mechanics and elasticity (Srigiriraju and Powers, 2005, 2006). Beyond swimming, bacteria have evolved a number of different motility methods (Jarrell and McBride, 2008). A relevant one in the context of helical propulsion is swarming (Kearns, 2010), in which swimming cells near nutrient-rich surfaces differentiate to a state where their cell body elongates, they grow more flagella and they move in a coordinated fashion (Darnton et al., 2010).

Historically, the fluid dynamics of helical propulsion was tackled by Chwang and Wu (1971) at the time when the nature of the helical wave used by bacteria (passive vs. active) was still unclear. A full hydrodynamic study was then proposed by Lighthill (1976), who took advantage of helical symmetry in order to calculate the value of the hydrodynamic force distribution along an infinite helix (Lighthill, 1996a). Full hydrodynamic interactions between the flagellar filament and the cell body were included by Higdon (1979b). Finally, the hydrodynamics of helical propulsion has been exploited to design artificial swimmers in the form of magnetised helices driven in rotation by an external rotating magnetic field (Ghosh and Fischer, 2009; Zhang et al., 2010), which offer force-free, but not torque-free, artificial swimmers up to a maximum step-out frequency (see Exercises). Other hydrodynamic aspects of the life of bacteria were reviewed by Lauga (2016), including topics covered in the subsequent chapters of this book.

Exercises

1. Consider a rigid body whose hydrodynamic resistance tensor is denoted by $\mathbf{R}$ when torques are expressed about point A on the body. Determine how the tensor $\mathbf{R}$ is modified when torques are computed about another point B on the body. Show that the hydrodynamic resistance matrix remains symmetric.

2. Using resistive-force theory, show that the helix angle (defined as in Fig. 8.1) maximising the propeller efficiency, $\bar{\varepsilon} = B^2/(4AD)$, is $\theta = 45°$.

3. Two rigid helices, H_1 and H_2, undergo relative rotation in a viscous fluid in which inertial effects are neglected. Helix H_1 is left-handed, of total length ℓ, helical radius a and helix angle $0 < \theta < \pi/2$. Helix H_2 has the same radius, a, and helix angle, θ, but is right-handed and has length $n\ell$, where $n \geqslant 0$ (not necessarily an integer). Both helices are aligned with their axes along $\mathbf{e}_z$ and linked in such a way that they cannot translate relative to each other but they do undergo a prescribed relative rotation of magnitude $\omega\mathbf{e}_z$ induced by a rotary motor of negligible hydrodynamic influence. The two-helix system is force- and torque-free. Compute the rotational speed of each helix and the total translational speed of the two-helix system. Interpret physically the results for the special cases $n = 0$, $n = 1$ and the limit $n \to \infty$.

4. An artificial swimmer in the shape of a rigid helix translates along, and rotates about, the z axis. The swimmer, equipped with a permanent magnetic torque of magnitude m_0 perpendicular to the z direction, is subject to an external magnetic

field $\mathbf{B}(t)$ of constant magnitude $\mathcal{B}_0$ rotating in the (x, y) plane with frequency ω. By considering the force and torque balances on the swimmer, show that below a critical 'step-out' frequency, ω_c, the swimmer rotates with the external magnetic field, but that it is unable to keep up with the field above ω_c. Furthermore, show that below ω_c there is a phase delay between the swimmer and the magnetic field that increases with ω.

$\left[\text{You may use the result } \int_0^{2\pi} (a - \sin x)^{-1}\, dx = 2\pi(a^2 - 1)^{-1/2}, a > 1.\right]$

5. The derivation of the maximum propulsion efficiency for a bacterium powered by a helical flagellar filament can be adapted to the case of the artificial swimmer from Exercise 4. Consider the efficiency for artificial swimming analogous to Eq. (8.43) and defined as

$$\hat{\varepsilon} = \frac{AU^2}{T\Omega},$$

where A is the resistance coefficient to translation, U and Ω the swimming speed and angular velocity of the artificial swimmer and T the external torque applied by the magnetic field. Show that

$$\hat{\varepsilon} = \frac{4\bar{\varepsilon}}{1 - 4\bar{\varepsilon}},$$

where $\bar{\varepsilon}$ is the optimal helical efficiency from Eq. (8.46), i.e. $\bar{\varepsilon} = B^2/(4AD)$.

6. Helical swimming can also be obtained by rotating flexible filaments. Consider a slender flexible filament of length L aligned with the x direction and subject to small-amplitude motion in the directions y and z. The end of the filament at $x = L$ is free, while the end at $x = 0$ is clamped along x at a point, M_0, undergoing circular motion $(y_0, z_0) = R_0(\cos(\omega t), \sin(\omega t))$. Derive the three-dimensional version of the linearised, passive elasto-hydrodynamic balance in Eq. (7.53) for $y(x, t)$ and $z(x, t)$ in the lab frame of reference and in the frame rotating with the point M_0. Assuming that the shape of the filament reaches a steady shape $(\tilde{y}(x), \tilde{z}(x))$ in the rotating frame, show that $\tilde{y}(x)$ satisfies

$$\frac{\partial^8 \tilde{y}}{\partial x^8} + \left(\frac{c_\perp \omega}{B}\right)^2 \tilde{y} = 0.$$

9

Flows and Stresses Induced by Cells

We have so far described the fluid dynamics relevant to the propulsion mechanisms of individual microorganisms, with a focus on the physical principles dictating cell locomotion. The length scales involved ranged from tens of nanometres to a few microns. In contrast, the dynamics of cell populations is characterised by much larger length scales, typically hundreds of microns and above. The flow disturbances induced by swimming cells on these large length scales play important biophysical roles, from governing the mixing and transport of nutrients to impacting the physical interactions of cells with their environment and the collective dynamics of populations. In this chapter we address the consequences of the force-free and torque-free swimming constraints of swimming cells on the flows they create. We develop the framework to describe these flow signatures mathematically on length scales larger than that of the organisms, compare them with experimental measurements, explain how to find better approximations near the cells and discuss implications for the hydrodynamic stresses induced by suspensions of swimming microorganisms.

9.1 Force-Free Swimming

In Chapter 6 we introduced the fundamental solution to the Stokes equations, the stokeslet, which characterises the flow due to a point force. We recall that in a Newtonian fluid of viscosity μ, the corresponding flow decays spatially as $1/r$ and may be written as

$$\mathbf{u}_F(\mathbf{r}; \mathbf{F}) = \frac{1}{8\pi\mu} \left(\frac{1}{r} + \frac{\mathbf{rr}}{r^3} \right) \cdot \mathbf{F}. \tag{9.1}$$

The stress field, σ_F, associated with a point force at the origin is the solution to

$$\nabla \cdot \sigma_F + \delta(\mathbf{r})\mathbf{F} = \mathbf{0}. \tag{9.2}$$

If we integrate this equation on a volume V bounded by a surface S that contains the origin and use the divergence theorem we obtain

$$\iint_S \sigma_F \cdot \mathbf{n} \, dS + \mathbf{F} = \mathbf{0}, \tag{9.3}$$

where $\mathbf{n}$ is the outward unit normal to S (i.e. pointing towards the fluid at infinity). The force on the fluid at infinity is therefore exactly equal to $\mathbf{F}$.

For the vast majority of small organisms, however, swimming occurs with no net force. Far from the cell there is therefore no stokeslet component of the flow, and the leading-order flow is a force dipole, decaying spatially as $1/r^2$. The physical origin of this force dipole is illustrated in Fig. 9.1 for three microorganisms, namely a swimming spermatozoon, a bacterium and a biflagellated alga. In all cases, the flagella produce local hydrodynamic forces (drawn with empty arrows) as they move relative to the surrounding fluid. This results in locomotion and therefore the moving cell exerts a force on the fluid in the opposite direction (filled arrows). Since only the flagella create propulsion but both the flagella and the cell body experience drag, the centre of drag and the centre of propulsion are physically separated, leading to a dipolar structure of the induced flow. This simple physical picture can be extended naturally to the case of more complex geometries, larger organisms and multiple flagella. In what follows, we develop the mathematical framework to describe such dipolar flows.

9.2 Force Dipoles

For notation convenience, we first rewrite the force in Eq. (9.1) as $\mathbf{F} = F\mathbf{e}$, where F is its magnitude and $\mathbf{e}$ is a unit vector, so that the stokeslet flow is

$$\mathbf{u}_F(\mathbf{r}; \mathbf{F}) = \frac{F}{8\pi\mu} \left[\frac{\mathbf{e}}{r} + \frac{(\mathbf{e} \cdot \mathbf{r})\mathbf{r}}{r^3} \right] \equiv F\mathbf{G}(\mathbf{r}; \mathbf{e}), \tag{9.4}$$

where the vector $\mathbf{G}$ is related to the Green's function of the Stokes equations. The schematic representation of this singularity using a single arrow (modelling the presence of the external force) is shown in Fig. 9.2(a).

To compute the flow due to a force dipole, consider a force $-F\mathbf{e}$ located at the origin and an equal and opposite force $F\mathbf{e}$ at a location $\epsilon\ell\mathbf{d}$ away, where $\mathbf{d}$ is a unit vector and ϵ a small dimensionless number (the vectors $\mathbf{e}$ and $\mathbf{d}$ do not have to be identical). By linearity of the Stokes equations, the total flow, $\mathbf{u}$, in the limit $\epsilon \ll 1$ is given by

$$\mathbf{u}(\mathbf{r}) = F\mathbf{G}(\mathbf{r} - \epsilon\ell\mathbf{d}; \mathbf{e}) - F\mathbf{G}(\mathbf{r}; \mathbf{e}) = -\epsilon\ell F(\mathbf{d} \cdot \nabla)\mathbf{G}(\mathbf{r}; \mathbf{e}) + O(\epsilon^2). \tag{9.5}$$

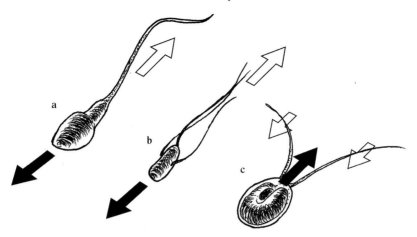

Figure 9.1 Schematic force dipoles for swimming microorganisms. Force-free swimming of a spermatozoon (a) and a bacterium (b). In both cases, the moving cell body induces a local leftward force on the surrounding fluid (drawn with filled arrow) while the flagellum induces a rightward propulsive force (empty arrow) leading to a force dipole. In contrast, for a biflagellated alga (c), the force dipole is reversed in sign and the cell is pulling itself into the fluid. Drawings courtesy of Jacques Lauga.

Defining $\mathcal{P} \equiv \epsilon\ell F$, with dimensions of force times length, the dipolar flow is the leading-order term in ϵ, namely

$$\mathbf{u}_{\text{dipole}}(\mathbf{r}; \mathbf{e}, \mathbf{d}) = -\mathcal{P}(\mathbf{d} \cdot \nabla)\mathbf{G}(\mathbf{r}; \mathbf{e}), \qquad (9.6)$$

and we call $\mathcal{P}$ the dipole strength.

From Eq. (9.4) and using standard results from vector calculus, we obtain

$$\nabla\mathbf{G}(\mathbf{r}; \mathbf{e}) = \frac{1}{8\pi\mu}\left[\frac{\mathbf{er} - \mathbf{re}}{r^3} + \frac{(\mathbf{e} \cdot \mathbf{r})\mathbf{1}}{r^3} - \frac{3(\mathbf{e} \cdot \mathbf{r})\mathbf{rr}}{r^5}\right], \qquad (9.7)$$

and thus

$$-\mathbf{d} \cdot \nabla\mathbf{G}(\mathbf{r}; \mathbf{e}) = \frac{1}{8\pi\mu}\left[\frac{(\mathbf{d} \cdot \mathbf{r})\mathbf{e} - (\mathbf{d} \cdot \mathbf{e})\mathbf{r}}{r^3} - \frac{(\mathbf{e} \cdot \mathbf{r})\mathbf{d}}{r^3} + \frac{3(\mathbf{e} \cdot \mathbf{r})(\mathbf{d} \cdot \mathbf{r})\mathbf{r}}{r^5}\right]. \qquad (9.8)$$

Finally, using the identity $(\mathbf{d} \times \mathbf{e}) \times \mathbf{r} = (\mathbf{d} \cdot \mathbf{r})\mathbf{e} - (\mathbf{e} \cdot \mathbf{r})\mathbf{d}$, we can write the dipolar flow from Eq. (9.6) explicitly as

$$\mathbf{u}_{\text{dipole}}(\mathbf{r}; \mathbf{e}, \mathbf{d}) = \frac{\mathcal{P}}{8\pi\mu}\left[\frac{(\mathbf{d} \times \mathbf{e}) \times \mathbf{r}}{r^3} - \frac{(\mathbf{d} \cdot \mathbf{e})\mathbf{r}}{r^3} + \frac{3(\mathbf{e} \cdot \mathbf{r})(\mathbf{d} \cdot \mathbf{r})\mathbf{r}}{r^5}\right]. \qquad (9.9)$$

This flow is often represented using two equal and opposite arrows, as shown in Fig. 9.2(b).

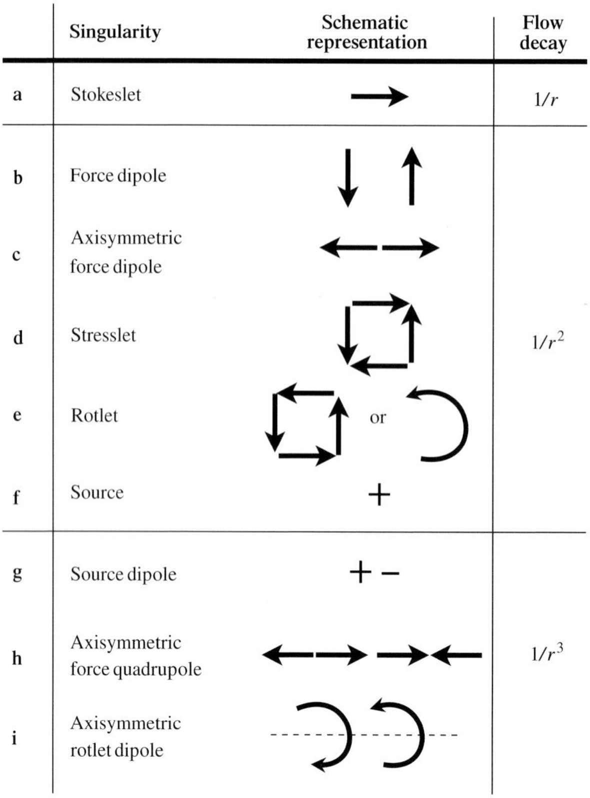

Singularity	Schematic representation	Flow decay
a	Stokeslet		$1/r$
b	Force dipole		
c	Axisymmetric force dipole		
d	Stresslet		$1/r^2$
e	Rotlet	or	
f	Source		
g	Source dipole		
h	Axisymmetric force quadrupole		$1/r^3$
i	Axisymmetric rotlet dipole		

Figure 9.2 Flow singularities with their schematic representation and spatial decay. The stokeslet flow (a), decaying as $1/r$, results from a point force acting on the fluid and is represented by a single arrow. Force dipoles due to equal and opposite forces (b) have flows decaying as $1/r^2$ and are represented by two arrows. The force dipole can be axisymmetric (c) or can be decomposed as the sum of a stresslet (symmetric part, d) and a rotlet (antisymmetric part, e). The point source (f), represented by a $+$, is an irrotational singularity decaying as $1/r^2$ and associated with fluid injected at the origin (see Chapter 6). Many flow singularities decay spatially as $1/r^3$ and the axisymmetric ones are: an irrotational source dipole (g) composed of a source ($+$) and a sink ($-$), a force quadrupole (h) and a rotlet dipole (i).

The flow in Eq. (9.9) decays spatially as $1/r^2$, and can be interpreted as the superposition of two fundamental solutions. The first term on the right-hand side of Eq. (9.9) is antisymmetric in $\mathbf{d}$ and $\mathbf{e}$ and thus disappears when $\mathbf{d} = \mathbf{e}$. In contrast, the other two terms are symmetric, and when $\mathbf{d} = \mathbf{e}$ they lead to the axisymmetric solution, illustrated in Fig. 9.2(c) using two aligned arrows. Introducing the symmetric/antisymmetric decomposition of the force dipole, we may write

$$\mathbf{u}_{\text{dipole}}(\mathbf{r}; \mathbf{e}, \mathbf{d}) = \mathbf{u}^{\text{sym}}_{\text{dipole}}(\mathbf{r}; \mathbf{e}, \mathbf{d}) + \mathbf{u}^{\text{antisym}}_{\text{dipole}}(\mathbf{r}; \mathbf{e}, \mathbf{d}), \qquad (9.10)$$

with

$$\mathbf{u}_{\text{dipole}}^{\text{sym}}(\mathbf{r}; \mathbf{e}, \mathbf{d}) = \frac{1}{2}\left[\mathbf{u}_{\text{dipole}}(\mathbf{r}; \mathbf{e}, \mathbf{d}) + \mathbf{u}_{\text{dipole}}(\mathbf{r}; \mathbf{d}, \mathbf{e})\right]$$

$$= \frac{\mathcal{P}}{8\pi\mu}\left[-\frac{(\mathbf{d}\cdot\mathbf{e})\mathbf{r}}{r^3} + \frac{3(\mathbf{e}\cdot\mathbf{r})(\mathbf{d}\cdot\mathbf{r})\mathbf{r}}{r^5}\right], \qquad (9.11\text{a})$$

$$\mathbf{u}_{\text{dipole}}^{\text{antisym}}(\mathbf{r}; \mathbf{e}, \mathbf{d}) = \frac{1}{2}\left[\mathbf{u}_{\text{dipole}}(\mathbf{r}; \mathbf{e}, \mathbf{d}) - \mathbf{u}_{\text{dipole}}(\mathbf{r}; \mathbf{d}, \mathbf{e})\right]$$

$$= \frac{\mathcal{P}}{8\pi\mu}\left[\frac{(\mathbf{d}\times\mathbf{e})\times\mathbf{r}}{r^3}\right]. \qquad (9.11\text{b})$$

In the following two sections, we interpret each of these two flows physically. The symmetric solution, Eq. (9.11a), is a particular case of the so-called stresslet, while the antisymmetric solution, Eq. (9.11b), is interpreted as due to a net torque applied to the fluid.

9.2.1 Stresslets

The symmetric part of the force dipole, Eq. (9.11a), is named a stresslet, a term coined by Batchelor (1970a) in the context of the fluid mechanics of suspensions. The corresponding flow is in the radial direction, $\mathbf{u} \sim \pm\mathbf{r}/(r^3)$. We illustrate this flow schematically in Fig. 9.2(d) using four straight arrows. The most general stresslet, as obtained by Batchelor, is

$$\mathbf{u}_S(\mathbf{r}; \mathbf{S}) = -\frac{3}{8\pi\mu}\frac{(\mathbf{r}\cdot\mathbf{S}\cdot\mathbf{r})\mathbf{r}}{r^5}, \qquad (9.12)$$

where the factor of -3 allows the stresslet tensor $\mathbf{S}$ in Eq. (9.12) to be interpreted physically as an extra stress generated by the swimmers (see Section 9.5). This tensor is symmetric and, in order to enforce incompressibility, has zero trace. Indeed, computing the volume flux due to the flow in Eq. (9.12) through a closed surface A with normal $\mathbf{n}$ leads to

$$\iint_A u_{S,i} n_i \, dS \propto S_{mk} \iint_A n_k n_m \, dS \propto S_{mm}, \qquad (9.13)$$

and thus $S_{mm} = 0$.

What is the stresslet tensor for the symmetric part of the force dipole? Using index notation, Eq. (9.12) becomes

$$u_{S,i}(\mathbf{r}; \mathbf{S}) = -\frac{3}{8\pi\mu}\frac{S_{jk}x_j x_k x_i}{r^5}, \qquad (9.14)$$

while the ith component of Eq. (9.11a) is given by

$$e_i \cdot \mathbf{u}_{\text{dipole}}^{\text{sym}}(\mathbf{r}; \mathbf{e}, \mathbf{d}) = \frac{\mathcal{P}}{8\pi\mu}\left[-\frac{d_j e_j x_i}{r^3} + \frac{3e_j x_j d_k x_k x_i}{r^5}\right], \qquad (9.15)$$

which can be made more obviously symmetric when written as

$$e_i \cdot \mathbf{u}_{\text{dipole}}^{\text{sym}}(\mathbf{r}; \mathbf{e}, \mathbf{d}) = \frac{\mathcal{P}}{8\pi\mu} \left[-\frac{d_j e_j x_i}{r^3} + \frac{3(e_j d_k + e_k d_j) x_j x_k x_i}{2r^5} \right]. \tag{9.16}$$

Comparing Eq. (9.14) with Eq. (9.16) we can identify the tensor **S** as

$$\mathbf{S} = \mathcal{P} \left[\frac{1}{3}(\mathbf{d} \cdot \mathbf{e})\mathbf{1} - \frac{1}{2}(\mathbf{ed} + \mathbf{de}) \right], \tag{9.17}$$

which is symmetric and trace-free. In the particular case where the unit vectors **d** and **e** are identical (case of an axisymmetric dipole, Fig. 9.2(c)) we have the simplification

$$\mathbf{S} = \mathcal{P} \left(\frac{1}{3}\mathbf{1} - \mathbf{ee} \right), \tag{9.18}$$

and the sign of $\mathcal{P}$ is interpreted physically in Section 9.3.

9.2.2 Rotlets

The antisymmetric part of the force dipole, Eq. (9.11b), is termed a rotlet,

$$\mathbf{u}_{\text{dipole}}^{\text{antisym}}(\mathbf{r}; \mathbf{e}, \mathbf{d}) = \frac{\mathcal{P}}{8\pi\mu} \left[\frac{(\mathbf{d} \times \mathbf{e}) \times \mathbf{r}}{r^3} \right]. \tag{9.19}$$

Physically, this flow results from a net external torque applied to the fluid, which can be seen in two different ways. First, we may use a classical solution of Stokes flow, namely the flow induced by the rotation of a rigid sphere of radius a (Leal, 2007). Using spherical coordinates centred at the origin of the sphere, the flow induced by a rotation at rate $\mathbf{\Omega}$ is given by

$$\mathbf{u} = \frac{a^3}{r^3} \mathbf{\Omega} \times \mathbf{r}, \tag{9.20}$$

and the net torque, **R**, applied to the fluid is given by (we use here the symbol **R** as in 'rotlet')

$$\mathbf{R} = 8\pi\mu a^3 \mathbf{\Omega}. \tag{9.21}$$

These two results can be combined to rewrite the flow as

$$\mathbf{u}_R(\mathbf{r}; \mathbf{R}) = \frac{1}{8\pi\mu} \frac{\mathbf{R} \times \mathbf{r}}{r^3}. \tag{9.22}$$

Comparing Eq. (9.19) and Eq. (9.22), the antisymmetric force dipole is seen to apply a net torque on the fluid, $\mathbf{R} = \mathcal{P}\mathbf{d} \times \mathbf{e}$.

Alternatively, we may compute the torque applied to the fluid directly from the Stokes equations. Following Eq. (9.6), for a force dipole the Stokes equations are forced by a dipolar density $\mathbf{\Lambda} = -\mathcal{P}\mathbf{d} \cdot \nabla[\delta(\mathbf{r})\mathbf{e}]$ as

$$\nabla \cdot \boldsymbol{\sigma} + \mathbf{\Lambda} = \mathbf{0}. \tag{9.23}$$

Taking the cross product with $\mathbf{r}$ and integrating over a closed fluid volume V enclosing the origin leads to

$$\iiint_V \mathbf{r} \times (\nabla \cdot \boldsymbol{\sigma}) \, dV + \iiint_V (\mathbf{r} \times \mathbf{\Lambda}) \, dV = \mathbf{0}. \tag{9.24}$$

The first term has the integrand

$$\mathbf{r} \times (\nabla \cdot \boldsymbol{\sigma}) = \epsilon_{ijk} r_j (\partial_m \sigma_{mk}) = \epsilon_{ijk} \partial_m (r_j \sigma_{mk}) - \epsilon_{ijk} \sigma_{jk}, \tag{9.25}$$

and the second term on the right-hand side of Eq. (9.25) is zero because ϵ is an antisymmetric tensor while σ is symmetric. Denoting by S the surface around V with outward normal $\mathbf{n}$, we can apply the divergence theorem on the first term to get

$$\iiint_V \mathbf{r} \times (\nabla \cdot \boldsymbol{\sigma}) \, dV = \iint_S \epsilon_{ijk} r_j \sigma_{mk} n_m \, dS = \iint_S \mathbf{r} \times (\boldsymbol{\sigma} \cdot \mathbf{n}) \, dS, \tag{9.26}$$

which is the torque applied on the surface by the fluid forces.

The second integral in Eq. (9.24) has the integrand

$$\mathbf{r} \times \mathbf{\Lambda} = \mathcal{P}[\mathbf{d} \cdot \nabla(\delta(\mathbf{r})\mathbf{e})] \times \mathbf{r} = \mathcal{P}\epsilon_{ijk} e_j d_m [\partial_m(\delta r_k) - \delta \cdot \partial_m r_k]. \tag{9.27}$$

The first term on the right-hand side of Eq. (9.27) can be evaluated using the divergence theorem, yielding no contribution, since the delta function is zero everywhere on S. For the second term we have $\partial_m r_k = 1$ if $m = k$ and 0 otherwise, and thus the volume integral is given by

$$\iiint_V (\mathbf{r} \times \mathbf{\Lambda}) \, dV = -\mathcal{P} \iiint_V \epsilon_{ijk} e_j d_k \delta(\mathbf{r}) \, dV = -\mathcal{P}\mathbf{e} \times \mathbf{d}. \tag{9.28}$$

So finally we get the equality

$$\iint_S \mathbf{r} \times (\boldsymbol{\sigma} \cdot \mathbf{n}) \, dS = \mathcal{P}\mathbf{e} \times \mathbf{d}, \tag{9.29}$$

and obtain the same physical interpretation as in Eq. (9.22), namely that a force dipole applies a net torque on the fluid equal to $\mathcal{P}\mathbf{e} \times \mathbf{d}$. As illustrated in Fig. 9.2(e), this singularity can be represented schematically using either four straight arrows or a singe curved arrow.

9.3 Leading-Order Flow Around Cells

9.3.1 Swimming Cells Are Stresslets

The vast majority of swimming cells are not only force-free, but also torque-free; their motion therefore applies zero net force and torque on the surrounding fluid. Given the result from the previous section, this means that the antisymmetric part of any force dipole cannot be part of the flow in the far field, which can thus only include its symmetric part. The most general instantaneous flow far from a swimming cell is thus a stresslet, given by Eq. (9.12). This flow is purely radial, with a magnitude that decays as $1/r^2$, and a direction that depends on the components of $\mathbf{S}$ in relation to the local position $\mathbf{r}$. Note that since the stresslet tensor $\mathbf{S}$ is symmetric and $\mathrm{tr}\,\mathbf{S} = 0$ due to the incompressibility constraint, it possesses only five independent components. Furthermore, the flow given by Eq. (9.12) is of this form in a lab frame that is centred instantaneously on the swimmer. It remains of this form for all times, with a stresslet tensor that, in general, is time-dependent, $\mathbf{S}(t)$.

A particularly illustrative case is that of an axisymmetric swimming cell with a constant stresslet (in the swimming frame). To a good approximation, this describes the flow created by most flagellated bacteria. Indeed, recall that a swimming bacterium is powered by a helical flagellum rotating quickly at about 100 Hz and a slow-rotating, approximately axisymmetric cell body (see Chapters 1 and 8). Since the relevant swimming timescale for the bacterium is on the order of a fraction of a second (typical length scale on the order of $5\,\mu$m and typical swimming speed on the order of $20\,\mu$m/s), we can average over many flagellar rotations on the order of the swimming time, and obtain an approximately axisymmetric flow.

In that case, if $\mathbf{e}$ denotes the axis of symmetry of a cell, the stresslet $\mathbf{S}$ is given by Eq. (9.18), $\mathbf{S} = \mathcal{P}\left(\frac{1}{3}\mathbf{1} - \mathbf{ee}\right)$, where $\mathcal{P}$ is constant. When written in the frame instantaneously centred on the swimmer, the flow field is then

$$\mathbf{u}_S\left(\mathbf{r}; \mathbf{S}\right) = \frac{\mathcal{P}}{8\pi\mu}\left[-\frac{1}{r^3} + \frac{3(\mathbf{e}\cdot\mathbf{r})^2}{r^5}\right]\mathbf{r}. \tag{9.30}$$

To gain further insight into this flow, let us use spherical coordinates (r, θ, ϕ) such that $\mathbf{e}$ denotes the polar axis $\theta = 0$, so that the flow may be rewritten as

$$\mathbf{u}_S\left(\mathbf{r}; \mathbf{S}\right) = \frac{\mathcal{P}}{8\pi\mu}\left(\frac{3\cos^2\theta - 1}{r^2}\right)\mathbf{e}_r. \tag{9.31}$$

This radial flow is axisymmetric and its magnitude changes sign at the critical angle such that $\cos^2\theta_c = 1/3$, corresponding to $\theta_c \approx 54.7°$ and $125.3°$. The radial flow closer to the polar axis than θ_c has the same sign as $\mathcal{P}$, while flow near the equator ($\theta = \pi/2$) has the sign opposite to that of $\mathcal{P}$. The structure of this flow is illustrated

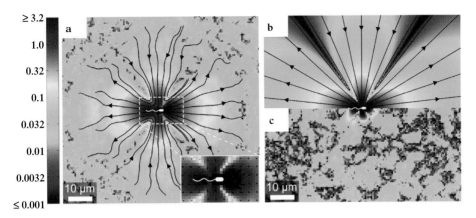

Figure 9.3 Flow induced by a swimming bacterium averaged over the timescale of rotation of its flagella. (a) Measurement for swimming *E. coli* (scale is in μm/s). (b) Theoretical prediction from force dipole, Eq. (9.31), with $\mathcal{P} \approx 0.8$ pN μm. (c) Difference between measurement and theoretical prediction. Reproduced from Drescher et al. (2011) with permission.

in Fig. 9.3(b) and described in further detail in Section 14.1, where we explore hydrodynamic interactions between microorganisms.

9.3.2 Pushers vs. Pullers

The sign of $\mathcal{P}$ plays an important biophysical role, as is further explored in the next chapters. Cells characterised by $\mathcal{P} > 0$ are called 'pushers', a category that includes spermatozoa (Fig. 9.1(a)) and the majority of flagellated bacteria (Fig. 9.1(b)). As their names indicate, pusher cells use their flagella to push on the surrounding viscous fluid, and thus swim body-first with their flagella behind the cell body, drawing in fluid from the sides. In contrast, cells with $\mathcal{P} < 0$ are called 'pullers', a denomination that includes the time-averaged flow of biflagellated algae (Fig. 9.1(c)). In that situation, the flagella are located in front of the cell body during swimming, and continuously drag the cell body behind them as they pull on the fluid, expelling fluid to the sides.

Notably, given the $\theta \leftrightarrow \pi - \theta$ symmetry, we see that the sign of $\mathcal{P}$ is not correlated with the direction of swimming. More generally, for an arbitrary cell the components of the instantaneous stresslet tensor, $\mathbf{S}$, are in fact unrelated to its instantaneous direction of swimming. For example, using the squirmer framework from Chapter 4, examples may be constructed where the swimming speed is fixed and the tensor $\mathbf{S}$ is arbitrary.

9.3.3 Comparison with Experiments: Flows Created by E. coli *and*
Chlamydomonas reinhardtii

How do these theoretical predictions compare with experiments? This question has been answered by measurements of the flow around two model organisms.

First, Drescher et al. (2011) characterised the flow field created by *E. coli* bacteria swimming in the field of view of a microscope in a viscous fluid seeded with small particles. The results are illustrated in Fig. 9.3. The measured flow field, averaged over many organisms, is displayed in Fig. 9.3(a) (the units in the figure are μm/s). In Fig. 9.3(b) we show the theoretical prediction from Eq. (9.31) with the best-fit value of the dipole strength, $\mathcal{P} \approx 0.8$ pN μm, with the difference between theory and measurements displayed in Fig. 9.3(c). The steady dipolar picture is seen to be very accurate and extends a few body lengths away from the cell, outside which the $1/r^2$-decaying flow field is drowned out by thermal noise.

While modelling the swimming cell as a steady dipole appears accurate in the case of *E. coli*, the flow created by other organisms can be strongly time-dependent. This is illustrated in Fig. 9.4 in the case of the biflagellate alga *Chlamydomonas reinhardtii*. Measurements by Guasto et al. (2010) for cells located in thin liquid films show that both the magnitude and the direction of the flow induced by the cell vary in time, and the far-field signature alternates between puller and pusher (Fig. 9.4(a)). When averaged over the periodic flagellar beats, the mean flow is that of a puller in the far field (Fig. 9.4(b)) (Drescher et al., 2010a). The entire mean flow structure is well approximated by a model comprising three balancing stokeslets, one for the cell body and one for each propelling flagellum (Fig. 9.4(c); see also Fig. 9.1(c)).

The stresslet flow induced far from low Reynolds number swimmers has important consequences for interactions between cells and their environment. For example, we can already anticipate that two pusher cells swimming parallel to one another have a tendency to attract each other while two pullers repel (see Exercises). Further work in Chapter 14 shows how the flow induced by cells reorient neighbouring organisms, and how this leads to hydrodynamic instabilities. Similarly, and as detailed in Chapter 11, a pusher cell swimming parallel to a surface is attracted by it, due to hydrodynamic images.

9.4 Other Relevant Flow Singularities

9.4.1 Flows at Higher Order

The stresslet is an accurate representation of the flow produced in the far field by many self-propelled organisms. However, in some cases, the stresslet created by an organism can be identically zero (see Exercises). Furthermore, even when a nonzero stresslet is created, the far-field approximation underlying the dipolar

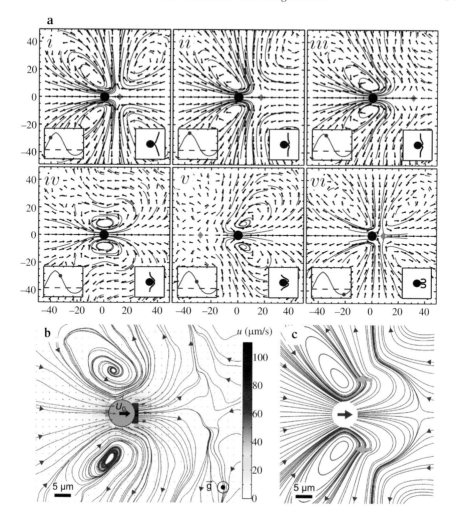

Figure 9.4 Flow induced by the biflagellate *Chlamydomonas reinhardtii* during locomotion. (a) Time dependence of the flow created by the cell located in a thin film during one period of flagellar beat with the cell swimming from left to right (axis labels are in μm) (reprinted figure with permission from Guasto et al. (2010), Copyright 2010 by the American Physical Society). (b) The mean flow averaged over many flagellar beats is a puller in the far field (reprinted figure with permission from Drescher et al. (2010a), Copyright 2010 by the American Physical Society). (c) A model comprising three stokeslets, one for the cell body and one for each flagellum so that forces balance, reproduces the main features of the mean flow.

flow is no longer valid very close to the cell, as is apparent in Figs. 9.3 and 9.4, and the detailed geometry of the organism and its propulsive appendages is important. These can be modelled using higher-order flow singularities decaying faster in space. Mathematically, these singularities are higher-order corrections in ϵ to the Taylor expansion of Eq. (9.5).

The next-order flow singularities decay spatially as $1/r^3$ and are of two types, namely irrotational source dipoles, which were already seen to characterise the flow near flagellar filaments in Chapter 6 (point sources of flow are represented schematically using '+' as shown in Fig. 9.2(f) while for source dipoles we use '+−'; see Fig. 9.2(g)), and force quadrupoles (i.e. flows obtained as spatial gradients of force dipoles). The two types of force quadrupoles relevant for an axisymmetric swimmer include the axisymmetric force quadrupole (Fig. 9.2(h)) and the rotlet dipole (see Fig. 9.2(i)).

9.4.2 Rotlet Dipoles

The rotlet dipole plays a particularly important role in nature, as it is the signature of helical propulsion by bacteria. Physically, a rotating helical flagellum applies locally a hydrodynamic torque to the surrounding fluid, which is cancelled out by the equal and opposite torque applied by the rotating cell body, leading to zero net torque (see Chapter 8). However, similar to the case of force dipoles, since the centre of propulsive torque is not the same as the centre of resistive (i.e. drag) torque, the torque distribution has in general a nonzero first moment, which corresponds to a torque dipole. So while the stresslet is a generic feature of all low Reynolds number swimmers, the rotlet dipole arises only for swimmers using rotational propulsion. Notably, for an axisymmetric swimmer, the rotlet dipole is the only flow component that induces azimuthal flow, which near a wall can lead to symmetry breaking in the trajectories of the cells (see Chapter 11).

Mathematically, the flow induced by a rotlet dipole may be derived using a method similar to the force dipole in Section 9.2. The axisymmetric rotlet dipole is obtained by placing a rotlet of strength $\mathbf{R} = R\mathbf{e}$ with flow given by Eq. (9.22) at position $\epsilon \ell \mathbf{e}$ from the origin, an opposite rotlet at the origin, and taking the limit $\epsilon \to 0$. Using $\mathcal{X} \equiv \epsilon \ell R$ to denote the magnitude of the dipole strength (dimensions of force times length squared), the resulting flow is given by

$$\mathbf{u}_{RD}(\mathbf{r}; \mathbf{e}, \mathcal{X}) = -\mathcal{X}(\mathbf{e} \cdot \nabla)\left(\frac{1}{8\pi\mu}\frac{\mathbf{e} \times \mathbf{r}}{r^3}\right) = \frac{3\mathcal{X}}{8\pi\mu}\frac{(\mathbf{e} \cdot \mathbf{r})(\mathbf{e} \times \mathbf{r})}{r^5}. \tag{9.32}$$

Note that for a swimming bacterium such as *E. coli*, flagellar filaments rotate behind the cell in a CCW direction while the cell body counter-rotates, so the total flow contains the singularity in Eq. (9.32) with $R > 0$ and $\mathcal{X} > 0$.

9.4.3 Swimming with a Net Force or Torque

We close by mentioning two notable exceptions to the force-free, torque-free motion for swimming cells. Some large organisms have a small density difference with the surrounding fluid, which means that when we look sufficiently far from the cell, a net stokeslet component dominates, proportional to the small density difference between the organism and the fluid (Drescher et al., 2010a). More common are cells that might be on average as dense as the fluid they swim in, but with non-uniform densities. This is the case for many algae species that are bottom-heavy (Pedley and Kessler, 1992) and thus experience net gravitational torques when they are not aligned with the vertical direction. This design characteristic allows the cells to bias their locomotion so that they swim upward, an important feature for oceanic organisms requiring light for photosynthesis and oxygen (for more details, see Chapter 10).

9.5 Average Stress Induced by Cells

9.5.1 Batchelor's Average Stress in a Suspension

The flow created by a self-propelled cell is, within a very good approximation, a stresslet, given instantaneously by Eq. (9.12). How can we interpret the components of the tensor $\mathbf{S}$ physically? Following Batchelor's classical calculation on the fluid dynamics of suspensions (Batchelor, 1970a), we now compute the mean mechanical stress in a collection of swimming cells and show that it is in fact related to the stresslet tensor $\mathbf{S}$.

In order to characterise the mean stress, we consider a volume V of fluid with a large number of swimmers in it, each identical and of volume $V_0 \ll V$ and surface S_0 (see sketch in Fig. 9.5). We denote by $\mathbf{u}$ and p the velocity and pressure fields in the Newtonian fluid (of dynamic viscosity μ) and call σ the mechanical stress throughout V (both in the fluid and in the swimmers). The mean stress in the suspension, Σ, is defined as the average of σ over the whole volume, hence

$$\Sigma_{ij} = \frac{1}{V} \iiint_V \sigma_{ij} \, dV. \tag{9.33}$$

To make progress, we separate Eq. (9.33) into an integral in the fluid domain (of volume denoted by $V - \sum V_0$ to indicate that all swimmers are excluded) and integrals inside the swimmers (each of volume V_0) as

$$\Sigma_{ij} = \frac{1}{V} \iiint_{V-\sum V_0} \left[-p\delta_{ij} + \mu \left(\frac{\partial u_i}{\partial x_j} + \frac{\partial u_j}{\partial x_i} \right) \right] dV + \frac{1}{V} \sum \iiint_{V_0} \sigma_{ij} \, dV, \tag{9.34}$$

where the sum in the second term is taken over all the swimmers in the suspension. We next transform the second volume integral into a surface integral by exploiting mechanical equilibrium. Indeed, we can write the stress tensor as

$$\sigma_{ij} = \frac{\partial}{\partial x_k} (\sigma_{ik} x_j) - \frac{\partial \sigma_{ik}}{\partial x_k} x_j, \tag{9.35}$$

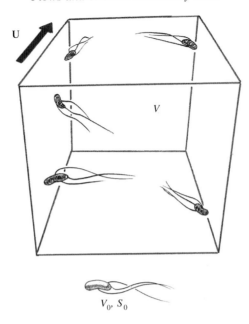

V_0, S_0

Figure 9.5 A population of swimming cells (here, flagellated bacteria). Each cell has a volume V_0 and surface S_0 while the volume of the whole population is V. The mean flow in the suspension is denoted $\mathbf{U}$. Drawings courtesy of Jacques Lauga.

where the second term in Eq. (9.35) is zero because in the absence of inertia and body forces, all surface forces need to balance (divergence of stress is zero). We can then use the divergence theorem to rewrite the integrals inside each swimmer as

$$\iiint_{V_0} \sigma_{ij}\, dV = \iiint_{V_0} \frac{\partial}{\partial x_k}(\sigma_{ik} x_j)\, dV = \iint_{S_0} \sigma_{ik} x_j n_k\, dS, \qquad (9.36)$$

where the normal vector $\mathbf{n}$ to S_0 is pointing into the fluid.

9.5.2 Mean Flow Gradient

We next have to evaluate the first term on the right-hand side of Eq. (9.34), which is the mean value of the stress inside the fluid. Using $\mathbf{U}$ to denote the mean flow in the suspension, we now show that the mean stress is related to the stress of the mean flow.

The mean velocity gradient inside the suspension is defined as

$$\frac{\partial U_i}{\partial x_j} = \frac{1}{V} \iiint_V \frac{\partial u_i}{\partial x_j}\, dV, \qquad (9.37)$$

which here again can be split into the fluid region and the swimmers as

$$\frac{\partial U_i}{\partial x_j} = \frac{1}{V} \iiint_{V-\sum V_0} \frac{\partial u_i}{\partial x_j} \, dV + \frac{1}{V} \sum \iiint_{V_0} \frac{\partial u_i}{\partial x_j} \, dV. \tag{9.38}$$

The second integral can be simplified using the divergence theorem

$$\iiint_{V_0} \frac{\partial u_i}{\partial x_j} \, dV = \iint_{S_0} u_i n_j \, dS, \tag{9.39}$$

leading to the equality

$$\frac{1}{V} \iiint_{V-\sum V_0} \frac{\partial u_i}{\partial x_j} \, dV = \frac{\partial U_i}{\partial x_j} - \frac{1}{V} \sum \iint_{S_0} u_i n_j \, dS. \tag{9.40}$$

9.5.3 Swimmer Stresses

Substituting Eqs. (9.36) and (9.40) into Eq. (9.34), we find that the average stress in the suspension is given by

$$\Sigma_{ij} = -\left(\frac{1}{V} \iiint_{V-\sum V_0} p \, dV \right) \delta_{ij} + \mu \left(\frac{\partial U_i}{\partial x_j} + \frac{\partial U_j}{\partial x_i} \right) + \Sigma_{ij}^s, \tag{9.41}$$

where an extra 'swimmer' stress, Σ_{ij}^s, has been introduced, given by the integrals

$$\Sigma_{ij}^s = \frac{1}{V} \sum \iint_{S_0} \left[\sigma_{ik} x_j n_k - \mu(u_i n_j + u_j n_i) \right] \, dS. \tag{9.42}$$

Since the extra stress in Eq. (9.42) might contain an isotropic term, we can remove it and write

$$\Sigma_{ij}^s = \frac{1}{3} \Sigma_{mm}^s \delta_{ij} + \Sigma_{ij}^{s,dev}, \tag{9.43}$$

where $\Sigma_{ij}^{s,dev}$ is a deviatoric (i.e. non-isotropic) swimmer stress. Combining the isotropic terms from Eqs. (9.41) and (9.43) allows us to define the mean pressure through the suspension, P, as

$$P = \left(\frac{1}{V} \iiint_{V-\sum V_0} p \, dV \right) - \frac{1}{3} \Sigma_{mm}^s, \tag{9.44}$$

and thus to obtain finally that

$$\Sigma_{ij} = -P\delta_{ij} + \mu \left(\frac{\partial U_i}{\partial x_j} + \frac{\partial U_j}{\partial x_i} \right) + \Sigma_{ij}^{s,dev}, \tag{9.45}$$

with

$$\Sigma_{ij}^{s,dev} = \frac{1}{V} \sum \iint_{S_0} \left[\sigma_{ik} x_j n_k - \frac{1}{3} \sigma_{mk} x_m n_k \delta_{ij} - \mu(u_i n_j + u_j n_i) \right] \, dS. \tag{9.46}$$

This is the fundamental result obtained by Batchelor. The mean stress in a suspension is equal to a mean Newtonian stress plus the sum of all the contributions from the individual constituents, all of which are written simply as surface integrals.

9.5.4 Stresslets Measure Mean Swimmer Stresses

The final step is to evaluate Eq. (9.46) for a self-propelled organism. To do so we first see that the value of each surface integral in Eq. (9.46) is unchanged if it is evaluated on any another surface $\tilde{S}_0$ containing S_0, with the fluid of viscosity μ located in between S_0 and $\tilde{S}_0$. Indeed, the difference between the two surface integrals may be evaluated using the divergence theorem and we obtain

$$\iint_{S_0-\tilde{S}_0} \left[\sigma_{ik}x_jn_k - \frac{1}{3}\sigma_{mk}x_mn_k\delta_{ij} - \mu(u_in_j + u_jn_i) \right] \mathrm{d}S$$

$$= \iiint_{\tilde{V}_0-V_0} \left[\sigma_{ij} + p\delta_{ij} - \mu \left(\frac{\partial u_i}{\partial x_j} + \frac{\partial u_j}{\partial x_i} \right) \right] \mathrm{d}V = 0. \tag{9.47}$$

For a sufficiently dilute suspension, we may then simply take the arbitrary surface $\tilde{S}_0$ to be a sphere S_R of radius R centred at the location of the swimmer. In the limit of large R, the flow is a stresslet, with velocity field $\mathbf{u}_S$ from Eq. (9.12) and stress field given by (see Exercises)

$$\sigma_{ij}(\mathbf{r}) = \frac{3S_{pq}}{4\pi} \left[-\frac{(x_j\delta_{pi} + x_i\delta_{pj})x_q}{r^5} + \frac{5x_px_qx_ix_j}{r^7} \right]. \tag{9.48}$$

With this, each term in Eq. (9.46) may then be calculated exactly after a radial rescaling as integrals on the surface S_1 as

$$\iint_{S_R} \sigma_{ik}x_jn_k \, \mathrm{d}S = \frac{3}{4\pi}S_{pq} \left(4\iint_{S_1} n_in_jn_pn_q\mathrm{d}S - \delta_{ip}\iint_{S_1} n_jn_q \, \mathrm{d}S \right), \tag{9.49}$$

$$\iint_{S_R} \frac{1}{3}\sigma_{mk}x_mn_k\delta_{ij} \, \mathrm{d}S = \frac{\delta_{ij}}{4\pi}S_{pq} \left(3\iint_{S_1} n_pn_q \, \mathrm{d}S \right), \tag{9.50}$$

and

$$\iint_{S_R} -\mu(u_in_j + u_jn_i) \, \mathrm{d}S = \frac{3}{4\pi}S_{pq}\iint_{S_1} n_in_jn_pn_q \, \mathrm{d}S. \tag{9.51}$$

Using the classical identities for isotropic second and fourth-rank tensors

$$\iint_{S_1} n_kn_m \, \mathrm{d}S = \frac{4\pi}{3}\delta_{km}, \tag{9.52a}$$

$$\iint_{S_1} n_in_jn_pn_q \, \mathrm{d}S = \frac{4\pi}{15}(\delta_{ij}\delta_{qp} + \delta_{pj}\delta_{qi} + \delta_{ip}\delta_{qj}), \tag{9.52b}$$

we can then compute the integral on the swimmer surface as

$$\iint_{S_0} \left[\sigma_{ik}x_jn_k - \frac{1}{3}\sigma_{mk}x_mn_k\delta_{ij} - \mu(u_in_j + u_jn_i) \right] dS = S_{ij}, \tag{9.53}$$

and thus the mean stress from Eq. (9.46) is given finally by

$$\Sigma_{ij}^{s,dev} = \frac{1}{V}\sum S_{ij}. \tag{9.54}$$

The formula in Eq. (9.54) provides us with the physical interpretation of the stresslet tensor, **S**. For each swimmer, **S** is the volume contribution to the mean stress in the suspension of cells. This result allows us to link the configuration of the swimmers (position and orientation) to the fluid motion in the suspension, and is further exploited in Chapter 14 where we propose a continuum model of swimmer suspensions.

Further Reading

As first seen in Chapter 6, superpositions of hydrodynamic singularities can be used to tackle a variety of low Reynolds number fluid dynamics problems (Chwang and Wu, 1975). Systematic Taylor expansions of such flow solutions in the far field form the basis of the so-called multipole expansion of Stokes flows (Kim and Karrila, 1991; Pozrikidis, 1992; Leal, 2007).

The leading-order flow around a swimming microorganism is a stresslet, and the instantaneous stresslet tensor for a monotrichous bacterium was determined computationally by Ishikawa et al. (2007). Further numerical work modelled the spatial structure of flow around swimming bacteria (Watari and Larson, 2010; Hu et al., 2015) and near a rotating flagellar filament (Spagnolie and Lauga, 2011). Beyond stresslets, fast-decaying force quadrupoles dominate velocity correlations between swimming bacteria due to the front-back symmetry of force dipoles (Liao et al., 2007).

The flows created by swimming cells have been shown experimentally to induce enhanced transport of molecules and particles above that due to Brownian motion (Wu and Libchaber, 2000; Kim and Breuer, 2004; Leptos et al., 2009), and theoretical models based on collisions between swimmers and passive tracers have led to predictions for effective (enhanced) diffusivities (Lin et al., 2011; Miño et al., 2013; Pushkin et al., 2013; Kasyap et al., 2014; Thiffeault, 2015).

Exercises

1. Verify that the flow for the most general trace-free stresslet $\mathbf{S}$, given by

$$\mathbf{u}_S(\mathbf{r}; \mathbf{S}) = -\frac{3}{8\pi\mu} \frac{(\mathbf{r} \cdot \mathbf{S} \cdot \mathbf{r})\mathbf{r}}{r^5},$$

   is incompressible.

2. For the general stresslet, whose flow is given by

$$\mathbf{u}_S(\mathbf{r}; \mathbf{S}) = -\frac{3}{8\pi\mu} \frac{(\mathbf{r} \cdot \mathbf{S} \cdot \mathbf{r})\mathbf{r}}{r^5},$$

   calculate the associated pressure field and deduce the stress field.

3. A swimmer, modelled as an axisymmetric stresslet of strength $\mathcal{P}$ with flow

$$\mathbf{u}(\mathbf{r}) = \frac{\mathcal{P}}{8\pi\mu} \left[-\frac{1}{r^3} + \frac{3(\mathbf{e} \cdot \mathbf{r})^2}{r^5} \right] \mathbf{r},$$

   undergoes circular motion of radius R and frequency ω. The swimmer also rotates its shape so that its long axis, $\mathbf{e}(t)$, always remains in the plane of the circle and tangent to it. By considering a point far from the circle, show that the flow induced by the rotating swimmer averaged over one period of rotation is also an axisymmetric stresslet, and derive its strength, $\tilde{\mathcal{P}}$, and direction, $\tilde{\mathbf{e}}$. Deduce that the time-averaged flow is a pusher if the original is a puller, and vice versa.

4. Two identical swimmers are described by an axisymmetric stresslet of strength $\mathcal{P}$ and direction $\mathbf{e}$ with flow given by

$$\mathbf{u}(\mathbf{r}) = \frac{\mathcal{P}}{8\pi\mu} \left[-\frac{1}{r^3} + \frac{3(\mathbf{e} \cdot \mathbf{r})^2}{r^5} \right] \mathbf{r}.$$

   If $\mathcal{P} > 0$, show that swimmers separated by a vector $\ell\mathbf{e}$ repel each other. Given an initial separation ℓ_0, compute the solution for $\ell(t)$.

5. What happens to the relative motion in Exercise 4 when the swimmers are instead separated by a vector $\ell\mathbf{j}$, where the unit vector $\mathbf{j}$ is perpendicular to $\mathbf{e}$?

6. In a simple model of biflagellate algae (such as the one in Fig. 9.1(c)), the flow induced by the cell in the surrounding fluid is instantaneously described by a force $F\mathbf{e}$ located at the origin and two forces $-\frac{1}{2}F\mathbf{e}$ located at $(\epsilon\ell, \pm\epsilon h, 0)$, with $h \neq 0$. By considering the flow far from the cell and thus taking the limit $\epsilon \ll 1$, what is the condition on ℓ for the instantaneous flow to be a force dipole? If that condition is not satisfied, what is the flow in the far field?

PART THREE

INTERACTIONS

10

Swimming Cells in Flows

So far, we have focused on the mathematical description of individual cells in infinite quiescent fluids. Now, in the third part of this book, we model how cells interact with their environment. In this chapter we begin by addressing the impact of external flows on cell locomotion, with two primary motivations. First, in many biological environments cells swim in moving fluids, for example bacteria in the circulatory system, spermatozoa in uterine peristaltic flow or plankton in oceanic turbulence. Second, one of the key features required to model the collective locomotion of swimming microorganisms is a quantitative understanding of the changes in position and orientation of one cell in the flow created by its neighbours, as is described in Chapter 14. We start by considering the dynamics of spherical swimmers in arbitrary external flows. In this case, the impact on cell translation and rotation can be obtained exactly (Faxén's laws), which we then use to address cell trajectories in simple canonical flows. We next examine the case of elongated swimmers, which may be analysed when the flow is linear, a limit relevant to many situations where the typical length scale over which the flow varies is much larger than the size of the organism. For slender swimmer shapes, we derive a simplified version of Jeffery's exact equation for ellipsoids in linear flows. Jeffery's equation is then used, in agreement with experiments, to characterise the angular dynamics of elongated bodies in shear flows (Jeffery's orbits) and to address the trajectories of elongated swimmers in elementary flows. We finally consider the case where swimmers have a preferential swimming direction, as relevant e.g. to cells swimming against gravity or towards a light source. The preferred swimming direction can be modelled by an additional external torque, which may then lead to cell trapping in high-shear regions and to hydrodynamic focusing.

10.1 Spherical Swimmers in Flows

10.1.1 Faxén's Laws

When cells are spherical in shape, the impact of arbitrary external flows can be calculated exactly in a series of results known as Faxén's laws (Leal, 2007). Consider
a spherical swimmer of radius a in a prescribed, unbounded flow given by $\mathbf{u}_\infty(\mathbf{x})$ in
the absence of the swimmer. Without the flow, the swimmer would translate with
its intrinsic swimming velocity $\mathbf{U}_s$ and rotate with angular velocity $\mathbf{\Omega}_s$. Due to the
linearity of the Stokes equations, we may invoke linear superposition so that the to-
tal rigid-body motion of the spherical swimmer ($\mathbf{U}$ and $\mathbf{\Omega}$) is equal instantaneously
to the sum of the swimming kinematics in the absence of flow ($\mathbf{U}_s$ and $\mathbf{\Omega}_s$) and the
kinematics due solely to the flow assuming that the swimmer is rigid (denoted $\mathbf{U}_f$
and $\mathbf{\Omega}_f$), i.e.

$$\mathbf{U} = \mathbf{U}_s + \mathbf{U}_f, \tag{10.1a}$$

$$\mathbf{\Omega} = \mathbf{\Omega}_s + \mathbf{\Omega}_f. \tag{10.1b}$$

The goal of this section is then to compute the values of $\mathbf{U}_f$ and $\mathbf{\Omega}_f$ as a function
of $\mathbf{u}_\infty(\mathbf{r})$ for a force- and torque-free rigid sphere. In what follows, we use $\mathbf{u}(\mathbf{r})$ to
denote the total flow field when the sphere is present, so that $\mathbf{u} \to \mathbf{u}_\infty$ far from the
sphere.

In order to derive Faxén's laws, we use the reciprocal theorem seen in Sec-
tion 2.6.2. For two incompressible and unbounded Stokes flows $\bar{\mathbf{u}}$ and $\hat{\mathbf{u}}$, with associated stress fields $\bar{\sigma}$ and $\hat{\sigma}$, with different boundary conditions acting on the
(same) instantaneous geometry S of the rigid sphere, we have the equality

$$\iint_S \bar{\mathbf{u}} \cdot \hat{\sigma} \cdot \mathbf{n}\, dS = \iint_S \hat{\mathbf{u}} \cdot \bar{\sigma} \cdot \mathbf{n}\, dS, \tag{10.2}$$

where $\mathbf{n}$ is the unit outward normal to the sphere. For the flow ($\hat{\mathbf{u}}$, $\hat{\sigma}$), we pick the
same drag problem as in Section 2.6.2, i.e. rigid-body motion with instantaneous
velocities $\hat{\mathbf{U}}$ and $\hat{\mathbf{\Omega}}$, resulting in the boundary condition on the swimmer ($\mathbf{r} = \mathbf{r}_s$)

$$\hat{\mathbf{u}}(\mathbf{r}_s) = \hat{\mathbf{U}} + \hat{\mathbf{\Omega}} \times \mathbf{r}_s. \tag{10.3}$$

For the other flow ($\bar{\mathbf{u}}$, $\bar{\sigma}$), we pick the perturbation to the external flow due to the
presence of the rigid sphere, i.e. $\bar{\mathbf{u}} = \mathbf{u} - \mathbf{u}_\infty$, which decays to zero at infinity and
is associated with boundary conditions on the sphere

$$\bar{\mathbf{u}}(\mathbf{r}_s) = \mathbf{U}_f + \mathbf{\Omega}_f \times \mathbf{r}_s - \mathbf{u}_\infty(\mathbf{r}_s). \tag{10.4}$$

Since $\hat{\mathbf{u}}$ is associated with constant translation and rotation on the surface of the swimmer, the right-hand side of Eq. (10.2) is

$$\iint_S \hat{\mathbf{u}} \cdot \bar{\sigma} \cdot \mathbf{n}\, dS = \hat{\mathbf{U}} \cdot \bar{\mathbf{F}} + \hat{\boldsymbol{\Omega}} \cdot \bar{\mathbf{T}}, \tag{10.5}$$

where $\bar{\mathbf{F}}$ and $\bar{\mathbf{T}}$ are the net forces and torques on the sphere in the $\bar{\mathbf{u}}$ flow (see also Eq. (2.31)). The rigid sphere is force- and torque-free and since the $\mathbf{u}_\infty$ flow does not lead to net forces or torques on the volume of fluid inside S (see Section 2.2), the flow field $(\bar{\mathbf{u}}, \bar{\sigma})$ does not impose a net force or torque on the sphere, and $\bar{\mathbf{F}} = \mathbf{0}$ and $\bar{\mathbf{T}} = \mathbf{0}$. The right-hand side of Eq. (10.5), and thus that of Eq. (10.2), are therefore identically zero.

In order to evaluate the left-hand side of Eq. (10.2) we use that, for a sphere in rigid-body motion,

$$\hat{\sigma} \cdot \mathbf{n} = -\frac{3\mu}{2a}\hat{\mathbf{U}} - 3\mu\hat{\boldsymbol{\Omega}} \times \mathbf{n}, \tag{10.6}$$

so that Eq. (10.2) becomes

$$\frac{3\mu}{2a}\hat{\mathbf{U}} \cdot \iint_S \bar{\mathbf{u}}\, dS + 3\mu\hat{\boldsymbol{\Omega}} \cdot \iint_S \mathbf{n} \times \bar{\mathbf{u}}\, dS = 0. \tag{10.7}$$

Since the velocity $\hat{\mathbf{U}}$ and angular velocity $\hat{\boldsymbol{\Omega}}$ are arbitrary, both integrals in Eq. (10.7) have to be zero, i.e.

$$\iint_S \bar{\mathbf{u}}\, dS = \mathbf{0}, \quad \iint_S \mathbf{n} \times \bar{\mathbf{u}}\, dS = \mathbf{0}. \tag{10.8}$$

Substituting the boundary conditions from Eq. (10.4) into the integrals of Eq. (10.8) and using the identities

$$\iint_S \mathbf{n}\, dS = \mathbf{0}, \quad \iint_S \mathbf{nn}\, dS = \frac{4}{3}\pi a^2 \mathbf{1}, \tag{10.9}$$

we obtain the results

$$\mathbf{U}_f = \frac{1}{4\pi a^2} \iint_S \mathbf{u}_\infty\, dS, \tag{10.10a}$$

$$\boldsymbol{\Omega}_f = \frac{3}{8\pi a^3} \iint_S \mathbf{n} \times \mathbf{u}_\infty\, dS. \tag{10.10b}$$

The kinematics of the rigid sphere are thus fully determined by the flow $\mathbf{u}_\infty$, and more specifically its mean value (and moment) on its surface. Importantly, $\mathbf{u}_\infty$ is the external flow in the absence of the sphere, so it is defined everywhere in space including inside the sphere.

In order to make further progress, we use vector calculus to relate the integrals in Eq. (10.10) to the value of $\mathbf{u}_\infty$ and its derivatives at the instantaneous centre of the sphere, which we denote by $\mathbf{0}$ (the origin). In order to determine $\mathbf{U}_f$, we use the Taylor expansion of the flow $\mathbf{u}_\infty$ around the centre of the sphere,

$$\mathbf{u}_\infty(\mathbf{r}_s) = \mathbf{u}_\infty|_0 + \mathbf{r}_s \cdot [\nabla \mathbf{u}_\infty]|_0 + \frac{1}{2}\mathbf{r}_s\mathbf{r}_s : [\nabla[\nabla \mathbf{u}_\infty]]|_0 + \cdots, \qquad (10.11)$$

where we use ':' to denote the trace of the product of two tensors. When substituted into the integral in Eq. (10.10a), all odd terms in $\mathbf{r}_s$ in Eq. (10.11) vanish by symmetry. The first term contributes $4\pi a^2 \mathbf{u}_\infty(\mathbf{0})$ to the value of the integral, while the second term leads to

$$\frac{a^2}{2}[\nabla[\nabla \mathbf{u}_\infty]]|_0 : \iint_S \mathbf{n}\mathbf{n}\,dS = \frac{2\pi}{3}a^4\nabla^2\mathbf{u}_\infty|_0, \qquad (10.12)$$

where we have used the fact that $\mathbf{r}_s = a\mathbf{n}$ on the sphere. All other even terms include higher-order Laplacians $\nabla^{2n}\mathbf{u}_\infty|_0$, which are all zero because $\nabla^4\mathbf{u}_\infty = \mathbf{0}$ for a Stokes flow (see Section 2.2). The result in Eq. (10.10a) becomes Faxén's law of velocity,

$$\mathbf{U}_f = \mathbf{u}_\infty|_0 + \frac{a^2}{6}\nabla^2\mathbf{u}_\infty|_0. \qquad (10.13)$$

This result, mathematically exact for a sphere, can be seen as a mean value theorem for the bi-Laplacian operator (Pozrikidis, 1992).

Physically, the result in Eq. (10.13) states that the velocity of the sphere is equal to the velocity of the fluid in the absence of the sphere at its centre, corrected by a term related to the finite size of the sphere and the 'curvature' of the velocity field (its second spatial derivative). Since the external flow is a solution to the Stokes equations, we know that its Laplacian is proportional to the local pressure gradient in the fluid (Eq. (2.9)), an equality that provides therefore a physical interpretation for the corrective term. The relative value of this correction term to the flow velocity is seen to be of order $O((a/L)^2)$, where L is the relevant length scale for a typical velocity gradient, and can thus be neglected for small swimmers in external flows varying sufficiently slowly.

In order to obtain the corresponding expression for the angular velocity of the sphere, Eq. (10.10b), we take advantage of the fact that the flow field $\mathbf{u}_\infty$ is defined everywhere in space in order to apply the divergence theorem, and write

$$\iint_S \mathbf{n} \times \mathbf{u}_\infty\,dS = \iiint_V \nabla \times \mathbf{u}_\infty\,dV, \qquad (10.14)$$

where V is the volume of the sphere. The integrand on the right-hand side of Eq. (10.14) is the flow vorticity, which we know is harmonic (see Section 2.2). We can make use of the mean value theorem for the Laplacian stating that the mean value of a harmonic function over a sphere is equal to value at its centre, and thus Eq. (10.10b) becomes Faxén's law of angular velocity,

$$\mathbf{\Omega}_f = \frac{1}{2}\nabla \times \mathbf{u}_\infty|_0. \tag{10.15}$$

The passive sphere rotates at exactly one half of the flow vorticity evaluated at the centre of the sphere in its absence. In particular, the sphere does not rotate if the flow is irrotational.

With these results, we can now summarise the total kinematics of spherical swimmers in unbounded flows. Consider a spherical cell located at position $\mathbf{X}(t)$ and oriented along a unit vector $\mathbf{p}(t)$, swimming with velocity $\mathbf{U}_s$ and rotating with angular velocity $\mathbf{\Omega}_s$ in the absence of flow. In that case, Eqs. (10.1), (10.13) and (10.15) mean that the instantaneous rates of change of position and orientation of the swimmer are given by

$$\dot{\mathbf{X}} = \mathbf{U}_s + \mathbf{u}_\infty|_\mathbf{X} + \frac{a^2}{6}\nabla^2\mathbf{u}_\infty|_\mathbf{X}, \tag{10.16a}$$

$$\dot{\mathbf{p}} = \left(\mathbf{\Omega}_s + \frac{1}{2}\nabla \times \mathbf{u}_\infty|_\mathbf{X}\right) \times \mathbf{p}, \tag{10.16b}$$

where we use dots to denote time derivatives.

In the next section, we analyse four prototypical flows where, for simplicity, we assume that the swimming speed is constant along the body-fixed orientation ($\mathbf{U}_s = U\mathbf{p}$) and ignore any intrinsic rotation ($\mathbf{\Omega}_s = \mathbf{0}$; see extension in Exercises).

10.1.2 Spherical Swimmer in a Vortex (Solid-Body Rotation)

We start by considering the trajectories of spherical swimmers in a simple viscous vortex, i.e. a solid-body rotation flow. The external flow is given by

$$\mathbf{u}_\infty(\mathbf{r}) = \mathbf{\Omega}_\infty \times \mathbf{r}, \tag{10.17}$$

where the rotation rate $\mathbf{\Omega}_\infty = \Omega_\infty \mathbf{e}_z$ is constant and prescribed. We assume that the swimmer starts and remains in the plane of the flow, (x, y), that it has initial position $\mathbf{X}_0 = (X_0, Y_0)$, and we place the origin of the lab frame at the centre of the vortex. Since the flow velocity varies linearly in space, the second term in Eq. (10.13) is identically zero while the flow vorticity is constant and given by $\nabla \times \mathbf{u}_\infty = 2\mathbf{\Omega}_\infty$. The trajectories of the swimmer, solutions to Eq. (10.16), therefore satisfy

$$\dot{\mathbf{X}} = U\mathbf{p} + \mathbf{\Omega}_\infty \times \mathbf{X}, \tag{10.18a}$$

$$\dot{\mathbf{p}} = \mathbf{\Omega}_\infty \times \mathbf{p}. \tag{10.18b}$$

Writing the swimmer orientation as $\mathbf{p}(t) = (\cos\phi(t), \sin\phi(t))$, Eq. (10.18b) leads to $\dot{\phi} = \Omega_\infty$ with solution $\phi = \Omega_\infty t + \phi_0$. Without loss of generality we may choose the (x, y) axis so that the swimmer is initially aligned with the x axis and thus pick $\phi_0 = 0$. Written in coordinate form, Eq. (10.18a) then becomes

$$\dot{X} = U \cos(\Omega_\infty t) - \Omega_\infty Y, \tag{10.19a}$$
$$\dot{Y} = U \sin(\Omega_\infty t) + \Omega_\infty X. \tag{10.19b}$$

This is best solved using complex variables. Upon defining $Z \equiv X + iY$, Eq. (10.19) becomes

$$\dot{Z} - i\Omega_\infty Z = U e^{i\Omega_\infty t}, \tag{10.20}$$

which indicates that resonance occurs as both cell rotation and translation are characterised by the same frequency. The solution to Eq. (10.20) is elementary and given by

$$Z(t) = (Z_0 + Ut)e^{i\Omega_\infty t}, \tag{10.21}$$

where $Z_0 = X_0 + iY_0$. The trajectories characterised by Eq. (10.21) are spirals, illustrated in Fig. 10.1(a) for the dimensionless parameters $X_0 = Y_0 = -2$, $\Omega_\infty = 1$ and $U = 5$. Interestingly, we see that at long times, i.e. when $t \gg |Z_0/U|$, the position of the swimmer is given by $(X(t) = Ut \cos(\Omega_\infty t),\ Y(t) = Ut \sin(\Omega_\infty t))$, i.e. it orbits the vortex with frequency Ω_∞ while the distance from the centre increases linearly with time. Due to this phase-locking between translation and rotation, the swimmer points radially outwards regardless of the parameter values. As we see below, if the swimmer is no longer spherical but elongated, we obtain exactly the same behaviour because the flow in Eq. (10.17) has no irrotational component.

10.1.3 Spherical Swimmer in Shear Flow

As a second example, we consider the kinematics of a spherical swimmer in a shear flow, for which the trajectories may be found analytically (Ten Hagen et al., 2011). We assume that the swimmer remains in the (x, y) plane and consider a linear shear flow as (Leal, 2007)

$$\mathbf{u}_\infty(\mathbf{r}) = \dot{\gamma} y \mathbf{e}_x, \tag{10.22}$$

where $\dot{\gamma} > 0$ is the constant shear rate.

We start by considering the orientation of the swimmer, satisfying

$$\dot{\mathbf{p}} = \frac{1}{2}(\nabla \times \mathbf{u}_\infty) \times \mathbf{p}. \tag{10.23}$$

The vorticity of the shear flow is constant and given by

$$\frac{1}{2}(\nabla \times \mathbf{u}_\infty) = -\frac{1}{2}\dot{\gamma}\mathbf{e}_z, \tag{10.24}$$

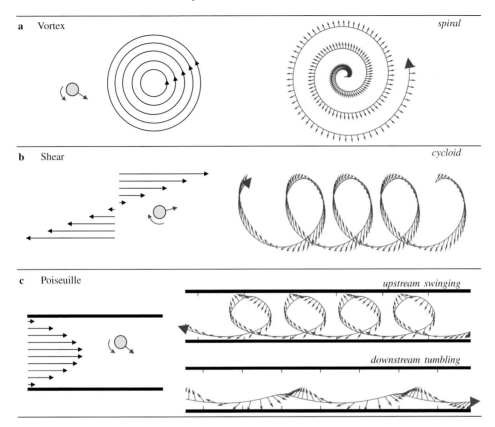

Figure 10.1 Illustration of the trajectories of spherical swimmers in three proto-typical flows (the small arrow shows the instantaneous orientation of the swimmer while the large arrow shows the direction of its trajectory). (a) Swimmer in a vortex (solid-body rotation); trajectory from Eq. (10.21) plotted for dimensionless parameters $X_0 = Y_0 = -2$, $\Omega_\infty = 1$ and $U = 5$. (b) Swimmer in simple shear flow; trajectory given by Eqs. (10.27c) and (10.30) shown with dimensionless parameters $X_0 = 0$, $Y_0 = 3$, $\phi_0 = 0.5$, $\dot{\gamma} = 1$ and $U = 3$. (c) Swimmer in pressure-driven Poiseuille flow; numerical solution to Eq. (10.33) with parameters $h = 1$, $U = 1$, $X_0 = 0$, $Y_0 = -0.6$, $\phi_0 = 1$ and flow speed $U_0 = 2.5$ (upstream swinging, top) and $U_0 = 10$ (downstream tumbling, bottom).

and writing $\mathbf{p}(t) = \left(\cos \phi(t), \sin \phi(t) \right)$, Eq. (10.23) becomes

$$\dot{\phi} = -\frac{1}{2}\dot{\gamma}, \tag{10.25}$$

with solution

$$\phi = \phi_0 - \frac{1}{2}\dot{\gamma}t, \tag{10.26}$$

where ϕ_0 is the initial angle. If the shear rate is positive, $\dot{\gamma} > 0$, the orientation of the swimmer undergoes clockwise (CW) rotation in the (x, y) plane.

In order to solve for the position of the swimmer, note again that the second term in Eq. (10.13) drops out because the flow is linear, and thus the trajectories are solutions to

$$\dot{X} = U \cos\left(\phi_0 - \frac{1}{2}\dot{\gamma}t\right) + \dot{\gamma}Y, \tag{10.27a}$$

$$\dot{Y} = U \sin\left(\phi_0 - \frac{1}{2}\dot{\gamma}t\right). \tag{10.27b}$$

We can first integrate Eq. (10.27b) to

$$Y(t) - Y_0 = \frac{2U}{\dot{\gamma}}\left[\cos\left(\phi_0 - \frac{1}{2}\dot{\gamma}t\right) - \cos\phi_0\right], \tag{10.28}$$

and use this solution to write Eq. (10.27a) as

$$\dot{X} = 3U \cos\left(\phi_0 - \frac{1}{2}\dot{\gamma}t\right) + \dot{\gamma}Y_0 - 2U \cos\phi_0, \tag{10.29}$$

which can be integrated exactly to

$$X(t) - X_0 = (\dot{\gamma}Y_0 - 2U \cos\phi_0)\, t - \frac{6U}{\dot{\gamma}}\left[\sin\left(\phi_0 - \frac{1}{2}\dot{\gamma}t\right) - \sin(\phi_0)\right]. \tag{10.30}$$

The trajectories given by Eqs. (10.27c) and (10.30) are cycloids, illustrated in Fig. 10.1(b) for the parameters $X_0 = 0$, $Y_0 = 3$, $\phi_0 = 0.5$, $\dot{\gamma} = 1$ and $U = 3$. Notably, the sign of $\dot{\gamma}Y_0 - 2U \cos\phi_0$ dictates whether the cell is moving towards the positive or negative x direction on average.

10.1.4 Spherical Swimmer in Poiseuille Flow

A more complex picture emerges for spherical swimmers in Poiseuille flow (Zöttl and Stark, 2012). Consider a steady pressure-driven flow along the x direction between two flat plates at $y = \pm h$. The flow in this case, a classical solution to the (Navier–)Stokes equations, is quadratic

$$\mathbf{u}_\infty(\mathbf{r}) = U_0\left(1 - \frac{y^2}{h^2}\right)\mathbf{e}_x, \tag{10.31}$$

where U_0 is the fluid speed at the centre of the channel (Leal, 2007). Since the flow is quadratic, the second term in Faxén's law of velocity, Eq. (10.13), should in general be kept. It can however be neglected provided the swimmer is small compared to the channel width, $a \ll h$, and remains far from the walls, which we assume here (this term merely contributes to an overall decrease in the swimming velocity in the x direction).

The flow vorticity is given by

$$\frac{1}{2}(\nabla \times \mathbf{u}_\infty) = U_0 \frac{y}{h^2}, \tag{10.32}$$

which now depends on the swimmer position across the channel. Focusing on swimmers remaining in the (x, y) plane and ignoring hydrodynamic interactions between the swimmer and the channel walls (see Chapter 11), Eq. (10.16) for the swimmer kinematics can now be written, using Eqs. (10.31) and (10.32), as

$$\dot{X} = U \cos \phi + U_0 \left(1 - \frac{Y^2}{h^2} \right), \tag{10.33a}$$

$$\dot{Y} = U \sin \phi, \tag{10.33b}$$

$$\dot{\phi} = U_0 \frac{Y}{h^2}. \tag{10.33c}$$

The autonomous system in Eqs. (10.33b)–(10.33c) can be solved first and then used to determine $X(t)$. Differentiating Eq. (10.33c) and using Eq. (10.33b) leads to

$$\ddot{\phi} = \frac{U_0 U}{h^2} \sin \phi, \tag{10.34}$$

which is the equation of a pendulum. It is characterised by one unstable fixed point at $\phi = 0$ (cell pointing downstream) and one stable fixed point at $\phi = \pi$ (cell pointing upstream) associated with a natural frequency $\left(U U_0/(h^2) \right)^{1/2}$. Furthermore, we note that the dynamical system in Eqs. (10.33b)–(10.33c) has a Hamiltonian structure. Indeed, if we define the Hamiltonian

$$H(Y, \phi) = \frac{U_0}{2} \frac{Y^2}{h^2} + U \cos \phi, \tag{10.35}$$

then Eqs. (10.33b)–(10.33c) can be rewritten in the usual Hamiltonian fashion as

$$\dot{Y} = -\frac{\partial H}{\partial \phi}, \tag{10.36a}$$

$$\dot{\phi} = \frac{\partial H}{\partial Y}. \tag{10.36b}$$

An examination of the contours of H reveals the presence of a separatrix between two domains of stable periodic oscillations with qualitatively different trajectories (Zöttl and Stark, 2012). These trajectories are illustrated in Fig. 10.1(c) for the (dimensionless) parameters $h = 1$, $U = 1$, $X_0 = 0$, $Y_0 = -0.6$ and $\phi_0 = 1$ by solving the system in Eq. (10.33) numerically. For a weak external flow, $U_0 = 2.5$, the cells swim upstream on a cycloid-like trajectory while their orientation oscillates about the upstream direction (upstream swinging, Fig. 10.1(c), top). In contrast, for a stronger external flow, $U_0 = 10$, the swimming cells are advected downstream while rotating periodically in all directions (downstream tumbling, Fig. 10.1(c), bottom).

10.1.5 Spherical Swimmer in Extensional Flow

The final situation we consider is that of a spherical swimmer in an extensional flow. In this case, the external velocity field is given by

$$\mathbf{u}_\infty(\mathbf{r}) = \dot{\epsilon}\left(x\mathbf{e}_x - y\mathbf{e}_y\right), \tag{10.37}$$

where $\dot{\epsilon} > 0$ is the constant extensional rate and we assume again that the cell remains in the plane of the flow.

Since the flow in Eq. (10.37) is irrotational, its vorticity is zero and thus the orientation of the cell does not change, $\dot{\phi} = 0$ i.e. $\phi = \phi_0$. In that case, the trajectory of the swimmer, found by integrating Eq. (10.16), is a solution to

$$\dot{X} = \dot{\epsilon}X + U\cos\phi_0, \tag{10.38a}$$
$$\dot{Y} = -\dot{\epsilon}Y + U\sin\phi_0. \tag{10.38b}$$

These two equations may be integrated exactly using exponential integrating factors and we obtain

$$X(t) = \left(X_0 + \frac{U\cos\phi_0}{\dot{\epsilon}}\right)e^{\dot{\epsilon}t} - \frac{U\cos\phi_0}{\dot{\epsilon}}, \tag{10.39a}$$

$$Y(t) = \left(Y_0 - \frac{U\sin\phi_0}{\dot{\epsilon}}\right)e^{-\dot{\epsilon}t} + \frac{U\sin\phi_0}{\dot{\epsilon}}. \tag{10.39b}$$

The swimmer is seen to asymptote exponentially to $|X| \rightarrow \infty$, $Y \rightarrow U\sin\phi_0/\dot{\epsilon}$, in a manner that depends on its initial conditions and its swimming speed.

10.2 Elongated Swimmers in Flows

If the shape of the swimming cell is not spherical, its orientation dynamics is in general more complex than a simple rotation due to the flow vorticity. In the special case where the swimmer has an elongated body, as is approximately the case for bacteria and spermatozoa, its dynamics in a special class of external flows (namely, linear flows) can be solved analytically.

10.2.1 Relevant Gradient Length Scales

Unlike the case of spherical swimmers, the dynamics of elongated swimmers cannot be derived analytically for an arbitrary external flow but may still be obtained when the flow varies linearly in space. Fortunately, provided that the swimmer is much smaller than the relevant flow gradient length scale, any external flow may be reasonably approximated as locally linear. This is the case in at least three biologically and environmentally important situations. First, swimming bacteria may enter the bloodstream and be subject to the flows present in the circulatory system. The relevant length scale in that case is the diameter of the blood vessel, which for all but the smallest capillaries is much larger than 10 μm and thus larger than swimming bacteria too (Pedley, 2000; Campbell and Reece, 2014). A second example concerns the locomotion of spermatozoa in the female reproductive tract during mammalian reproduction. As the cells make their way through the female uterus, they are subject to peristaltic flows with characteristic length scales of centimetres, again much larger than the cells (Eytan and Elad, 1999; Myers and Elad, 2017). As a final example, we mention the locomotion of plankton (including algae and bacteria) in turbulent oceanic flows. The smallest length scale in fully developed turbulence is the Kolmogorov scale (Tennekes and Lumley, 1972). Applied to oceanic turbulence, this scale is associated with turbulent eddies no smaller than a few millimetres, again safely larger than most microscopic organisms (Lazier and Mann, 1989; Guasto et al., 2012).

10.2.2 Resistive-Force Theory in an External Flow

In order to characterise the dynamics of elongated swimmers in flows, we use resistive-force theory from Eq. (6.40). To account for the presence of an external flow, $\mathbf{u}_\infty(\mathbf{r})$, we now interpret the velocity on the right-hand side of Eq. (6.40) as the difference between the velocity of the centreline of the filament and that of the background fluid at the location of the centreline. For a filament centreline denoted instantaneously by $\mathbf{X}(s,t)$ moving with velocity $\mathbf{u}(s,t)$ where s is the arclength and t the time, we use $\mathbf{u}_\infty(s,t)$ to denote the velocity $\mathbf{u}_\infty(\mathbf{X}(s,t))$. The hydrodynamic force density along the elongated body, $\mathbf{f}$, is now written as

$$\mathbf{f}(s,t) = -\left[c_\|\mathbf{tt} + c_\perp(\mathbf{1} - \mathbf{tt})\right] \cdot \left[\mathbf{u}(s,t) - \mathbf{u}_\infty(s,t)\right], \qquad (10.40)$$

where we recall that $\mathbf{t}$ is the unit tangent vector along the body centreline.

10.2.3 Self-Propelled Rods in a Linear Flow

Following Hohenegger and Shelley (2011), we derive in this section the equivalent to Eqs. (10.13) and (10.15) for non-spherical, slender swimmers in linear flows. We model the cell as a force- and torque-free self-propelled rod swimming along its elongated direction measured by the unit vector $\mathbf{p}$. As we did in Section 10.1.1, we may exploit the linearity of Stokes flows and so can write that the elongated

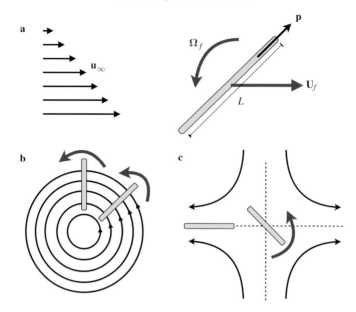

Figure 10.2 Dynamics of a passive rigid rod in an external linear flow. (a) The force- and torque-free rod, of length L, is aligned with the unit vector $\mathbf{p}(t)$ and, when located in the linear flow $\mathbf{u}_\infty$, is advected at velocity $\mathbf{U}_f$ and rotates with angular velocity $\mathbf{\Omega}_f$. (b) In a flow with constant vorticity, the rod rotates at a constant rate. (c) In an extensional flow, the rod aligns with the principal axis of strain and rotates at a rate that depends on its orientation.

swimmer moves with the instantaneous sum of its swimming kinematics in a quiescent fluid and the rigid-body kinematics imposed by the external flow with no-slip boundary conditions, i.e. Eq. (10.1). As above, we focus on the second part here and aim to compute the values of $\mathbf{U}_f$ and $\mathbf{\Omega}_f$.

In Fig. 10.2(a) we show a schematic representation of the rod, which has length L and whose centreline position is measured by its arclength s, ranging from $s = -L/2$ to $s = +L/2$. Taking $\mathbf{U}_f$ to denote the velocity of the centre of the rod (located at $s = 0$), the velocity of a point at arclength s along the rod is obtained as

$$\mathbf{u}_{\mathrm{rod}}(s, t) = \mathbf{U}_f + \mathbf{\Omega}_f \times (s\mathbf{p}). \tag{10.41}$$

Alternatively, using $\mathbf{X}(t)$ to denote the position of the centre of the swimmer, points along the centreline of the rod are located at $\mathbf{X}(t) + s\mathbf{p}(t)$ and their velocity may thus be written as

$$\mathbf{u}_{\mathrm{rod}}(s, t) = \dot{\mathbf{X}} + s\dot{\mathbf{p}}. \tag{10.42}$$

The purpose of the calculation below is to derive the values of $\dot{\mathbf{X}}$ ($= \mathbf{U}_f$) and $\dot{\mathbf{p}}$ ($= \mathbf{\Omega}_f \times \mathbf{p}$).

The external flow is assumed to vary linearly in space,

$$\mathbf{u}_\infty(\mathbf{r}) = \mathbf{A} \cdot \mathbf{r}, \tag{10.43}$$

where $\mathbf{A}$ is a constant, trace-free second-rank tensor. Its value at the location of each point along the rod is thus given by

$$\mathbf{u}_\infty(\mathbf{X} + s\mathbf{p}) = \mathbf{A} \cdot (\mathbf{X} + s\mathbf{p}). \tag{10.44}$$

In order to calculate the distribution of hydrodynamic forces along the elongated rod, $\mathbf{f}$, we make use of resistive-force theory from Eq. (10.40), applicable to slender shapes and leading to the force density

$$\mathbf{f}(s, t) = -[c_\| \mathbf{pp} + c_\perp (1 - \mathbf{pp})] \cdot [\mathbf{u}_{\mathrm{rod}}(s, t) - \mathbf{u}_\infty(\mathbf{X} + s\mathbf{p})], \tag{10.45}$$

where the unit tangent vector $\mathbf{t}$ appearing in Eq. (10.40) is now the unit vector $\mathbf{p}$ for the rod. Using Eqs. (10.42) and (10.44), we note that along the rod the relative velocity is given by

$$\mathbf{u}_{\mathrm{rod}}(s, t) - \mathbf{u}_\infty(\mathbf{X} + s\mathbf{p}) = \dot{\mathbf{X}} - \mathbf{A} \cdot \mathbf{X} + s(\dot{\mathbf{p}} - \mathbf{A} \cdot \mathbf{p}). \tag{10.46}$$

Exploiting that $c_\|/c_\perp = 1/2$ (see Section 6.3) and the tensor equality

$$\left(1 - \frac{1}{2}\mathbf{pp}\right) \cdot (1 + \mathbf{pp}) = 1, \tag{10.47}$$

we may invert Eq. (10.45) as

$$\mathbf{u}_{\mathrm{rod}} - \mathbf{u}_\infty = -\frac{1}{c_\perp}(1 + \mathbf{pp}) \cdot \mathbf{f}. \tag{10.48}$$

In order to determine the distribution of forces and the kinematics of the cell, we need to enforce that the swimmer is instantaneously force- and torque-free. Starting with force balance, we note that the tensor in Eq. (10.45) is constant and thus the constraint of zero net force becomes

$$\int_{-L/2}^{L/2} \mathbf{f} \, \mathrm{d}s = \mathbf{0} \rightarrow \int_{-L/2}^{L/2} [\mathbf{u}_{\mathrm{rod}}(s, t) - \mathbf{u}_\infty(\mathbf{X} + s\mathbf{p}, t)] \, \mathrm{d}s = \mathbf{0}. \tag{10.49}$$

In Eq. (10.46), we see that the relative velocity between the rod and the fluid includes a constant plus a term varying linearly in s. This linear term integrates to zero on the symmetric interval $-L/2 \leqslant s \leqslant L/2$ and therefore the constant term has to be identically zero, i.e.

$$\mathbf{U}_f = \dot{\mathbf{X}} = \mathbf{A} \cdot \mathbf{X}. \tag{10.50}$$

The centre of the slender swimmer moves therefore with the instantaneous velocity of the fluid in its absence evaluated at its position. Note that Eq. (10.50) may also be rewritten as

$$\mathbf{U}_f = \mathbf{u}_\infty|_{\mathbf{x}}, \tag{10.51}$$

which is the same result as for a sphere in linear flow.

Substituting Eq. (10.50) into Eq. (10.46), we see that the relative velocity between the rod and the fluid varies linearly along the rod as

$$\mathbf{u}_{\mathrm{rod}} - \mathbf{u}_\infty = s\left(\dot{\mathbf{p}} - \mathbf{A} \cdot \mathbf{p}\right). \tag{10.52}$$

Therefore the force density is linear in s and may be written as $\mathbf{f}(s,t) = s\tilde{\mathbf{f}}(t)$. To derive the value of $\tilde{\mathbf{f}}$ we invoke the constraint of torque-free swimming for the cell. The balance of torques about the centre of the rod is

$$\mathbf{0} = \int_{-L/2}^{L/2} s\mathbf{p}(t) \times s\tilde{\mathbf{f}}(t)\,\mathrm{d}s = \mathbf{p}(t) \times \tilde{\mathbf{f}}(t) \int_{-L/2}^{L/2} s^2\,\mathrm{d}s. \tag{10.53}$$

The vectors $\mathbf{p}$ and $\tilde{\mathbf{f}}$ are therefore parallel and we may write $\tilde{\mathbf{f}}(t) = \Gamma(t)\mathbf{p}(t)$. In order to determine the function $\Gamma(t)$, we return to Eq. (10.48) and enforce that it is true everywhere along the rod. Since we have $\mathbf{f}(s,t) = s\tilde{\mathbf{f}}(t) = s\Gamma(t)\mathbf{p}(t)$, then Eq. (10.48) becomes

$$s\left(\dot{\mathbf{p}} - \mathbf{A} \cdot \mathbf{p}\right) = -\frac{s\Gamma}{c_\perp}[\mathbf{1} + \mathbf{pp}] \cdot \mathbf{p} = -\frac{2s\Gamma}{c_\perp}\mathbf{p}, \tag{10.54}$$

and therefore

$$\dot{\mathbf{p}} - \mathbf{A} \cdot \mathbf{p} = -\frac{2\Gamma}{c_\perp}\mathbf{p}. \tag{10.55}$$

The value Γ is obtained by remembering that the vector $\mathbf{p}$ is of unit norm and therefore $\dot{\mathbf{p}} \cdot \mathbf{p} = 0$, so taking the scalar product of Eq. (10.55) with $\mathbf{p}$ gives

$$\mathbf{p} \cdot \mathbf{A} \cdot \mathbf{p} = \frac{2\Gamma}{c_\perp}. \tag{10.56}$$

As a side note, we remark that the term $\mathbf{p} \cdot \mathbf{A} \cdot \mathbf{p}$ is sometimes found in the literature written as $\mathbf{p}^T \cdot \mathbf{A} \cdot \mathbf{p}$ or $\mathbf{A} : \mathbf{pp}$. Using Eqs. (10.55)–(10.56), the evolution equation for the orientation of the cell, $\mathbf{p}$, is finally obtained as

$$\dot{\mathbf{p}} = \mathbf{A} \cdot \mathbf{p} - (\mathbf{p} \cdot \mathbf{A} \cdot \mathbf{p})\mathbf{p} = (\mathbf{1} - \mathbf{pp}) \cdot (\mathbf{A} \cdot \mathbf{p}). \tag{10.57}$$

A simple geometrical interpretation of this orientation evolution equation was offered by Doi and Edwards (1988), who noted that the rod rotates with the projection perpendicular to the vector $\mathbf{p}$ (operator $\mathbf{1} - \mathbf{pp}$) of the linear velocity induced along the direction of the rod (term $\mathbf{A} \cdot \mathbf{p}$).

The equation for $\dot{\mathbf{p}}$, Eq. (10.57), may be transformed to obtain an explicit expression for the angular velocity of the rod, $\boldsymbol{\Omega}_f$. If we note that

$$[\mathbf{p} \times (\mathbf{A} \cdot \mathbf{p})] \times \mathbf{p} = \mathbf{A} \cdot \mathbf{p} - (\mathbf{p} \cdot \mathbf{A} \cdot \mathbf{p})\mathbf{p}, \tag{10.58}$$

we can deduce

$$\dot{\mathbf{p}} = \boldsymbol{\Omega}_f \times \mathbf{p}, \quad \boldsymbol{\Omega}_f = \mathbf{p} \times (\mathbf{A} \cdot \mathbf{p}). \tag{10.59}$$

The expression for $\boldsymbol{\Omega}_f$ may be further simplified by splitting the flow into its symmetric and antisymmetric parts. Writing Eq. (10.43) using components we see that $u_{\infty,i} = A_{ij}x_j$ and thus $\mathbf{A} = (\nabla \mathbf{u}_\infty)^T$. Classically, velocity gradients may be written as $\nabla \mathbf{u}_\infty = \mathbf{E} + \frac{1}{2}\boldsymbol{\epsilon} \cdot \boldsymbol{\omega}$, where $\mathbf{E}$ is the symmetric rate of strain tensor, $\boldsymbol{\omega}$ is the flow vorticity and $\boldsymbol{\epsilon}$ is the third-rank permutation (or Levi-Civita) tensor (Batchelor, 1967), and thus

$$\mathbf{A} = (\nabla \mathbf{u}_\infty)^T = \mathbf{E} - \frac{1}{2}\boldsymbol{\epsilon} \cdot \boldsymbol{\omega}. \tag{10.60}$$

The second term in Eq. (10.60) may be substituted in Eq. (10.59) and we obtain, using index notation,

$$\left[\mathbf{p} \times \left(-\frac{1}{2}\boldsymbol{\epsilon} \cdot \boldsymbol{\omega}\right) \cdot \mathbf{p}\right]_i = -\frac{1}{2}\epsilon_{kij}\epsilon_{kmn}\omega_n p_m p_j = -\frac{1}{2}(\omega_n p_i p_n - \omega_i p_m p_m), \tag{10.61}$$

and therefore

$$\mathbf{p} \times \left[\left(-\frac{1}{2}\boldsymbol{\epsilon} \cdot \boldsymbol{\omega}\right) \cdot \mathbf{p}\right] = \frac{1}{2}[\boldsymbol{\omega} - (\boldsymbol{\omega} \cdot \mathbf{p})\mathbf{p}]. \tag{10.62}$$

Since the rotation rate impacts the rate of change of $\mathbf{p}$ through a cross product with $\mathbf{p}$, the second term on the right-hand side of Eq. (10.62) does not contribute to the value of $\dot{\mathbf{p}}$ and thus the final equations for the rotation rate of the rod are

$$\dot{\mathbf{p}} = \boldsymbol{\Omega}_f \times \mathbf{p}, \quad \boldsymbol{\Omega}_f = \frac{1}{2}\boldsymbol{\omega} + \mathbf{p} \times (\mathbf{E} \cdot \mathbf{p}), \tag{10.63}$$

where we recall that since the flow is linear, both $\boldsymbol{\omega}$ and $\mathbf{E}$ are constant.

Physically, the rotation rate of the slender rod arises from two flow contributions. The first term in Eq. (10.63) indicates a constant rotation with half of the flow vorticity, analogously to Faxén's law for a sphere, Eq. (10.15), and illustrated in Fig. 10.2(b) in the case of solid-body rotation. The second term in Eq. (10.63) is rotation of the rod due to the irrotational component of the flow, since the rate of strain tensor is symmetric. It indicates a rotation rate that depends on the rod orientation and acts to align it with the principal axis of strain (i.e. when the vectors $\mathbf{E} \cdot \mathbf{p}$ and $\mathbf{p}$ are parallel), as illustrated in Fig. 10.2(c) in the case of an extensional flow. As a result of the presence of these two different terms, the angular velocity of the rod is in general time-dependent.

Note that in the case of solid-body rotation, $\mathbf{E} = \mathbf{0}$, so the second term for the rotation rate in Eq. (10.63) is identically zero, and therefore elongated swimmers in a viscous vortex (solid-body rotation) have the same trajectories as spherical swimmers (Section 10.1.2).

10.2.4 Jeffery's Equation

The result derived in Eq. (10.63) is a special case of Jeffery's equation (Jeffery, 1922) in the limit of asymptotically slender rods. Jeffery's classical calculation provides the exact value of the angular velocity, $\mathbf{\Omega}_f$, for spheroids in linear flows (see Further Reading). Using the same notation as above, Jeffery's result is written as

$$\dot{\mathbf{p}} = \mathbf{\Omega}_f \times \mathbf{p}, \quad \mathbf{\Omega}_f = \frac{1}{2}\omega + B\mathbf{p} \times (\mathbf{E} \cdot \mathbf{p}), \tag{10.64}$$

or alternatively as

$$\dot{\mathbf{p}} = \frac{1}{2}\omega \times \mathbf{p} + B[\mathbf{E} \cdot \mathbf{p} - (\mathbf{p} \cdot \mathbf{E} \cdot \mathbf{p})\mathbf{p}], \tag{10.65}$$

where, for a prolate spheroid, $0 \leqslant B < 1$ is a shape factor equal to

$$B = \left(\frac{r^2 - 1}{r^2 + 1}\right), \tag{10.66}$$

with $r \geqslant 1$ the aspect ratio of the ellipsoid. The cases treated above can be recovered as special limits. If the spheroid is a sphere, then $r = 1$, $B = 0$ and we recover Faxén's result, Eq. (10.15). In the opposite limit, if the spheroid is slender, then $r \gg 1$, $B \approx 1$ and we obtain the result of Eq. (10.63). Note that for a rod with an aspect ratio of 10, which can be used as the effective aspect ratio of a flagellated bacterium, the slender result differs by less than 2% from the exact one.

Note also that in order to avoid the use of the cross product, some studies introduce the tensor $\mathbf{W} = -\frac{1}{2}\boldsymbol{\epsilon} \cdot \omega = \mathbf{E} - \nabla\mathbf{u}_\infty$ to write

$$\frac{1}{2}\omega \times \mathbf{p} = [\mathbf{p} \times (\mathbf{W} \cdot \mathbf{p})] \times \mathbf{p} = (\mathbf{1} - \mathbf{pp}) \cdot \mathbf{W} \cdot \mathbf{p}, \tag{10.67}$$

so that Eq. (10.65) can be written under the tensorial form

$$\dot{\mathbf{p}} = (\mathbf{1} - \mathbf{pp}) \cdot (\mathbf{W} + B\mathbf{E}) \cdot \mathbf{p}. \tag{10.68}$$

10.2.5 Jeffery's Orbits and Elongated Swimmers in Shear Flow

For elongated swimmers in a linear flow, the evolution equations for the position
($\mathbf{X}$) and orientation of the swimmer ($\mathbf{p}$) are written as

$$\dot{\mathbf{X}} = \mathbf{U}_s + \mathbf{A} \cdot \mathbf{X}, \tag{10.69a}$$

$$\dot{\mathbf{p}} = \mathbf{\Omega} \times \mathbf{p}, \quad \mathbf{\Omega} = \mathbf{\Omega}_s + \frac{1}{2}\boldsymbol{\omega} + B\mathbf{p} \times (\mathbf{E} \cdot \mathbf{p}). \tag{10.69b}$$

In what follows, we assume again for simplicity that the swimmer has no intrin-
sic rotational velocity ($\mathbf{\Omega}_s = \mathbf{0}$) and that swimming occurs along the body-fixed
direction at constant speed ($\mathbf{U}_s = U\mathbf{p}$).

Focusing on the case of a shear flow as in Eq. (10.22), $\mathbf{u}_\infty(\mathbf{r}) = \dot{\gamma}y\mathbf{e}_x$ with
constant shear rate $\dot{\gamma} > 0$, and assuming that the elongated swimmer remains in the
(x, y) plane of flow, we may write its orientation as $\mathbf{p} = (\cos\phi(t), \sin\phi(t))$. The
instantaneous velocities of the swimmer satisfy Eq. (10.69a), which in components
reads

$$\dot{X} = \dot{\gamma}Y + U\cos\phi(t), \tag{10.70a}$$

$$\dot{Y} = U\sin\phi(t). \tag{10.70b}$$

In order to solve for the trajectories of the swimmer, we need to derive the angular dynamics, which is done by calculating the components of Eq. (10.64). The
vorticity of the shear flow has already been computed in Eq. (10.24), while the rate
of strain tensor in the (x, y) plane is given by

$$\mathbf{E} = \frac{\dot{\gamma}}{2}\begin{pmatrix} 0 & 1 \\ 1 & 0 \end{pmatrix}. \tag{10.71}$$

The rotation rate of the swimmer follows from Eq. (10.69b), which becomes

$$\mathbf{\Omega} = \Omega\mathbf{e}_z, \quad \Omega = -\frac{1}{2}\dot{\gamma} + \frac{1}{2}B\dot{\gamma}(\cos^2\phi - \sin^2\phi). \tag{10.72}$$

Using Eq. (10.66) and the identity $1 = \cos^2\phi + \sin^2\phi$, we obtain that the rate of
change of the swimmer orientation, $\dot{\phi} = \Omega$, obeys

$$\dot{\phi} = -\frac{\dot{\gamma}}{r^2 + 1}\left(\cos^2\phi + r^2\sin^2\phi\right). \tag{10.73}$$

This equation is the same as for passive rods ($U = 0$), and the resulting angular
dynamics may be solved exactly (Kim and Karrila, 1991). Dividing Eq. (10.73)
by $\cos^2\phi$ and recognising the derivative of $\tan\phi$, we may then use the substitution
$f = r\tan\phi$ to obtain a simplified differential equation,

$$\frac{\dot{f}}{1 + f^2} = -\frac{r\dot{\gamma}}{r^2 + 1}, \tag{10.74}$$

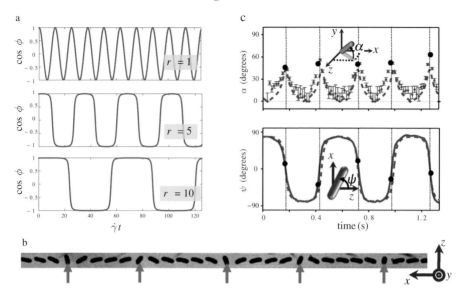

Figure 10.3 Jeffery's orbits. (a) Numerical illustration of angular dynamics; $\cos\phi$ as a function of dimensionless time, $\dot{\gamma}t$, for three rod aspect ratios, $r = 1, 5$ and 10, where ϕ is the angle between the rod and the flow direction. (b) Modified Jeffery's orbits of non-flagellated *E. coli* cells in a Poiseuille flow created by a microfluidic device (reprinted figure with permission from Kaya and Koser (2009), Copyright 2009 by the American Physical Society); at each height above the surface, the flow behaves locally like a shear flow, denoted $\mathbf{u}_\infty(\mathbf{r}) \approx \dot{\gamma}y\mathbf{e}_x$ using axes (x, y, z) defined to be consistent with the text. (c) For a cell with aspect ratio $r = 2.96$ and subject to shear rate $\dot{\gamma} = 65.6/\text{s}$, comparison between the experimental time dependence of the angles α and ψ (symbols and solid line) and the theoretical prediction from Jeffery's three-dimensional orbits (dashed line); here α denotes the acute 'pitch' angle between the axis of symmetry of the cell and the (x, z) plane, while ψ is the acute 'yaw' angle between the z axis and the projection of the cell axis onto the (x, z) plane. The black circles correspond to the five snapshots indicated by arrows in (b).

where again we can spot the derivative of the inverse tangent function. The exact solution to Eq. (10.73) is then given by

$$\tan\phi(t) = -\frac{1}{r}\tan\left(\frac{(t - t_0)\dot{\gamma}}{r + r^{-1}}\right),\tag{10.75}$$

where the rod is aligned with the x direction at $t = t_0$. The result in Eq. (10.75) indicates periodic rotation of the rod with period

$$T = \frac{2\pi}{\dot{\gamma}}\left(r + \frac{1}{r}\right),\tag{10.76}$$

which remains unchanged if the rod is initially tilted with respect to the plane of the flow (see Exercises). The angular motion of the rod, classically referred to as Jeffery's orbits, is illustrated in Fig. 10.3(a) for three different aspect ratios of the rod ($r = 1, 5, 10$). We can see the period increasing with aspect ratio, as predicted in Eq. (10.76). Furthermore, while a spherical particle ($r = 1$) rotates at a constant rate, elongated rods spend much time aligned with the flow direction ($\cos \phi = \pm 1$) followed by quick reorientation events. Note that the limit $r \to \infty$ is in fact singular since in that case slender rods are predicted to rotate only until they align permanently with the flow direction.

A comparison with experiments is illustrated in Fig. 10.3(b–c) in the case of non-flagellated *E. coli* bacteria advected by a plane Poiseuille flow created in a microfluidic device (Kaya and Koser, 2009). With a distance between the top and bottom plates in the device of $30\ \mu$m, the typical flow length scale is much larger than the bacterial cell bodies and the velocity field around each cell is approximately locally linear, $\mathbf{u}_\infty(\mathbf{r}) \approx \dot{\gamma} y \mathbf{e}_x$. As a result, the non-motile bacteria undergo Jeffery's orbits while being advected by the flow, as shown in Fig. 10.3(b). In Fig. 10.3(c), we display Jeffery's theoretical predictions (dashed line) along with the experimental results (symbols and solid line) using the angles defined by Kaya and Koser (2009), where α is the acute 'pitch' angle between the axis of symmetry of the cell and the (x, z) plane while ψ is the acute 'yaw' angle between the z axis and the projection of the cell axis onto the (x, z) plane. The theoretical results derived for prolate spheroids are able to reproduce the experimental measurements with excellent agreement.

For swimming cells in simple shear, Eq. (10.70) may be solved numerically using, for the angular dynamics, the solution to Eq. (10.75) that is continuous in time, and we obtain oscillating trajectories with distorted shapes analogous to the cycloids derived in Section 10.1.3. Similar to spherical swimmers, the cells undergo net motion in either the $+x$ or $-x$ direction as a function of the shear rate and their initial position, aspect ratio and swimming speed.

10.2.6 Elongated Swimmer in Poiseuille Flow

The dynamics of elongated swimmers in Poiseuille flow may be tackled by assuming that the swimmer is smaller than the width of the channel and locally linearising the flow in Eq. (10.31) around the instantaneous location of the swimmer. The resulting system has to be integrated numerically (Zöttl and Stark, 2013). Unlike the case of spherical swimmers, the dynamics for elongated swimmers no longer has a Hamiltonian structure, but there is still a conservative dynamical system with a constant of motion. The swimmer trajectories are very similar to the case of spheres, with either upstream swinging around the centreline of the channel or downstream tumbling, depending on the initial location and orientation of the swimmer, its geometry, swimming speed and the flow strength.

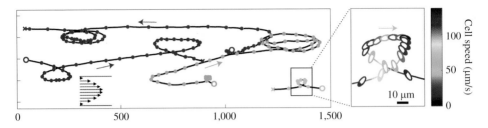

Figure 10.4 Swimming trajectories of four smooth-swimming *Bacillus subtilis* cells in Poiseuille flow (direction of flow indicated by sketch) displaying both upstream- and downstream-moving motion as indicated by arrows (reprinted by permission from Springer Nature: Rusconi et al. (2014)). Units on the *x*-axis are in μm.

The role of Brownian noise on these results can be illustrated by examining the dynamics of swimming cells in microfluidic experiments. We show in Fig. 10.4 the trajectories of smooth-swimming *Bacillus subtilis* bacteria in (approximately) plane Poiseuille flow inside a microchannel (Rusconi et al., 2014). We observe both upstream- and downstream-moving cells, with noise allowing the cells to transition from one type of trajectory to the next. At long times, the interaction between fluid shear and motility leads to strong heterogeneity in the spatial distribution of elongated bacteria, which tend to accumulate in high-shear regions.

10.2.7 Elongated Swimmer in Extensional Flow

As a final example, we consider the dynamics of an elongated swimmer in an irrotational extensional flow. With the external flow velocity from Eq. (10.37) characterised by the constant extensional rate $\dot{\epsilon} > 0$, the trajectories of the swimmer follow

$$\dot{X} = \dot{\epsilon} X + U \cos \phi(t), \qquad (10.77a)$$

$$\dot{Y} = -\dot{\epsilon} Y + U \sin \phi(t), \qquad (10.77b)$$

with the equation for $\dot{\phi}$ obtained by solving Eq. (10.64). Since the flow has no vorticity, the only relevant component arises from the rate of strain tensor

$$\mathbf{E} = \dot{\epsilon} \begin{pmatrix} 1 & 0 \\ 0 & -1 \end{pmatrix}, \qquad (10.78)$$

leading to

$$\dot{\phi} = -2B\dot{\epsilon} \sin \phi \cos \phi, \qquad (10.79)$$

with the exact solution

$$\tan \phi(t) = \tan \phi_0 e^{-2B\dot{\epsilon} t}. \qquad (10.80)$$

Independently of its initial location or orientation, the swimmer becomes aligned with the principal axis of strain. The exact solution of Eq. (10.77) for the position of the swimmer involves hypergeometric functions and is not very enlightening. The swimmer is always eventually advected by the flow like a passive particle. However, at early times $t \ll \dot{\epsilon}^{-1}$, the dynamics of the swimmer is dominated by self-propulsion and thus the quadrant in which the swimmer is eventually trapped and advected away is not necessarily the one it started from.

10.3 Biased Swimmers in Flows

So far the orientation of swimmers has been set by a combination of intrinsic rotation and flow-induced changes in direction. In several biological settings, self-propelled organisms also interact with an effective external field and as a result are biased in favour of swimming in a particular direction. The resulting dynamics was first studied in the context of alga cells swimming in a gravitational field (Kessler, 1985) with a mathematical treatment that turned out to be applicable to a wide variety of other situations.

10.3.1 Biased Locomotion and Taxis

Consider a fixed lab frame vector $\mathbf{k}$, which indicates the preferred swimming direction of the organism. The standard modelling approach to capture the tendency of the swimmer to reorient along $\mathbf{k}$ consists in assuming that the cell is subject to an external torque of magnitude

$$\mathbf{T} = T_0 \mathbf{p} \times \mathbf{k}, \qquad (10.81)$$

where the unit vector $\mathbf{p}$ denotes the direction of swimmer. Clearly, $\mathbf{T}$ is zero when $\mathbf{p} \parallel \mathbf{k}$ and maximal when the two vectors are perpendicular to each other. This modelling assumption turns out to be exact mathematically in at least three cases, and is used as an empirical modelling approach in other situations.

First, there exists a class of bacteria that are magnetic due to the presence of intracellular magnetic organelles (Blakemore, 1982). When subject to an external magnetic field, $\mathbf{B} = B\mathbf{k}$, these cells experience a net magnetic torque between the direction of the field and that of the cellular dipole, as in Eq. (10.81), of magnitude $T_0 = \mu_0 B$ where μ_0 is the strength of the cellular dipole (Hall and Busenberg, 1969). This gives rise classically to *magnetotaxis*, where bacteria swim by following magnetic field lines (Blakemore, 1982).

A second situation where the external torque from Eq. (10.81) is correct quantitatively applies to bottom-heavy algae, a class of marine microorganisms with heavy organelles located in the posterior part of their cell body. The centre of mass of these organisms is located below their centre of buoyancy, offset by a distance h, and every time the posterior–anterior axis of such a cell is not oriented vertically, it is subject to a gravitational torque as in Eq. (10.81) with $T_0 = hmg$, where m is the mass of the cell and $\mathbf{g} = -g\mathbf{k}$ is the acceleration due to gravity. This is known to give rise to *gravitaxis*, the tendency of these cells to swim upward (Kessler, 1985; Pedley and Kessler, 1992), a clever physical mechanism allowing cells to have access to light for photosynthesis and to oxygen passively.

The final configuration in which an external torque of the form of Eq. (10.81) is induced is discussed in Section 11.3, and arises for cells swimming near a surface in a shear flow. In that case, the torque originates from the fore–aft asymmetry of the cell and it leads to a reorientation of the swimmer along (or against) the flow direction.

Beyond these situations where the external torque is exact, a term of the form of Eq. (10.81) has also been proposed as an empirical modelling tool to describe the tendency of some cells to reorient towards light sources (Williams and Bees, 2011), leading to *phototaxis*. A similar term could also be used to model *chemotaxis*, the tendency of some cells to swim towards (or away from) specific chemicals (Berg and Brown, 1972).

In all cases, balancing the external torque in Eq. (10.81) with a viscous torque gives rise to an additional term in the angular velocity of the swimming cell in Eq. (10.64) of the form

$$\Omega = \Omega_0 \mathbf{p} \times \mathbf{k}, \tag{10.82}$$

so that the reorientation equation now becomes

$$\dot{\mathbf{p}} = \Omega \times \mathbf{p}, \quad \Omega = \frac{1}{2}\omega + B\mathbf{p} \times (\mathbf{E} \cdot \mathbf{p}) + \Omega_0 \mathbf{p} \times \mathbf{k}, \tag{10.83}$$

or alternatively Eq. (10.65) is now

$$\dot{\mathbf{p}} = \frac{1}{2}\omega \times \mathbf{p} + B[\mathbf{E} \cdot \mathbf{p} - (\mathbf{p} \cdot \mathbf{E} \cdot \mathbf{p})\mathbf{p}] + \Omega_0[\mathbf{k} - (\mathbf{k} \cdot \mathbf{p})\mathbf{p}]. \tag{10.84}$$

10.3.2 Biased Swimming under Shear

We demonstrated in Section 10.2.5 that cells in shear flows always reorient periodically. The presence of a bias in the swimming direction can now lead to a balance between the viscous torque exerted by the shear flow and the restoring torque towards the preferred direction, resulting in a fixed orientation of the swimming cells (Pedley and Kessler, 1987, 1992).

Consider a shear flow perpendicular to the preferred swimming direction of the organism, here chosen to be $\mathbf{k} = \mathbf{e}_z$, and assume for simplicity that the swimmer is spherical. The flow is written as $\mathbf{u}_\infty = \dot{\gamma}z\mathbf{e}_y$ in the (x, y, z) frame (see notation in

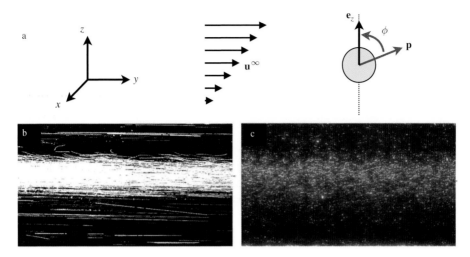

Figure 10.5 Biased swimming under shear. (a) Schematic representation of spherical swimmer in a shear flow oriented at angle ϕ from its preferred swimming direction ($\mathbf{e}_z$). (b–c) Experimental illustration of gyrotactic trapping in a laboratory
setup with vertically varying shear rates. Populations of algae *Chlamydomonas
nivalis* (b) and *Heterosigma akashiwo* (c) in high-shear regions where the torques
arising from the external shear are larger than the restoring torques from bottom-
heaviness so the cells are trapped in layers perpendicular to their preferred direction (Durham et al. (2009) reproduced with permission).

Fig. 10.5(a)). We denote by ϕ the angle between the direction of the swimmer and
$\mathbf{e}_z$ so that $\mathbf{p} = \cos\phi\,\mathbf{e}_z + \sin\phi\,\mathbf{e}_y$. The rotation of the cell from Eq. (10.83) occurs
about the x axis perpendicular to the plane of shear, which leads to the so-called
forced Adler equation for the angular dynamics,

$$\dot{\phi} = \frac{1}{2}\dot{\gamma} - \Omega_0 \sin\phi. \tag{10.85}$$

Note that the change of sign compared to Eq. (10.25) for the shear contribution is
merely due to the change of definition of ϕ.

   The result in Eq. (10.85) shows that the balance between shear and restoring
torque can lead to a constant orientation for the swimmer. The fixed points $\hat{\phi}$, which
are solutions to $\dot{\phi} = 0$, satisfy

$$\sin\hat{\phi} = \frac{\dot{\gamma}}{2\Omega_0}. \tag{10.86}$$

If $\dot{\gamma}/(2\Omega_0) < 1$, then Eq. (10.86) has two solutions. One equilibrium is stable (the
solution with $\cos\hat{\phi} > 0$ where the cell is swimming towards $\mathbf{e}_z$ at an angle) while
the other one is unstable (the solution with $\cos\hat{\phi} < 0$ where the cell is swimming
against $\mathbf{e}_z$).

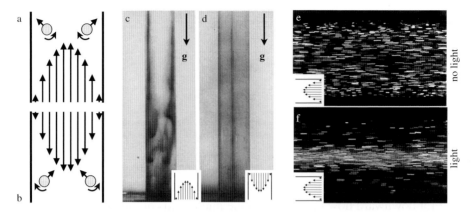

Figure 10.6 Hydrodynamic focusing of biased swimmers. (a–b) Focusing mechanism for spherical cells swimming upward in a Poiseuille flow. In an upward flow, the cells are reoriented towards the side walls (a) while in a downward flow they are rotated towards the centre of the flow (b). (c–d) Gyrotactic focusing of upward-swimming bottom-heavy algae in Poiseuille flow showing accumulation on the side walls for upward flow (c) and focusing for downward flow (d) (reprinted by permission from Springer Nature: Kessler (1985)). (e–f) Phototactic focusing of swimming algae showing that cells distributed uniformly in the absence of light (e) focus in the centre of the flow when a light source is turned on in the direction opposite to that of the flow (f) (reprinted figure with permission from Garcia et al. (2013), Copyright 2013 by the American Physical Society).

If the shear rate increases sufficiently, we might reach a regime where $\dot{\gamma}/(2\Omega_0) > 1$ in which case Eq. (10.86) has no solution. In that situation, the viscous torques from shear dominates the bias of the swimmer, and the cell reorients periodically without the ability to escape. For bottom-heavy marine organisms with the tendency to swim upward, this leads to the phenomenon of *gyrotactic trapping*. In regions of the ocean subject to strong horizontal shear flows, cells accumulate at specific depths in the regions where the local shear rate is large enough to satisfy $\dot{\gamma}/(2\Omega_0) > 1$. Gyrotactic trapping is illustrated experimentally in Fig. 10.5 for populations of the green alga *Chlamydomonas nivalis* (Fig. 10.5(b)) and toxic alga *Heterosigma akashiwo* (Fig. 10.5(c)) constrained to remain in high-shear regions in a laboratory setup with vertically varying shear rates (Durham et al., 2009).

10.3.3 Hydrodynamic Focusing

The combination of a directional bias with hydrodynamic reorientation can also lead to the phenomenon of hydrodynamic focusing in flows going against the preferred swimming direction. The physical principle of this focusing is best understood by considering spherical cells swimming in a Poiseuille flow directed along, or against, the preferred swimming direction of the organism. It is illustrated in

Figs. 10.6(a–b), where we assume that the cells are biased in favour of swimming upwards. If the flow is also oriented upward, the sign of the flow vorticity is such that the cells are tilted towards the side walls of the pipe (Fig. 10.6(a)). In contrast, if the fluid flows against the preferred swimming direction, the vorticity reorients the cells towards the centre of the pipe, and since they are self-propelled they accumulate there (Fig. 10.6(b)).

For bottom-heavy swimmers, this phenomenon is termed *gyrotactic focusing* and was discovered by Kessler (1985). It is illustrated experimentally in Figs. 10.6(c–d) in the case of pipe flow of swimming algae. Upward-swimming cells migrate to the walls of the pipe in the case of an upward flow (Fig. 10.6(c)) while for downward flow they focus on the middle axis of the pipe (Fig. 10.6(d)).

A similar focusing phenomenon is observed in the case of phototactic algae, as illustrated experimentally in Figs. 10.6(e–f). In the absence of light and therefore of any bias in their swimming direction, the swimming algae are distributed uniformly in a Poiseuille flow (Fig. 10.6(e)). In contrast, when a light source attracting the cells against the flow is added, the swimming algae focus in the centre of the pipe (Fig. 10.6(f)) (Garcia et al., 2013).

Further Reading

The mathematics for the motion of a spherical swimmer in a three-dimensional pressure-driven pipe flow is similar to that in a Poiseuille flow between two plates. In that case, the cell dynamics is characterised by two constants of motion, leading to helical trajectories (both left-handed and right-handed chiralities are possible) analogous to the periodic swinging and tumbling seen in two dimensions (Zöttl and Stark, 2012; Zhu et al., 2013; Zöttl and Stark, 2013). The analysis may also be carried out in the case of a non-Newtonian mean flow with additional nonlinear effects (Mathijssen et al., 2016).

The derivation of Jeffery's result for the rotation rate of a rigid prolate spheroid in a linear flow, Eq. (10.64), may be found in the textbook by Kim and Karrila (1991). It involves computing the resistance matrix of spheroids and the stresslet they generate when subject to an external linear flow, generalising some of the arguments from Chapter 5. The predictions from Jeffery's equation have been verified experimentally for the flow between two concentric cylinders (Trevelyan and Mason, 1951) and can in fact be extended to essentially all bodies of revolution (Bretherton, 1962).

The ability of cells to respond to flow is termed *rheotaxis*, and in this chapter we have focused on the passive advection and rotation of organisms with fixed ellipsoidal shapes. When the complex details of the cell shape and its possible variation in time are taken into account, additional effects come into play. For example, a helix or ribbon subject to shear experiences hydrodynamic forces perpendicular to its axis (Makino and Doi, 2005). As a result, bacteria with helical flagellar filaments drift across streamlines in shear flows (Marcos et al., 2012). Similarly, the time-varying flagellar beating of spermatozoa impacts the ability of cells to respond to shear (Ishimoto and Gaffney, 2015).

The work in this chapter was limited to steady linear flows such as solid-body rotation, simple shear or extensional flows. When swimming in complex, time-varying or even turbulent flows, microorganisms display new dynamical behaviour (Rusconi and Stocker, 2015). For instance, in chaotic flows, the ability to swim can reduce the net rate of transport (Khurana et al., 2011), while deterministic arrays of vortices lead to clustered aggregations of gyrotactic organisms (Durham et al., 2011). The coupling between motility and shear in turbulent flows can induce strong patchiness in the distribution of phytoplankton (Durham et al., 2013; De Lillo et al., 2014), modify the stirring and mixing of nutrients and impact the chemotactic behaviour of swimming cells (Taylor and Stocker, 2012).

Exercises

1. A straight rigid rod is located in the plane of a two-dimensional viscous flow. Derive the equations governing the dynamics of the rod in: (i) a solid-body rotation (vortex) flow, $\mathbf{u}_\infty(\mathbf{r}) = \Omega_\infty \mathbf{e}_z \times \mathbf{r}$; (ii) a shear flow, $\mathbf{u}_\infty(x, y, z) = \dot{\gamma} y \mathbf{e}_x$; and (iii) an extensional flow, $\mathbf{u}_\infty(x, y, z) = \dot{\epsilon}(x\mathbf{e}_x - y\mathbf{e}_y)$, where Ω_∞, $\dot{\gamma}$ and $\dot{\epsilon}$ are constants. Identify the fixed points of the rod orientation and their stability.

2. The scallop theorem in Chapter 2 implied that a free two-sphere dumbbell could not swim (see also Exercise 7, Chapter 2). However, if both spheres in the dumbbell are actuated independently, a net force can be induced on the fluid. Consider two rigid spheres of radii a_1 and a_2 displaced periodically along the x axis in an infinite Newtonian fluid at zero Reynolds number. The positions of the spheres are prescribed as $x_1(t) = \delta \cos(\omega t)$ and $x_2(t) = \ell + \delta \cos(\omega t + \phi)$, where ω, δ, ϕ and ℓ are constants. Denote by F_1 the instantaneous force applied by the first sphere on the fluid. In the limit $a_i \ll \ell$ where hydrodynamic interactions can be modelled by interacting stokeslets, compute the leading-order value of the time-averaged force $\langle F_1 \rangle$ in powers of $1/\ell$. Interpret the values of ϕ for which the force averages to zero. Are these results in contradiction to the scallop theorem?

3. A spherical body swims with constant speed U in the (x, y) plane of a two-dimensional viscous flow. In addition to its translation, the swimmer rotates with intrinsic constant angular velocity ω_0 in the z direction. Compute the trajectory of the swimmer in: (i) a solid-body rotation flow (vortex),

$\mathbf{u}_\infty(\mathbf{r}) = \Omega_\infty \mathbf{e}_z \times \mathbf{r}$; (ii) a shear flow, $\mathbf{u}_\infty(x, y, z) = \dot{\gamma} y \mathbf{e}_x$; and (iii) an extensional flow, $\mathbf{u}_\infty(x, y, z) = \dot{\epsilon}(x\mathbf{e}_x - y\mathbf{e}_y)$, where Ω_∞, $\dot{\gamma}$ and $\dot{\epsilon}$ are constants.

4. A spherical swimmer is located in a solid-body rotation flow (vortex), $\mathbf{u}_\infty(\mathbf{r}) = \Omega_\infty \mathbf{e}_z \times \mathbf{r}$, where Ω_∞ is constant. The swimming velocity $\mathbf{U}$, of constant magnitude U, is oriented initially at an angle θ_0 from the z axis (i.e. $\mathbf{U}(t = 0) \cdot \mathbf{e}_z = U \cos \theta_0$). Calculate the position of the swimmer for all times, $(X(t), Y(t), Z(t))$, and describe the resulting trajectories.

5. A straight rigid rod is made to rotate by a shear flow, $\mathbf{u}_\infty(x, y, z) = \dot{\gamma} y \mathbf{e}_x$, where $\dot{\gamma}$ is the constant shear rate. As opposed to the situation in Section 10.2.5, the rod does not remain in the (x, y) plane of the flow but is tilted by an angle θ from the z direction. Using spherical coordinates (θ, ϕ), show that the dynamics of the azimuthal angle ϕ in the plane of flow is unchanged from the case $\theta = 0$, i.e. Eq. (10.75), and that the angle θ satisfies

$$\tan \theta = \frac{K}{\left[1 + (r^2 - 1) \sin^2 \phi\right]^{1/2}}.$$

6. An elongated swimmer (shape factor, $B > 0$) is biased in favour of swimming upwards and thus subject to an external torque, $\mathbf{\Omega} = \Omega_0 \mathbf{p} \times \mathbf{e}_z$, where $\mathbf{e}_z$ is the vertical unit vector and $\mathbf{p}$ the unit vector along the swimming direction. The swimmer is located in the plane of a shear flow, $\mathbf{u}_\infty(x, y, z) = \dot{\gamma} z \mathbf{e}_y$, where $\dot{\gamma}$ is constant. Show that two fixed points of the swimmer orientation exist provided that $(1 - B)\dot{\gamma} \leqslant 2\Omega_0$, and show that only one of them is stable.

7. A swimmer is composed of three aligned rigid spheres of identical radii a and linked by mechanisms with negligible hydrodynamic effect. The distance between the first and second spheres is denoted by $\ell_1(t)$ while that between the second and third spheres is $\ell_2(t)$, both of which are prescribed to vary periodically in time. In the limit $a \ll \ell_i$ where hydrodynamic interactions can be modelled by interacting stokeslets, show that, if the three-sphere swimmer is force-free at all times, the mean swimming speed V_0 is given at leading order in a/ℓ_i by

$$V_0 = \frac{a}{6}\left\langle \frac{\dot{\ell}_2 - \dot{\ell}_1}{\ell_1 + \ell_2} \right\rangle + \frac{a}{3}\left\langle \frac{\dot{\ell}_1}{\ell_2} - \frac{\dot{\ell}_2}{\ell_1} \right\rangle.$$

11

Self-Propulsion and Surfaces

In all models considered thus far, swimming cells are located in a fluid with no surrounding boundaries. However, the presence of surfaces near a swimmer can impact dramatically its ability to generate propulsive forces for locomotion. In this chapter, we use a series of far- and near-field models to review the various ways in which boundaries influence cellular propulsion from a hydrodynamic standpoint. At the cellular level, both the distribution of cells and their swimming kinematics are affected. On smaller length scales, boundaries govern the ability of appendages such as cilia to produce net forces and flow. We first consider length scales much larger than those of the cells. We show how the method of images for hydrodynamic singularities can be used to demonstrate that long-range hydrodynamic interactions lead to the attraction of swimming cells by boundaries and to a change in the swimming kinematics of bacteria from straight to circular. We then examine the dynamics of swimming cells in shear flows and explain how the presence of a surface leads to cell reorientation and upstream swimming. Focusing on swimmers close to surfaces, we next revisit the waving sheet model in the presence of a boundary to show how increased friction impacts locomotion kinematics. We finish by zooming in to the sub-cellular level and addressing the role played by surfaces on the generation of forces and flows by individual cilia.

11.1 Hydrodynamic Attraction by Surfaces

11.1.1 Observations

It has long been known that swimming cells are attracted by surfaces. Lord Rothschild demonstrated the effect by measuring the steady-state distribution of swimming bull spermatozoa between two rigid surfaces (Rothschild, 1963). His results, reproduced in Fig. 11.1 (left), show a strong tendency of the cells to swim near the surfaces, with a five-fold increase in cell concentration near walls

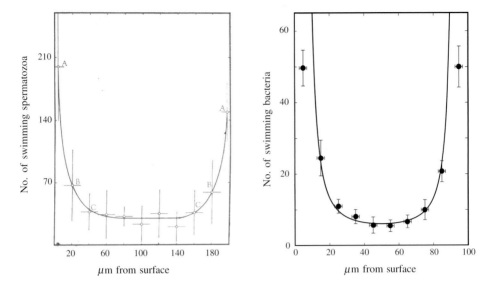

Figure 11.1 Left: distribution of swimming bull spermatozoa between two rigid surfaces demonstrating attraction (symbols) (reprinted by permission from Springer Nature: Rothschild (1963)); the line is a guide to the eye. Right: distribution of swimming bacteria between two glass coverslips (symbols) showing the same phenomenon (Berke et al., 2008); the line shows the theoretical model balancing hydrodynamic attraction from both surfaces (see Eq. (11.14) for one surface) with diffusion.

compared to its bulk value. Detailed measurements for smooth-swimming (i.e. non-tumbling) *E. coli* bacteria, reproduced in Fig. 11.1 (right), reveal the same phenomenon (Berke et al., 2008). Though there is still a debate on the relative importance of different physical processes in the measured results, it is clear that hydrodynamics does contribute to trapping of cells near walls (Berke et al., 2008; Giacché et al., 2010; Drescher et al., 2011; Bianchi et al., 2017). Here we use the method of images in order to derive a model of this hydrodynamic attraction.

11.1.2 Image Flow Singularities

Method of Images

A number of fundamental flow singularities were introduced in Chapter 6, and we saw in Chapter 9 how they may be used in order to construct far-field hydrodynamic models of swimming cells. While all singularities were derived in the case of an unbounded bulk fluid, they may be modified to account for the presence of a surface using a method of images similar to the one used classically to solve Laplace's equation (e.g. for problems in electrostatics). The application of this method to Stokes flow was pioneered by Blake (1971a), whose results have allowed

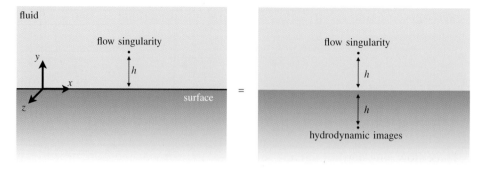

Figure 11.2 A flow singularity placed at distance h above an infinite flat surface (left) creates a velocity field equal to the flow induced by the original singularity plus a finite number of hydrodynamic image singularities located outside the fluid domain on the other side of the surface (right). The surface is either a rigid no-slip surface or an undeformed free surface.

the research community to reinterpret previous work in colloidal hydrodynamics near surfaces (Happel and Brenner, 1965) using intuitive physical ideas.

Consider a flow singularity placed at distance h above an infinite flat surface located at $y = 0$ (see notation in Fig. 11.2, left). The surface can either be a rigid no-slip boundary or an undeformed free surface (i.e. one unable to sustain any shear stress). The velocity field generated by the original singularity located at $\mathbf{r}_0$ and evaluated at a point $\mathbf{r}$ in space may be written as $\mathbf{u}^\infty(\mathbf{r} - \mathbf{r}_0)$. For any such singularity, as we show below, we may write the total flow field created by the singularity near the wall at any point $\mathbf{r}$ in space, $\mathbf{u}(\mathbf{r}; h)$ as a superposition of the original flow singularity plus an image flow field generated by a finite set of flow singularities located on the other side of the surface, i.e.

$$\mathbf{u}(\mathbf{r}; h) = \mathbf{u}^\infty(\mathbf{r} - h\mathbf{e}_y) + \mathbf{u}^{im}(\mathbf{r} + h\mathbf{e}_y; h). \qquad (11.1)$$

In other words, the hydrodynamic presence of a wall is equivalent mathematically to an appropriate linear superposition of additional hydrodynamic singularities located outside the fluid domain, as illustrated in Fig. 11.2 (right). The first term in Eq. (11.1) depends on the value of h only through the location of the singularity (which is at position $h\mathbf{e}_y$). In contrast, the second term depends on h through both the location of the image singularities (which are at $-h\mathbf{e}_y$, mirror image from the original singularity) and the strengths of some of these singularities (hence the notation used to indicate an additional parametric dependence on h in Eq. (11.1)). Note that the simplicity of the result in Eq. (11.1) and the fact that all image singularities are located at the same (mirror) image point are only true for plane boundaries. In what follows, we use tensor notation whenever possible in order to remain rigorous, although in some sections we find it more intuitive to use Cartesian coordinates.

Flow Singularities

All calculations carried out in this chapter can be deduced from the image system for the stokeslet (point force). Furthermore, by linearity, we can decompose the problem of a general stokeslet near a surface into the separate cases of a stokeslet either parallel or perpendicular to the boundary.

Three flow singularities already seen in the previous chapters are necessary to obtain our results. First we use the stokeslet, first seen in Eq. (6.3) and illustrated in Fig. 9.2(a). The stokeslet flow field allows us to define the stokeslet's Green's function tensor, $\mathbf{G}_F$, as

$$\mathbf{u}_F(\mathbf{r}) = \frac{1}{8\pi\mu}\left(\frac{1}{r} + \frac{\mathbf{rr}}{r^3}\right)\cdot\mathbf{F} \equiv \mathbf{G}_F(\mathbf{r})\cdot\mathbf{F}. \tag{11.2}$$

Similarly, we use the potential source dipole defined in Eq. (6.4) and illustrated in Fig. 9.2(g). We define its Green's function tensor, $\mathbf{G}_{SD}$, such that the velocity field for a source dipole of strength $\mathbf{M}$ is written

$$\mathbf{u}_{SD}(\mathbf{r}) = \frac{1}{4\pi}\left(\frac{1}{r^3} - \frac{3\mathbf{rr}}{r^5}\right)\cdot\mathbf{M} \equiv \mathbf{G}_{SD}(\mathbf{r})\cdot\mathbf{M}. \tag{11.3}$$

Finally, we use the general force dipole seen in Eq. (9.9) and illustrated in Fig. 9.2(b–c). Recall that this was obtained as the flow induced by a force $-F\mathbf{e}$ located at the origin and an equal and opposite force $F\mathbf{e}$ at a location $\epsilon\ell\mathbf{d}$ away, leading to the velocity field in the limit $\epsilon \to 0$

$$\mathbf{u}_{\text{dipole}}(\mathbf{r};\mathbf{e},\mathbf{d}) = \frac{\mathcal{P}}{8\pi\mu}\left[\frac{(\mathbf{d}\times\mathbf{e})\times\mathbf{r}}{r^3} - \frac{(\mathbf{d}\cdot\mathbf{e})\mathbf{r}}{r^3} + \frac{3(\mathbf{e}\cdot\mathbf{r})(\mathbf{d}\cdot\mathbf{r})\mathbf{r}}{r^5}\right]$$
$$\equiv \mathcal{P}\mathbf{G}_{FD}(\mathbf{r};\mathbf{e},\mathbf{d}), \tag{11.4}$$

which defines the force dipole's Green's function vector, $\mathbf{G}_{FD}$, where $\mathcal{P}$ denotes the dipole strength, $\mathcal{P} = \epsilon\ell F$. The notation for $\mathbf{G}_{FD}$ in Eq. (11.4) indicates both the spatial dependence of the flow ($\mathbf{r}$) and the parametric (bilinear) dependence on the vectors $\mathbf{e}$ and $\mathbf{d}$.

Stokeslet near a Free Surface

The case of a free surface, relevant to organisms swimming near air-water interfaces, is straightforward mathematically and leads to many of the same physical features as no-slip surfaces. For an infinite, flat free surface at $y = 0$, the boundary conditions are those of no-penetration and no-shear, i.e.

$$u_y(x, 0, z) = 0, \qquad \left.\frac{\partial u_x}{\partial y}\right|_{(x,0,z)} = 0, \qquad \left.\frac{\partial u_z}{\partial y}\right|_{(x,0,z)} = 0. \tag{11.5}$$

The conditions in Eq. (11.5) are satisfied automatically by a flow that is mirror-symmetric with respect to the surface at $y = 0$, i.e. a flow satisfying

$$u_x(x, -y, z) = u_x(x, y, z), \tag{11.6a}$$

$$u_y(x, -y, z) = -u_y(x, y, z), \tag{11.6b}$$

$$u_z(x, -y, z) = u_z(x, y, z). \tag{11.6c}$$

This suggests that the image system in the case of a free surface may be obtained simply by placing at the symmetric location below the free surface (i.e. at $y = -h$) a singularity equal to the mirror image of the original flow singularity in the fluid. By uniqueness of Stokes flows, this image system is the only solution to the problem. Note that in that case, the strength of the image singularities is independent of h and the flow depends on h only through the location of the singularity.

Specifically, in the case of a stokeslet induced by a force $\mathbf{F}_\parallel$ in a direction parallel to the surface, the original flow is given by $\mathbf{u}_\parallel^\infty(\mathbf{r}) = \mathbf{G}_F(\mathbf{r}) \cdot \mathbf{F}_\parallel$ and therefore the mirror image flow is $\mathbf{u}_\parallel^{im}(\mathbf{r}) = \mathbf{G}_F(\mathbf{r}) \cdot \mathbf{F}_\parallel$. Similarly, for a stokeslet induced by a point force perpendicular to the surface, $\mathbf{u}_\perp^\infty(\mathbf{r}) = \mathbf{G}_F(\mathbf{r}) \cdot \mathbf{F}_\perp$ with $\mathbf{F}_\perp = F_\perp \mathbf{e}_y$, the mirror image of the original singularity leads to the image flow as $\mathbf{u}_\perp^{im}(\mathbf{r}) = -\mathbf{G}_F(\mathbf{r}) \cdot \mathbf{F}_\perp$. In both cases, the total flow is given by Eq. (11.1). Obviously, the same mirror-symmetry principle holds for any flow singularity.

Stokeslet near a No-Slip Surface

The case of a stokeslet above a rigid surface, associated with the no-slip boundary condition, $\mathbf{u}(x, 0, z) = \mathbf{0}$, involves a more complex system of images made up of three different flow singularities. The image system in that case was first derived by Blake (1971a) using Fourier transforms (see also Blake and Chwang, 1974).

The simplest situation is that of a stokeslet perpendicular to the surface, $\mathbf{u}_\perp^\infty(\mathbf{r}) = \mathbf{G}_F(\mathbf{r}) \cdot \mathbf{F}_\perp$, since the resulting flow is axisymmetric. The full image system is given in that case by

$$\mathbf{u}_\perp^{im}(\mathbf{r}; h) = -\mathbf{G}_F(\mathbf{r}) \cdot \mathbf{F}_\perp - 2hF_\perp \mathbf{G}_{FD}(\mathbf{r}; \mathbf{e}_y, \mathbf{e}_y) - \frac{h^2}{\mu} \mathbf{G}_{SD}(\mathbf{r}) \cdot \mathbf{F}_\perp, \tag{11.7}$$

and involves three distinct singularities: a stokeslet of strength opposite to the original singularity, a vertical force dipole of vertical forces, and a source dipole. The full flow, obtained through the superposition of the original stokeslet with all images from Eq. (11.1), has streamlines illustrated in Fig. 11.3(a).

In the case where the stokeslet is parallel to the surface, the original singularity is $\mathbf{u}_\parallel^\infty(\mathbf{r}) = \mathbf{G}_F(\mathbf{r}) \cdot \mathbf{F}_\parallel$ while the image system is given exactly by

$$\mathbf{u}_\parallel^{im}(\mathbf{r}; h) = -\mathbf{G}_F(\mathbf{r}) \cdot \mathbf{F}_\parallel + 2hF_\parallel \mathbf{G}_{FD}(\mathbf{r}; \mathbf{e}_y, \mathbf{e}_x) + \frac{h^2}{\mu} \mathbf{G}_{SD}(\mathbf{r}) \cdot \mathbf{F}_\parallel. \tag{11.8}$$

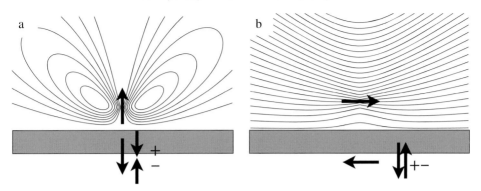

Figure 11.3 Illustration of hydrodynamic images and flow streamlines for a point force perpendicular (a) and parallel (b) to a flat no-slip surface. In both cases, the image system involves a stokeslet, a force dipole and a source dipole.

It is again made up of three terms: a stokeslet opposite to the original one, a horizontal force dipole of vertical forces, and a source dipole. The total flow, again obtained from Eq. (11.1), has streamlines illustrated in Fig. 11.3(b).

It is important to note that, in Eqs. (11.7) and (11.8), both the force dipole and the source dipole in the image systems have strengths that depend on the value of h, a feature that is important below when we manipulate the images of force dipoles. It is also to be noted that the flow far from the original singularity in both cases is no longer a stokeslet, but has a higher-order decay; this is sometimes referred to as hydrodynamic screening, and arises from the fact that both image systems include an equal and opposite force cancelling out the original stokeslet.

11.1.3 Far-Field Attraction by Surfaces

Equipped with the image system for the stokeslet, we may now propose a far-field model for the hydrodynamic attraction by surfaces discussed in Section 11.1.1. Consider a swimming cell whose distance to the closest surface, h, is much larger than its typical size, a. In that case, we may model the flow created by the swimmer as that due to a flow singularity. Specifically, following the results from Chapter 9, the swimmer is modelled as an axisymmetric force dipole (stresslet). Assuming that the cell is oriented parallel to the surface, we now show that the wall always attracts it hydrodynamically.

To compute the hydrodynamic effect of the wall, we consider the image system for the force dipole near the surface, and focus on the case of a no-slip boundary (see Exercises for the simpler case of a free surface). Superimposing the image system for two opposite point forces from the previous section, we deduce that the images for a force dipole near a no-slip surface are composed of an opposite force

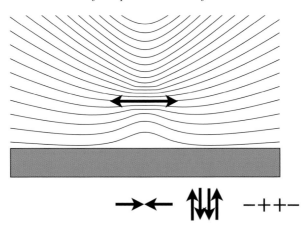

Figure 11.4 Illustration of hydrodynamic images and flow streamlines for a swimmer modelled as a force dipole parallel to a no-slip surface. The image system involves a force dipole, a force quadrupole and a source quadrupole. Image courtesy of Debasish Das.

dipole, a force quadrupole and a source quadrupole, with streamlines illustrated in Fig. 11.4.

The hydrodynamic impact of the wall on the cell can then be understood in the following intuitive manner. The cell creates a dipolar flow; this flow is modified by hydrodynamic images due to the presence of the wall and as a result the cell is advected (and potentially rotated) by these image flows. Using the methodology from Chapter 10, the flows from the hydrodynamic images, $\mathbf{u}^{im}$, provide, therefore, the external flow, $\mathbf{u}_\infty$, which moves the swimmer. If the cell is spherical in shape, the second term in Eq. (10.13) can be neglected because h is the relevant length scale for the gradients in the image flows and we are in the far-field limit, $a \ll h$. If instead the cell is elongated, Eq. (10.51) applies because the flow is locally linear, also as a result of the limit $a \ll h$. In both cases, the far-field approach results in the swimming cell being advected with a wall-induced velocity, $\mathbf{u}_{wall}$, such that

$$\mathbf{u}_{wall} = \mathbf{u}^{im}(2h\mathbf{e}_y; h), \qquad (11.9)$$

since the swimmer is located vertically above the hydrodynamic images at a distance $2h$ from them. By symmetry, $u_{wall,x} = u_{wall,z} = 0$, and the wall can at most induce a nonzero vertical advection velocity, which may be directed either away from or towards the surface as dictated by the sign of $u_{wall,y}$. Note also that due to the front-back symmetry of the force dipole, dipoles parallel to the surface experience no hydrodynamic reorientation from the image flows.

To compute the value of $u_{wall,y}$, we adapt the derivation of the dipolar flows in Chapter 9. Specifically, in order to discuss force dipoles we took the limit of two equal and opposite forces coming closer together. The same methodology may be used for the image flow. Consider an axisymmetric dipole obtained as the limit where $\epsilon \to 0$ of a force $-F_{\|}\mathbf{e}_x$ located at $(0, 0, h)$ and the opposite force, $F_{\|}\mathbf{e}_x$, at a location $\epsilon\ell\mathbf{e}_x$ away from it. During the limiting process, it is important to note that both forces remain parallel to the surface at height h.

For the original singularity, the flow is the symmetric force dipole with velocity field given in Eq. (9.30) as

$$\mathbf{u}_S(\mathbf{r}; \mathbf{S}) = \frac{\mathcal{P}}{8\pi\mu}\left[-\frac{1}{r^3} + \frac{3(\mathbf{e}_x \cdot \mathbf{r})^2}{r^5}\right]\mathbf{r}, \qquad (11.10)$$

which is a pusher if $\mathcal{P} > 0$ and a puller otherwise (alternatively, this can be seen as stresslet of strength $\mathbf{S}$ given by Eq. (9.18)). For the images, by linearity of the Stokes equations, the total image flow at leading order in ϵ is given by

$$\mathbf{u}_{\|}^{im}(\mathbf{r} + h\mathbf{e}_y - \epsilon\ell\mathbf{e}_x; h) - \mathbf{u}_{\|}^{im}(\mathbf{r} + h\mathbf{e}_y; h) \approx \mathcal{L}_x\left(\mathbf{u}_{\|}^{im}\right)\Big|_{(2h\mathbf{e}_y;h)}, \qquad (11.11)$$

where we have defined the differential operator

$$\mathcal{L}_x \equiv -\epsilon\ell\frac{\partial}{\partial x}, \qquad (11.12)$$

with of course all other variables (including h) being kept constant. This operator $\mathcal{L}_x$ is used repeatedly in what follows.

In order to evaluate the right-hand side of Eq. (11.11) we use the no-slip image system from Eq. (11.8). As we aim to compute the y component of the dipole images and take its derivatives along x, we first evaluate the stokeslet image flow at height h above the wall as

$$u_{\|,y}^{im}(x, 2h, 0; h) = \frac{3F_{\|}xh^3}{2\pi\mu r^5}, \quad r^2 = x^2 + (2h)^2. \qquad (11.13)$$

We may then take the x derivative and, using the definition $\mathcal{P} = \epsilon\ell F_{\|}$ (see Chapter 9), obtain the wall-induced velocity as

$$u_{\text{wall},y} = \mathcal{L}_x\left(u_{\|,y}^{im}\right)\Big|_{(2h\mathbf{e}_y;h)} = -\frac{3\mathcal{P}}{64\pi\mu h^2}. \qquad (11.14)$$

The result in Eq. (11.14) shows that a pusher cell (i.e one for which $\mathcal{P} > 0$) is attracted by the wall ($u_{\text{wall},y} < 0$) while a puller cell is repelled hydrodynamically (Berke et al., 2008). As seen in Chapter 9, most swimming cells are pushers, including spermatozoa and bacteria, and therefore with hydrodynamics alone we can explain the systematic wall accumulation seen in Fig. 11.1. Furthermore, in the case of bacteria, Berke et al. (2008) showed that balancing this wall attraction with diffusion leads to a density profile for the swimming cells in quantitative agreement with observations (Fig. 11.1, right; see also Chapter 13, Exercises). We note that the scaling of this wall effect with distance, $u_{\text{wall},y} \sim 1/h^2$, is a consequence of the dipolar nature of the flow induced by the swimmer.

If the surface is not a no-slip wall but instead a free surface, the image system for the dipole swimmer is an identical dipole on the other side of the surface, and we obtain a result similar to that in Eq. (11.14) with a different prefactor (specifically, wall attraction by a no-slip surface is 50% stronger than by a free surface; see Exercises).

11.1.4 Far-Field Reorientation by Surfaces

In the previous section, we saw that if a swimming cell, modelled at the level of a pusher force dipole, swims parallel to a surface then it is hydrodynamically attracted to it. What happens, however, if the cell is no longer parallel to the surface? The results in Chapter 10 quantified how swimming cells are affected by external flows. While the attraction in Eq. (11.14) is a consequence of the advection of the swimmer in the flow of its images, in general these images also lead to a hydrodynamic reorientation of the swimmer. Using a far-field model, we now show that hydrodynamics reorients pusher cells systematically towards the configuration where they are parallel to the surface, thereby always leading eventually to their attraction. We treat the case of a no-slip surface here, while the case of a free surface is one of the exercises.

Setup

Staying in the far field, we consider a swimmer of size a much smaller than its distance to the surface, $a \ll h$. The flows from the hydrodynamic images vary therefore on length scales much larger than the typical swimmer scale, and thus they may be approximated as locally linear around the cells. We can thus apply Jeffery's equation, Eq. (10.64), to characterise the angular velocity of the swimmer in the flow created by its images.

Consider a swimmer tilted by an angle θ away from the direction parallel to the surface, so that its direction of swimming is characterised by the unit vector $\mathbf{p} = \cos\theta \mathbf{e}_x + \sin\theta \mathbf{e}_y$ (we use the same Cartesian notation as in the previous section; see illustration in Fig. 11.5). In order to evaluate Jeffery's equation and calculate the reorientation rate of the swimmer, we need two ingredients, namely: (i) all hydrodynamic images for a tilted force dipole and (ii) the vorticity, ω, and symmetric rate of strain tensor, $\mathbf{E}$, of the flow from these images. By symmetry, it is clear that the only relevant component of the angular velocity of the swimmer is

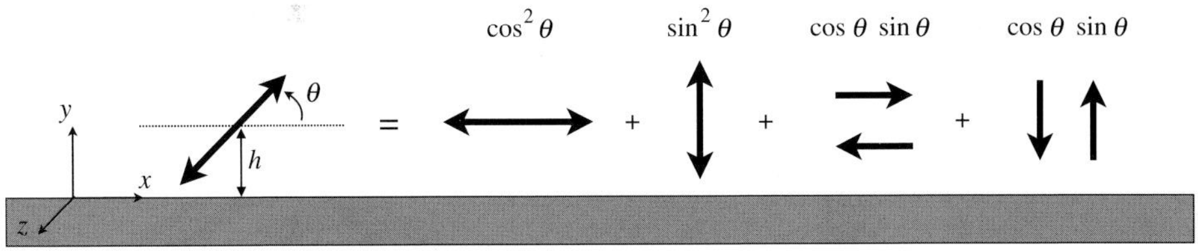

Figure 11.5 Notation for a force dipole at a distance h above the surface and tilted by an angle θ away from the horizontal direction. This dipole may be decomposed as the sum of four force dipoles whose directions are either parallel or perpendicular to the surface and with magnitude that depends on the value of θ as indicated above each singularity (Eq. (11.16)).

in the z direction. Evaluating Jeffery's equation, Eq. (10.64), the component Ω_z of the rotation rate is obtained as

$$\Omega_z = \frac{1}{2}\omega_z + B\cos\theta\sin\theta(E_{yy} - E_{xx}) + B(\cos^2\theta - \sin^2\theta)E_{xy}, \tag{11.15}$$

and thus we have to compute one component of the vorticity (ω_z) and three components of the rate of strain tensor (E_{xx}, E_{yy}, E_{xy}).

Decomposition of a Tilted Dipole

In order to carry out the calculation, we decompose the tilted dipole into simpler dipoles directed either parallel or perpendicular the surface, as illustrated graphically in Fig. 11.5. Given the bilinear dependence of the force dipole in Eq. (11.4) on the vectors **e** and **d**, the pusher of magnitude $\mathcal{P}$ tilted by an angle θ away from the direction parallel to the surface can be written as the sum of four dipoles, namely (i) a parallel axisymmetric dipole ($\leftrightarrow$) of strength $\mathcal{P}\cos^2\theta$, (ii) a perpendicular axisymmetric dipole ($\updownarrow$) of magnitude $\mathcal{P}\sin^2\theta$ and (iii) two xy dipoles ($\rightleftarrows + \downarrow\uparrow$), each of strength $p\cos\theta\sin\theta$ (i.e. an xy stresslet). As a consequence, we may write the cell rotation rate formally as the sum

$$\Omega_z = \cos^2\theta\,\Omega_{\leftrightarrow} + \sin^2\theta\,\Omega_{\updownarrow} + \cos\theta\sin\theta\,\Omega_{\rightleftarrows} + \cos\theta\sin\theta\,\Omega_{\downarrow\uparrow}, \tag{11.16}$$

and evaluate Eq. (11.15) for each term separately.

Obtaining Dipoles by Taking Derivatives

In Eq. (11.16) we notice that there are two types of force dipoles. In the first type, the flows are obtained by taking derivatives in the direction parallel to the surface of point forces and their images ($\leftrightarrow$ and $\downarrow\uparrow$) and thus by applying the operator $\mathcal{L}_x$ as in the previous section (see Eq. (11.12)). A more complicated situation arises for the two dipoles for which derivatives have to be taken in the direction perpendicular to the surface ($\updownarrow$ and $\rightleftarrows$) because the strengths of some of the hydrodynamic images also depend on the distance to the surface in that case. This situation may be addressed in the following manner. Consider the general situation of a hydrodynamic singularity whose total flow is given by Eq. (11.1). A force dipole in the direction perpendicular to the surface may be obtained by placing the original singularity at $y = h + \epsilon\ell$ and a singularity of opposite sign at $y = h$, and evaluating the flow in the limit $\epsilon \to 0$. In that case, the image flow is obtained at leading order in ϵ as

$$\mathbf{u}^{\mathrm{im}}_{\mathrm{dipole}}(\mathbf{r}; h) \approx \mathbf{u}^{\mathrm{im}}(\mathbf{r} + (h + \epsilon\ell)\mathbf{e}_y; h + \epsilon\ell) - \mathbf{u}^{\mathrm{im}}(\mathbf{r} + h\mathbf{e}_y; h)$$

$$\approx \mathcal{L}_y\left(\mathbf{u}^{\mathrm{im}}\right)\bigg|_{(\mathbf{r}+h\mathbf{e}_y;h)}, \tag{11.17}$$

where we have defined the new differential operator

$$\mathcal{L}_y \equiv \epsilon\ell\left(\frac{\partial}{\partial y} + \frac{\partial}{\partial h}\right). \tag{11.18}$$

The operator $\mathcal{L}_y$ plays the same role for dipoles perpendicular to the surface as the operator $\mathcal{L}_x$ for dipoles parallel to the surface, and we now use both to compute the reorientation of the swimmer. Note that in the first partial derivative in Eq. (11.18), both x and h are kept constant while in the second one both x and y are kept constant.

Dipoles Involving Forces Parallel to the Surface

Two of the dipoles in the decomposition from Fig. 11.5 are obtained by considering forces parallel to the surface ($\leftrightarrow$ and $\rightleftarrows$). Writing this parallel force as $\mathbf{F}_\parallel = F_\parallel\mathbf{e}_x$, the relevant flow components of its image system near a no-slip surface are written, following from Eq. (11.8) and using $\mathbf{r} = (x, y, z)$, as

$$u^{\mathrm{im}}_{\parallel,x}(\mathbf{r}; h) = \frac{F_\parallel}{8\pi\mu}\left(-\frac{1}{r} - \frac{x^2}{r^3} + \frac{2h^2}{r^3} - \frac{2hy}{r^3} - \frac{6h^2x^2}{r^5} + \frac{6hx^2y}{r^5}\right), \tag{11.19a}$$

$$u^{\mathrm{im}}_{\parallel,y}(\mathbf{r}; h) = \frac{F_\parallel}{8\pi\mu}\left(-\frac{xy}{r^3} + \frac{2hx}{r^3} + \frac{6hxy^2}{r^5} - \frac{6h^2xy}{r^5}\right). \tag{11.19b}$$

This image has z vorticity

$$\frac{1}{2}\omega^{\mathrm{im}}_{\parallel,z}(\mathbf{r}; h) = \mathbf{e}_z \cdot \left(\nabla \times \mathbf{u}^{\mathrm{im}}_\parallel\right) = \frac{F_\parallel}{8\pi\mu}\left(\frac{2h}{r^3} - \frac{y}{r^3} - \frac{6hx^2}{r^5}\right), \tag{11.20}$$

and its rate of strain tensor has components

$$E^{\mathrm{im}}_{\parallel,xy}(\mathbf{r}; h) = \frac{F_\parallel}{8\pi\mu}\left(\frac{3yx^2}{r^5} - \frac{6h^2y}{r^5} + \frac{6hy^2}{r^5} + \frac{30h^2x^2y}{r^7} - \frac{30hx^2y^2}{r^7}\right) \tag{11.21}$$

and

$$\left(E^{im}_{\|,yy} - E^{im}_{\|,xx}\right)(\mathbf{r};h) = \frac{F_\|}{8\pi\mu}\left(\frac{3xy^2}{r^5} - \frac{3x^3}{r^5} + \frac{12h^2x}{r^5} - \frac{12hxy}{r^5}\right.$$

$$\left. + \frac{30hx^3y}{r^7} + \frac{30h^2xy^2}{r^7} - \frac{30hxy^3}{r^7} - \frac{30h^2x^3}{r^7}\right). \quad (11.22)$$

Using Eq. (11.15) and the definition of the dipole strength, $\mathcal{P} = \epsilon\ell F_\|$, we obtain the angular velocities for the two dipoles as

$$\Omega_{\leftrightarrow} = B\cos\theta\sin\theta\mathcal{L}_x\left(E^{im}_{\|,yy} - E^{im}_{\|,xx}\right)\bigg|_{(2he_y;h)} = \frac{15\mathcal{P}}{128\pi\mu h^3}B\cos\theta\sin\theta \quad (11.23)$$

and

$$\Omega_{\rightleftarrows} = \mathcal{L}_y\left(\frac{1}{2}\omega^{im}_{\|,z} + B(\cos^2\theta - \sin^2\theta)E^{im}_{\|,xy}\right)\bigg|_{(2he_y;h)}$$

$$= \frac{\mathcal{P}}{64\pi\mu h^3} - \frac{3\mathcal{P}}{64\pi\mu h^3}B(\cos^2\theta - \sin^2\theta). \quad (11.24)$$

Dipoles Involving Forces Perpendicular to the Surface

The other two dipoles from the decomposition in Fig. 11.5 are the result of taking gradients for forces perpendicular to the surface ($\downarrow\uparrow$ and $\updownarrow$). Writing the perpendicular force as $\mathbf{F}_\perp = F_\perp\mathbf{e}_y$, the relevant components of its image system in the presence of a no-slip surface, Eq. (11.7), are given by

$$u^{im}_{\perp,x}(\mathbf{r};h) = \frac{F_\perp}{8\pi\mu}\left(-\frac{xy}{r^3} + \frac{2hx}{r^3} - \frac{6hxy^2}{r^5} + \frac{6h^2xy}{r^5}\right), \quad (11.25a)$$

$$u^{im}_{\perp,y}(\mathbf{r};h) = \frac{F_\perp}{8\pi\mu}\left(-\frac{1}{r} - \frac{y^2}{r^3} + \frac{2hy}{r^3} - \frac{2h^2}{r^3} - \frac{6hy^3}{r^5} + \frac{6h^2y^2}{r^5}\right). \quad (11.25b)$$

These lead to the z component of the image vorticity

$$\frac{1}{2}\omega^{im}_{\perp,z}(\mathbf{r};h) = \frac{F_\perp}{8\pi\mu}\left(\frac{x}{r^3} + \frac{6hxy}{r^5}\right), \quad (11.26)$$

and to the components of the image rate of strain tensor

$$E^{im}_{\perp,xy}(\mathbf{r};h) = \frac{F_\perp}{8\pi\mu}\left(\frac{3xy^2}{r^5} - \frac{12hxy}{r^5} + \frac{6h^2x}{r^5} + \frac{30hxy^3}{r^7} - \frac{30h^2xy^2}{r^7}\right) \quad (11.27)$$

and

$$\left(E^{\text{im}}_{\perp,yy} - E^{\text{im}}_{\perp,xx}\right)(\mathbf{r};h) = \frac{F_\perp}{8\pi\mu}\left(-\frac{3x^2y}{r^5} + \frac{6hx^2}{r^5} + \frac{12h^2y}{r^5}\right.$$

$$-\frac{18hy^2}{r^5} + \frac{3y^3}{r^5} + \frac{30h^2x^2y}{r^7}$$

$$\left.+\frac{30hy^4}{r^7} - \frac{30h^2y^3}{r^7} - \frac{30hx^2y^2}{r^7}\right). \tag{11.28}$$

Using the dipole definition, $\mathcal{P} = \epsilon\ell F_\perp$, along with Jeffery's equations, Eq. (11.15), we obtain the angular velocity induced by each dipole as

$$\Omega_{\downarrow\uparrow} = \mathcal{L}_x\left(\frac{1}{2}\omega^{\text{im}}_{\perp,z} + B(\cos^2\theta - \sin^2\theta)E^{\text{im}}_{\perp,xy}\right)\Bigg|_{(2he_y;h)} \tag{11.29}$$

$$= -\frac{\mathcal{P}}{16\pi\mu h^3} - \frac{3\mathcal{P}}{32\pi\mu h^3}B(\cos^2\theta - \sin^2\theta)$$

and

$$\Omega_{\updownarrow} = B\cos\theta\sin\theta\mathcal{L}_y\left(E^{\text{im}}_{\perp,yy} - E^{\text{im}}_{\perp,xx}\right)\Bigg|_{(2he_y;h)} = -\frac{3\mathcal{P}}{16\pi\mu h^3}B\cos\theta\sin\theta. \tag{11.30}$$

Hydrodynamic Reorientation of Swimmer by Surface

With the computed reorientation of the four fundamental dipoles in Eqs. (11.23), (11.24), (11.29) and (11.30), we may calculate the angular velocity of the tilted swimmer using Eq. (11.16), which leads to the value

$$\Omega_z = -\frac{3\mathcal{P}\cos\theta\sin\theta}{64\pi\mu h^3}\left[1 + \frac{B}{2}\left(2 - \cos^2\theta\right)\right], \tag{11.31}$$

as was computed by Berke et al. (2008) (note the different definition of θ used in that article).

The term in the bracket on the right-hand side of Eq. (11.31) is always positive since $0 \leqslant B < 1$, and therefore the angular velocity, $\Omega_z = \dot{\theta}$, has the same sign as $-\mathcal{P}\cos\theta\sin\theta$. It follows that here again pushers and pullers behave differently. For pusher cells with $\mathcal{P} > 0$, the deviation from the direction parallel to the surface follows the dynamics $\dot{\theta} \sim -\cos\theta\sin\theta$ and acts to align the swimmer parallel to the surface ($\theta = 0,\pi$ are the stable fixed points). If pushers ever deviate from being aligned with the nearby wall, they are reoriented parallel to it hydrodynamically, always leading to the attraction obtained in Eq. (11.14). In contrast, for puller swimmers with $\mathcal{P} < 0$, the angular dynamics obeys $\dot{\theta} \sim \cos\theta\sin\theta$. In that case, the surface reorients the cell hydrodynamically in the direction perpendicular to the wall, and the swimmer ends up either swimming straight into the surface or away from it (the stable fixed points are now at $\theta = \pm\pi/2$). This difference in angular dynamics between pushers and pullers may be used to explain the stark contrast between the forward and backward swimming of polar bacteria, which alternate the direction of rotation of their rotary motor (Magariyama et al., 2005). Note that here again, if the no-slip wall were to be replaced by a free surface, the reorientation dynamics would follow the same physics but with a different magnitude (see Exercises).

11.2 Circular Swimming near Surfaces

11.2.1 Observations

Following our investigation of the attraction and reorientation of cells by surfaces in the previous section, we now examine the impact of these surfaces on the trajectories of organisms. One of the most remarkable wall-induced changes in swimming kinematics is observed for flagellated bacteria. As reported by Berg and Turner (1990), and confirmed using cell tracking by Frymier et al. (1995), flagellated bacteria such as *E. coli* swim in straight lines in bulk fluids but their trajectories become circular near a surface. Furthermore, the nature of the surface affects the direction of rotation of the cells. This is illustrated in Fig. 11.6(a), where swimming *E. coli* cells undergo clockwise (CW) circular motion (when measured from the bulk of the fluid above the surface) above a no-slip surface (Lauga et al., 2006). For the same organism, swimming near a free surface changes the direction of rotation to counter-clockwise (CCW), as shown in Fig. 11.6(b) (Di Leonardo et al., 2011). These circular trajectories are the result of hydrodynamic interactions with the surfaces and may be rationalised using our far-field model.

11.2.2 Singularity Model and Images

Model

Consider a cell swimming parallel to a surface (either a free or a rigid surface). Modelled at the level of a force dipole, we saw that the swimming cell is attracted by the surface, but by symmetry a force dipole cannot be rotated in the direction perpendicular to the surface. Surface rotation must therefore result from the impact of surfaces on higher-order singularities in the flow created by the swimming cells. In Section 9.4 we listed the flow singularities decaying spatially as $1/r^3$ that are contained in the flow created by any swimmer, namely irrotational source dipoles, force quadrupoles and rotlet dipoles.

Given the fast rotation of the flagella ($\sim$ 100 Hz) compared to the timescale for the circular motion (on the order of seconds), we can furthermore make the assumption that the flow induced by a swimming bacterium averaged over a rotation of its flagellum is approximately axisymmetric. The singularities associated with

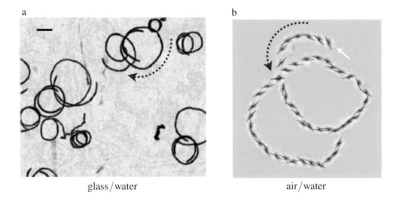

Figure 11.6 Circular swimming of *E. coli* bacteria in water near surfaces as
visualised from inside the fluid. (a) Swimming near a glass surface leads to CW
circular motion (Lauga et al. (2006) adapted with permission) (scale bar: 20 μm).
(b) Swimming near a free surface changes the direction of rotation to CCW
(reprinted figure with permission from Di Leonardo et al. (2011), Copyright 2011
by the American Physical Society).

axisymmetric flow are illustrated in Fig. 9.2(g–i). Both the source dipole
(Fig. 9.2(g)) and force quadrupole (Fig. 9.2(h)) induce flows that are right-left symmetric in the frame of the swimmer. In other words, using notation from Fig. 11.2,
if the swimmer is oriented along x and if y denotes the direction normal to the
surface, the flows from these two singularities are mirror-image symmetric with
respect to the (x, y) plane passing through the cell. A circular trajectory, resulting
from a y component of the cell rotation rate, therefore cannot be induced by the
source dipole nor by the force quadrupole.

In contrast, the rotlet dipole, illustrated in Fig. 9.2(i), does not obey this symmetry. As we now show, that singularity leads to circular swimming consistent with
experimental observations. We thus model a cell, swimming along the direction $\mathbf{e}_x$
parallel to the surface, as a rotlet dipole with flow given in Eq. (9.32), namely

$$\mathbf{u}_{RD}(\mathbf{r}; \mathbf{e}, \mathcal{X}) = \frac{3\mathcal{X}}{8\pi\mu} \frac{(\mathbf{e}_x \cdot \mathbf{r})(\mathbf{e}_x \times \mathbf{r})}{r^5}, \tag{11.32}$$

with dipole strength $\mathcal{X} > 0$ for *E. coli*.

Rotlet and Rotlet Dipole Parallel to Free Surface

Analogous to the point force and force dipole, the image system for a rotlet and a
rotlet dipole near a free surface is simple and only involves mirror-image singularities. The total flow still has the form given by Eq. (11.1), where $\mathbf{u}^\infty$ is the rotlet
of strength $\mathbf{R}$ of velocity given by Eq. (9.22). In the case where $\mathbf{R}$ is parallel to

the free surface, the hydrodynamic image, $\mathbf{u}_R^{\text{im}}$, is a rotlet with flow rotating in the opposite direction,

$$\mathbf{u}_R^{\text{im}}(\mathbf{r}) = \mathbf{u}_R(\mathbf{r}; -\mathbf{R}). \tag{11.33}$$

Similarly, for a rotlet dipole as in Eq. (11.32) parallel to the free surface, the image flow is a rotlet dipole of opposite strength,

$$\mathbf{u}_{RD}^{\text{im}}(\mathbf{r}) = \mathbf{u}_{RD}(\mathbf{r}; \mathbf{e}, -\mathcal{X}). \tag{11.34}$$

Rotlet Parallel to No-Slip Surface

In the case where the rotlet is parallel to a no-slip surface, the image system is more complex and involves three flow singularities. Consider a rotlet of strength $R\mathbf{e}_x$ located at $y = h$. The image flow is given in this case by

$$\mathbf{u}_R^{\text{im}}(\mathbf{r}; h) = \mathbf{u}_R(\mathbf{r}, -R\mathbf{e}_x) + R\left[\mathbf{G}_{FD}(\mathbf{r}; \mathbf{e}_y, \mathbf{e}_z) + \mathbf{G}_{FD}(\mathbf{r}; \mathbf{e}_z, \mathbf{e}_y)\right]$$
$$+ \frac{hR}{\mu}\mathbf{G}_{SD}(\mathbf{r}) \cdot \mathbf{e}_z, \tag{11.35}$$

and consists of (i) an opposite rotlet, (ii) a yz stresslet of strength R (i.e. 2R times the symmetric part of a yz force dipole) and (iii) a source dipole of strength $hR\mathbf{e}_z/\mu$ directed along the surface and perpendicular to the direction of the rotlet (Blake and Chwang, 1974).

11.2.3 Wall-Induced Hydrodynamic Reorientation

With these hydrodynamic images, we may now propose a model for the wall-induced reorientation of the swimming bacteria along the surface. Assume that the swimmer is oriented parallel to the surface along the x direction. We evaluate Eq. (10.64) in order to obtain the y component of the rotation rate, Ω_y, as

$$\Omega_y = \frac{1}{2}\omega_y^{\text{im}} - BE_{xz}^{\text{im}}. \tag{11.36}$$

We see that we only need to compute one component of the image vorticity, ω_y^{im}, and one component of the rate of strain tensor, E_{xz}^{im}.

Swimmer Parallel to Free Surface

The hydrodynamic image for the rotlet dipole parallel to a free surface induces the flow given by Eq. (11.34), with components in Cartesian coordinates

$$\mathbf{u}_{RD}^{\text{im}}(\mathbf{r}) = -\frac{3\mathcal{X}}{8\pi\mu}\left(0, -\frac{xz}{r^5}, \frac{xy}{r^5}\right). \tag{11.37}$$

Since the velocity field in Eq. (11.37) has no x component, both flow gradient terms in Eq. (11.36) are proportional to each other, and the cell rotation rate is given by

$$\Omega_y = -\frac{1}{2}(1+B)\frac{\partial u_{RD,z}^{im}}{\partial x}\bigg|_{2h\mathbf{e}_y}. \tag{11.38}$$

This result can be evaluated directly using Eq. (11.37), leading to the reorientation rate perpendicular to the free surface,

$$\Omega_y = \frac{3}{256\pi}\frac{X}{\mu h^4}(1+B). \tag{11.39}$$

Since $0 \leqslant B < 1$, the rotation rate satisfies $\Omega_y > 0$, and the swimming bacterium always rotates along the surface in a CCW fashion when viewed from above both the cell and the surface, as observed experimentally (see Fig. 11.6(b)). The scaling of the effect with the distance to the surface, $\Omega_y \sim 1/h^4$, is consistent with the flow gradients of a rotlet dipole, and indicates that the effect is strongly localised near the surface with a fast decay.

Swimmer Parallel to No-Slip Surface

In the situation where the rotlet dipole is parallel to a no-slip surface, we use the image system in Eq. (11.35) for the rotlet, to which we apply the differential operator in Eq. (11.12) to obtain the result applicable to the dipole. For a rotlet of strength $R\mathbf{e}_x$, the relevant velocity components of the image system are

$$u_{R,x}^{im}(\mathbf{r};h) = \frac{R}{8\pi\mu}\left(\frac{6xyz}{r^5} - \frac{6xhz}{r^5}\right), \tag{11.40a}$$

$$u_{R,z}^{im}(\mathbf{r};h) = \frac{R}{8\pi\mu}\left(\frac{2h}{r^3} - \frac{y}{r^3} + \frac{6yz^2}{r^5} - \frac{6hz^2}{r^5}\right), \tag{11.40b}$$

which lead to the vorticity

$$\frac{1}{2}\omega_y^{im}(\mathbf{r};h) = \frac{3R}{16\pi\mu}\left(\frac{xy}{r^5}\right) \tag{11.41}$$

and the rate of strain

$$E_{xz}^{im}(\mathbf{r};h) = \frac{3R}{16\pi\mu}\left(\frac{20xhz^2}{r^7} - \frac{20xyz^2}{r^7} - \frac{4xh}{r^5} + \frac{3xy}{r^5}\right). \tag{11.42}$$

With these results, we may compute the rotation rate of the rotlet dipole as

$$\Omega_y = \mathcal{L}_x\left(\frac{1}{2}\omega_y^{im} - BE_{xz}^{im}\right)\bigg|_{(2h\mathbf{e}_y;h)}, \tag{11.43}$$

which, when introducing the dipole strength $X = \epsilon \ell R$, is evaluated as

$$\Omega_y = -\frac{3}{256\pi}\frac{X}{\mu h^4}(1 - B). \tag{11.44}$$

Since $0 \leqslant B < 1$, the rotation rate in Eq. (11.44) is always negative. As observed experimentally, flagellated bacteria therefore swim along CW circles near no-slip surfaces (see Fig. 11.6(a)). While the order of magnitude of the rotation rate is the same for the free surfaces and the no-slip surfaces (in particular, both have the same spatial decay), the rotations are identical in magnitude for a spherical swimmer ($B = 0$) only. For elongated swimmers ($B > 0$) the rotation rate is larger near a free surface compared to a no-slip surface, leading to circles of larger radii on rigid surfaces.

11.3 Upstream Swimming

So far in this chapter, interactions with boundaries have resulted from the flows created by the cells themselves. However, important surface phenomena can also arise in externally generated flows. Sufficiently close to a no-slip surface, any external flow is approximately linear, i.e. it is locally a simple shear flow. When an external shear flow is imposed on a suspension of cells swimming near a wall, it has been observed that cells reorient to swim upstream against the flow. This behaviour is a special case of *rheotaxis*, the ability of organisms to reorient and change their behaviour in response to hydrodynamic forces. Here we combine the results of Chapter 10 with hydrodynamic images to derive a mathematical model for it.

11.3.1 Observations

Observations of upstream swimming were first reported for two species of bacteria, *E. coli* (Hill et al., 2007) and *B. subtilis* (Cisneros et al., 2007). Subsequent measurements for *E. coli* by Kaya and Koser (2012) showed that in the absence of flow bacteria swam in circles near the surfaces (Fig. 11.7(a)) but when flow was turned on, the organisms transitioned to direct upstream motility (Fig. 11.7(b)). Upstream swimming was also observed for larger eukaryotic cells, namely spermatozoa. Kantsler et al. (2014) reported on the upstream swimming of bull spermatozoa along spiral trajectories. Further work showed that bull spermatozoa swim in circles near walls as a result of the three-dimensionality of their flagellar beat (Fig. 11.7(c)). When an external flow is added with a sufficiently large shear rate, the cells transition to upstream swimming along straighter trajectories (Fig. 11.7(d)).

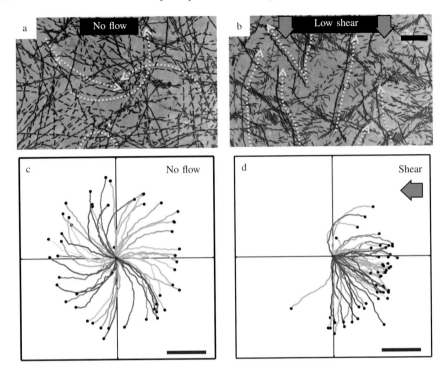

Figure 11.7 Upstream swimming of cells in shear flows near surfaces. Top: *E. coli* bacteria swim in circles near a rigid surface in the absence of external flow (a) but transition to upstream swimming when a shear flow is added (b) (adapted from Hill et al., 2007); scale bar: 10 μm. Bottom: bull spermatozoa swim in circles near walls due to their three-dimensional flagellar beat (c) but swim upstream when a sufficiently strong external shear is applied (d) ((c) and (d) reprinted with permission from Tung et al., (2015), Copyright 2015 by the American Physical Society.); scale bar: 100 μm.

11.3.2 Model

The original studies on *E. coli* made the argument that upstream motility resulted from a hydrodynamic reorientation of swimming cells in moderate shear resulting in stable mechanical equilibrium of the cell, where the swimmer points upstream and thus swims against the flow (Hill et al., 2007; Kaya and Koser, 2012). This model was further refined by Tung et al. (2015). Here we revisit these models using the far-field approach developed earlier in the chapter.

Motion of a Sphere in a Shear Flow near a Wall

The physics of upstream swimming lies in the three-way interactions between the swimmer, the surface and the shear flow. In order to understand the phenomenon, the key physical problem to understand is the motion of a passive sphere in a shear flow near a wall. Using standard notation with a no-slip surface at $y = 0$, we consider an external shear flow given by $\mathbf{u}_\infty(\mathbf{r}) = \dot{\gamma}y\mathbf{e}_x$. In an unbounded fluid, a passive sphere of radius a would simply translate with the flow (Eq. (10.13)). The presence of the wall modifies the perturbation to the flow due to the presence of the sphere, which in turns modifies the free-moving speed in a manner that depends on the size of the sphere and its distance to the wall.

The perturbation flow created by a sphere in an unbounded shear flow is a classical exact solution of Stokes flows (Leal, 2007). Decomposing the shear flow as $\mathbf{u}_\infty(\mathbf{r}) = \mathbf{E} \cdot \mathbf{r} + \boldsymbol{\Omega} \times \mathbf{r}$, where $\mathbf{E}$ is the rate of strain tensor and $\boldsymbol{\Omega}$ the rotational component of the flow, the perturbation flow $\bar{\mathbf{u}}$ decaying to zero at infinity is given exactly by

$$\bar{\mathbf{u}}(\mathbf{r}) = -\frac{1}{2}(\boldsymbol{\Omega} \times \mathbf{r})\frac{a^3}{r^3} - \left(\frac{5a^3}{2r^5} - \frac{5a^5}{2r^7}\right)(\mathbf{r} \cdot \mathbf{E} \cdot \mathbf{r})\mathbf{r} - (\mathbf{E} \cdot \mathbf{r})\frac{a^5}{r^5}. \tag{11.45}$$

This flow is composed of a rotlet (first term), a stresslet (second term) and two higher-order flow singularities, both decaying as $1/r^4$ (last two terms). Far from the sphere, $r \gg a$, the leading-order flow is therefore given by the stresslet and the rotlet, both of which decay spatially as $1/r^2$.

When the sphere is at a distance h above the no-slip surface, the first hydrodynamic effect of the wall is due to the images of both slowly decaying singularities, the stresslet and the rotlet. This leads to a change in the translational speed of the sphere parallel to the surface, $U_x = h\dot{\gamma} + \delta U_x$, where the speed perturbation, δU_x, can be evaluated analytically (the sphere does not move perpendicular to the surface due to time-reversibility). Inspecting Eq. (11.45), we see that the rotlet strength is $\mathbf{R} = -4\pi\mu a^3\boldsymbol{\Omega}$ and parallel to the wall. Using Eq. (11.40b), we get that the rotlet image does not lead to any velocity in the flow direction at the location of the sphere, and therefore the wall effect at leading order is due solely to the stresslet.

The stresslet in Eq. (11.45) has strength $\mathbf{S} = 20\pi\mu a^3\mathbf{E}/3$, using notation from Chapter 9. Since for the shear flow we have $\mathbf{E} = \dot{\gamma}(\mathbf{e}_x\mathbf{e}_y + \mathbf{e}_y\mathbf{e}_x)/2$, the velocity field, $\mathbf{u}$, associated with the stresslet can be written using our notation of force dipoles as

$$\mathbf{u}(\mathbf{r}) = \mathbf{u}_{\rightleftarrows}(\mathbf{r}) + \mathbf{u}_{\downarrow\uparrow}(\mathbf{r}), \tag{11.46}$$

with

$$\mathbf{u}_{\rightleftarrows} = \frac{\mathcal{P}}{2}\mathbf{G}_{FD}(\mathbf{r}; \mathbf{e}_x, \mathbf{e}_y), \quad \mathbf{u}_{\downarrow\uparrow} = \frac{\mathcal{P}}{2}\mathbf{G}_{FD}(\mathbf{r}; \mathbf{e}_y, \mathbf{e}_x) \tag{11.47}$$

and

$$\mathcal{P} = -\frac{20}{3}\pi\mu a^3\dot{\gamma}. \tag{11.48}$$

The x component of the first term in Eq. (11.46) can be obtained by applying the $\mathcal{L}_y$ operator to the x component of the stokeslet flow in Eq. (11.19a), $u_{\parallel,x}^{\text{im}}$, leading to the dipole image velocity

$$u_{\rightleftarrows} = \mathcal{L}_y(u_{\|,x}^{im})\Big|_{(2h\mathbf{e}_y;h)} = \frac{3\mathcal{P}}{64\pi\mu h^2}. \tag{11.49}$$

Similarly, the second term in Eq. (11.46) has a velocity along x whose magnitude is obtained by applying the $\mathcal{L}_x$ operator to the velocity component $u_{\perp,x}^{im}$ from Eq. (11.25a), also leading to the value

$$u_{\downarrow\uparrow} = \mathcal{L}_x\left(u_{\perp,x}^{im}\right)\Big|_{(2h\mathbf{e}_y;h)} = \frac{3\mathcal{P}}{64\pi\mu h^2}. \tag{11.50}$$

Adding the results from Eq. (11.49) and (11.50) and substituting the dipole strength from Eq. (11.48), we obtain the perturbation to the velocity of the free sphere parallel to the wall as

$$\delta U_x = u_{\rightleftarrows} + u_{\downarrow\uparrow} = -\frac{5a^3\dot{\gamma}}{16h^2}, \tag{11.51}$$

so that its total velocity can finally be expressed as

$$U_x = h\dot{\gamma} + \delta U_x = h\dot{\gamma}\left(1 - \frac{5}{16}\frac{a^3}{h^3}\right). \tag{11.52}$$

This result, valid in the far-field limit $a \ll h$, was obtained by Goldman et al. (1967) using a different perturbative method.

Upstream Reorientation of Asymmetric Bodies

The most important results predicted by Eq. (11.52) are that (i) the presence of a surface always slows down the sphere and that (ii) the effect is more pronounced if the sphere is larger or is closer to the wall.

To appreciate the consequences of this result for upstream swimming, let us now consider the case of an asymmetric body in the shape of a dumbbell composed of two spheres of different sizes. Assume the dumbbell is parallel to the surface (so both spheres are at the same height h above the wall) and subject to a shear flow. The result in Eq. (11.52) states that both spheres move at a speed smaller than that of the flow, with the larger sphere being slower than the smaller one. The difference in size leads thus to a difference in velocity, so that the dumbbell gets progressively reoriented by the flow with the larger sphere (of larger resistance) located upstream of the smaller sphere. Note that, although we derived this result in the far field, the same physics is at play in strong confinement (Brotto et al., 2013).

This hydrodynamic reorientation is very general since the front and back of any asymmetric body would be expected to move at different speeds, so we would always expect such a body to reorient. This is the physical mechanism responsible for upstream swimming. Cells such as spermatozoa and bacteria all have asymmetric shapes, with a cell body larger in width than its propelling flagellum. Therefore, in the presence of a shear flow, they become reoriented with the portion of their body with the largest resistance (cell body) located upstream of the portion with smallest resistance (flagellum). Since these cells are pushers and are propelled from the back by their flagellum, an upstream reorientation results in upstream swimming.

Mathematically, let us denote by θ the angle that the cell, of typical length L, makes with the upstream (i.e. $-x$) direction. The difference in size between the front and the back of the cell leads to a difference in velocity ΔU, inducing a rotation of the cell along the direction perpendicular to the planar surface. The magnitude of the cell rotation rate, $\Omega_\perp$, is obtained by projecting the velocity difference in the direction perpendicular to the cell direction, i.e. $L\Omega_\perp \sim -\Delta U \sin \theta$. The cell therefore reorients around the normal to the surface with angular velocity

$$\Omega_\perp = -v\dot{\gamma}\sin\theta, \tag{11.53}$$

where $v > 0$ is a dimensionless parameter related to the shape asymmetry of the organism. The value of $\Omega_\perp$ is proportional to the shear rate, as expected from the linearity of Stokes flow. Note that the angular dependence of Eq. (11.53) is similar to that obtained by a torque of the form in Eq. (10.81) applied in the direction normal to the surface, and thus upstream swimming is yet another example of biased locomotion.

Upstream Swimming of Asymmetric Swimmers

The transition to upstream swimming needs also to take into account the fact that the cells swim in circles in the absence of shear flows. We explored earlier in the chapter the rotation of bacteria near walls. Similarly, the spermatozoa in the experiments of Tung et al. (2015) have circular trajectories near surfaces as a result of their flagella deforming with a three-dimensional chiral beat. If we use ω to denote the intrinsic circular rotation rate of a cell near the wall in the absence of flow, the evolution equation for the orientation of the cell in the plane of the surface, $\theta(t)$, is thus obtained as the sum of the intrinsic rotation plus the reorientation from Eq. (11.53), i.e.

$$\dot{\theta} = -\omega + \Omega_\perp = -\omega - v\dot{\gamma}\sin\theta. \tag{11.54}$$

Here again we obtain the same forced Adler equation as Eq. (10.85) where, instead of trapping, shear rates now play the opposite role of 'freeing' the cells from their circular orbits.

The orientation of the cells in Eq. (11.54) undergoes a saddle-node bifurcation at a critical value of the shear rate $\dot{\gamma}_c = \omega/v$. The cells swim in circular orbits along the surface for $\dot{\gamma} < \dot{\gamma}_c$ but a transition to straight swimming occurs for $\dot{\gamma} > \dot{\gamma}_c$ with two equilibria, $\hat{\theta}$, satisfying $\sin\hat{\theta} = -\dot{\gamma}_c/\dot{\gamma}$. The stable equilibrium corresponds to

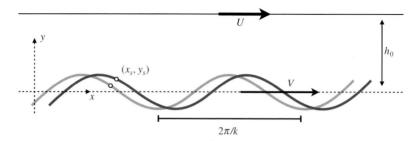

Figure 11.8 Two-dimensional waving sheet swimming at a mean distance h_0 from a no-slip surface. The deformation of the sheet is described with the same notation as Fig. 3.1, i.e. wavelength $2\pi/k$ and frequency ω. In the swimming frame (x, y) the rigid surface translates with velocity $U\mathbf{e}_x$.

the swimming cell pointing upstream $(\cos\hat{\theta} > 0)$ while the one pointing downstream $(\cos\hat{\theta} < 0)$ is unstable. Asymmetric cells with large cell bodies pushed from behind by flagella are thus expected to display a generic transition to stable upstream swimming in the presence of a wall-bound shear flow.

11.4 Impact of Surfaces on Swimming Speeds

The modelling approach developed in the previous sections allowed us to rationalise the hydrodynamic attraction and reorientation of swimming cells by surfaces. Once an organism swims near a wall, how is its locomotion speed affected? This problem is relevant, for example, to the locomotion of mammalian spermatozoa that migrate by following the walls of the female reproductive tract (Fauci and Dillon, 2006; Denissenko et al., 2012). One modelling option to capture the role of surfaces on flagellar beating consists of deriving the changes to drag coefficients for filaments translating very close to walls (Katz et al., 1975; Brennen and Winet, 1977). Here we instead adapt the waving sheet model first introduced in Chapter 3 to address the impact of walls on locomotion speeds. We compute the swimming speed of the sheet in two limits, first assuming that the waving occurs at small amplitude, and then by considering the situation in which the wavelength of the deformation is large compared to the distance to the surface.

11.4.1 Small-Amplitude Model

The calculation for the swimming speed near a wall of a waving sheet at small amplitude was carried out by Reynolds (1965), and we present it here using the same notation and methodology as in Chapter 3.

Setup

The two-dimensional sheet has wavelength $2\pi/k$, frequency ω and wave speed $V = \omega/k$, and is located at a mean distance h_0 from a no-slip surface (see Fig. 11.8). Assuming the sheet to be swimming with velocity $-U\mathbf{e}_x$, we solve the problem in the swimming frame (x, y), in which the no-slip surface moves with velocity $U\mathbf{e}_x$. For simplicity we focus on the case where the sheet undergoes only transverse deformation with amplitude B, so that the material points (x_s, y_s) on the sheets are described by Eq. (3.1) with $A = 0$, i.e.

$$x_s = x, \tag{11.55a}$$
$$y_s = B\sin(kx - \omega t). \tag{11.55b}$$

Non-dimensionalising lengths by k^{-1} and times by ω^{-1}, the positions of the material points are now given by Eq. (3.6) with $a = 0$, i.e.

$$x_s = x, \tag{11.56a}$$
$$y_s = \epsilon b\sin(x - t). \tag{11.56b}$$

Note that the use of both ϵ and b in Eq. (11.56) is redundant since there is only one dimensionless amplitude, but it allows the notation to be consistent with that of Chapter 3 and the rest of this book.

Similar to Section 3.3, the flow problem for the fluid located between the sheet and the surface is solved using a streamfunction, ψ, solution to the biharmonic equation

$$\nabla^4\psi = 0. \tag{11.57}$$

The boundary conditions for the flow on the sheet are the same as in Eq. (3.9) with $a = 0$, i.e.

$$u_x(x_s, y_s) = \left.\frac{\partial\psi}{\partial y}\right|_{(x_s,y_s)} = 0, \tag{11.58a}$$
$$u_y(x_s, y_s) = -\left.\frac{\partial\psi}{\partial x}\right|_{(x_s,y_s)} = -\epsilon b\cos(x - t). \tag{11.58b}$$

Defining $d \equiv h_0 k$ as the mean dimensionless distance between the waving sheet and the wall, the swimming-frame boundary conditions on the wall are given by

$$\left.\frac{\partial\psi}{\partial y}\right|_{(x,d)} = U, \tag{11.59a}$$
$$\left.\frac{\partial\psi}{\partial x}\right|_{(x,d)} = 0, \tag{11.59b}$$

and the goal is to compute U for given values of ϵb and d.

As in Section 3.3, we solve here the problem as a regular perturbation expansion in the small parameter ϵ:

$$\psi = \epsilon\psi_1 + \epsilon^2\psi_2 + \cdots, \tag{11.60a}$$

$$U = \epsilon U_1 + \epsilon^2 U_2 + \cdots. \tag{11.60b}$$

The expansions in the boundary conditions in powers of ϵ are identical to Eq. (3.15) with $a = 0$, i.e.

$$\epsilon\frac{\partial\psi_1}{\partial y}\bigg|_{(x,0)} + \epsilon^2\frac{\partial\psi_2}{\partial y}\bigg|_{(x,0)} + \epsilon^2 b\sin(x-t)\frac{\partial^2\psi_1}{\partial y^2}\bigg|_{(x,0)} = 0, \tag{11.61a}$$

$$\epsilon\frac{\partial\psi_1}{\partial x}\bigg|_{(x,0)} + \epsilon^2\frac{\partial\psi_2}{\partial x}\bigg|_{(x,0)} + \epsilon^2 b\sin(x-t)\frac{\partial^2\psi_1}{\partial x\partial y}\bigg|_{(x,0)} = \epsilon b\cos(x-t). \tag{11.61b}$$

In contrast, since the position of the wall is fixed, the boundary conditions on it, Eq. (11.59), remain of the same form at all orders in the expansion.

The general unit-speed 2π-periodic solution to Eq. (3.8) is given using complex notation in Eq. (3.17) and may be rewritten alternatively as

$$\psi = \psi_0 + Ky + Ly^2 + My^3 \tag{11.62}$$

$$+ \mathcal{R}\left\{\sum_{n\geqslant 1}\left[(G_n + yH_n)\cosh\left(n(y-d)\right)\right]e^{in(x-t)}\right\}$$

$$+ \mathcal{R}\left\{\sum_{n\geqslant 1}\left[(I_n + yJ_n)\sinh\left(n(y-d)\right)\right]e^{in(x-t)}\right\},$$

where capital letters denote undetermined constants and where we have used $\mathcal{R}$ to denote the real part of a complex number (in what follows we imply taking the real part in each instance where we use complex notation, as allowed for linear operations). At each order in the perturbation expansion, the value of the constant ψ_0 is irrelevant since the velocity is defined as a gradient of the streamfunction. Furthermore, both $L = 0$ and $M = 0$ at each order since they are associated with a mean shear and mean pressure gradient, respectively.

Solution at Order ϵ

The first non-trivial flow occurs at order ϵ. The streamfunction ψ_1 is a solution to the biharmonic equation subject to the boundary conditions

$$\left.\frac{\partial \psi_1}{\partial y}\right|_{(x,0)} = 0, \tag{11.63a}$$

$$\left.\frac{\partial \psi_1}{\partial x}\right|_{(x,0)} = be^{i(x-t)}, \tag{11.63b}$$

$$\left.\frac{\partial \psi_1}{\partial y}\right|_{(x,d)} = U_1, \tag{11.63c}$$

$$\left.\frac{\partial \psi_1}{\partial x}\right|_{(x,d)} = 0. \tag{11.63d}$$

Given the form of the boundary conditions in Eq. (11.63), only the $n = 1$ mode in Eq. (11.62) is nonzero and we may write the solution as

$$\psi_1 = K_1 y + [(G_1 + yH_1)\cosh(y - d) + (I_1 + yJ_1)\sinh(y - d)]\,e^{i(x-t)}, \tag{11.64}$$

with five constants to determine, K_1, G_1, H_1, I_1, J_1. To calculate their values, we first enforce the boundary conditions on the wall, Eqs. (11.63d)–(11.63c), and obtain $K_1 = U_1, G_1 = -dH_1$ and $I_1 = -H_1 - dJ_1$, so the first-order streamfunction now takes the form

$$\psi_1 = U_1 y + H_1 \left[(y - d)\cosh(y - d) - \sinh(y - d)\right] e^{i(x-t)}$$
$$+ J_1(y - d)\sinh(y - d)e^{i(x-t)}. \tag{11.65}$$

The boundary conditions on the sheet in Eqs. (11.63a)–(11.63b) then lead to $U_1 = 0$ (no swimming at first order, as expected by symmetry) and

$$H_1 d \sinh d - J_1(d \cosh d + \sinh d) = 0, \tag{11.66a}$$
$$H_1(\sinh d - d \cosh d) + J_1 d \sinh d = -ib, \tag{11.66b}$$

and thus we obtain the constants as

$$H_1 = -ib\left(\frac{\sinh d + d \cosh d}{\sinh^2 d - d^2}\right), \quad J_1 = -ib\left(\frac{d \sinh d}{\sinh^2 d - d^2}\right). \tag{11.67}$$

Solution at Order ϵ^2

To compute the value of the swimming speed, we need to consider the problem at the next order. At order ϵ^2, the streamfunction ψ_2 satisfies the biharmonic equation, Eq. (11.57), with boundary conditions on the sheet obtained from the Taylor expansion in Eq. (11.61), which now become, in the presence of the wall,

$$\left.\frac{\partial \psi_2}{\partial y}\right|_{(x,0)} = -b \sin(x - t)\left.\frac{\partial^2 \psi_1}{\partial y^2}\right|_{(x,0)}, \tag{11.68a}$$

$$\left.\frac{\partial \psi_2}{\partial x}\right|_{(x,0)} = -b \sin(x - t)\left.\frac{\partial^2 \psi_1}{\partial x \partial y}\right|_{(x,0)}, \tag{11.68b}$$

$$\left.\frac{\partial \psi_2}{\partial y}\right|_{(x,d)} = U_2, \tag{11.68c}$$

$$\left.\frac{\partial \psi_2}{\partial x}\right|_{(x,d)} = 0. \tag{11.68d}$$

As in Section 3.3, the value of U_2 can be obtained without having to solve for the full problem at $O(\epsilon^2)$. The solution for ψ_2 is of the general form shown in Eq. (11.62) and therefore contains both a general travelling wave solution plus a term of the form $K_2 y$. We first compute the average of Eq. (11.68a) on the mean position of the sheet, which gives us access to the value of K_2. In order to evaluate Eq. (11.68a), we calculate the derivative

$$\left.\frac{\partial^2 \psi_1}{\partial y^2}\right|_{(x,0)} = [H_1\left(-d \cosh d - \sinh d\right) + J_1\left(2 \cosh d + d \sinh d\right)] e^{i(x-t)}.$$

$$\tag{11.69}$$

Using the values for the constants in Eq. (11.67), we take the real part of Eq. (11.69), multiply it by $\sin(x - t)$ and compute its average over one wavelength of the sheet to determine the constant K_2. We next notice that, as a result of Eq. (11.68c), we have equality $U_2 = K_2$, which allows us to obtain the leading-order swimming speed as

$$U_2 = \frac{1}{2}b^2 \left(\frac{\sinh^2 d + d^2}{\sinh^2 d - d^2}\right). \tag{11.70}$$

Discussion

The result in Eq. (11.70) is consistent with the calculations from Chapter 3 for swimming in an unbounded fluid. Indeed, in the limit $d \gg 1$, the velocity becomes $U_2 \approx \frac{1}{2}b^2$, identical to the result in Eq. (3.30) for $a = 0$.

Returning to dimensional quantities, the waving sheet swims at leading order with velocity $-U\mathbf{e}_x$ of magnitude

$$U = \frac{\omega k}{2} B^2 \left(\frac{\sinh^2(kh_0) + (kh_0)^2}{\sinh^2(kh_0) - (kh_0)^2}\right), \tag{11.71}$$

a result valid in the limit of small amplitude, $B \ll h_0, k^{-1}$. Therefore, for fixed characteristics of the wave, the presence of the wall always increases the swimming speed. Indeed, it is straightforward to compute the rate of change of U with h_0 from

Eq. (11.71) and to show that $\partial U/\partial h_0 < 0$. The closer the swimmer is to the wall, the larger the swimming speed. An interesting limit of this result emerges when $kh_0 \ll 1$. Since we assumed that the waving amplitude was small, this therefore corresponds to the limit $B \ll h_0 \ll k^{-1}$. In that case, using the Taylor expansion $\sinh kh_0 \approx kh_0 + (kh_0)^3/6$, Eq. (11.70) leads to the approximate result

$$U_2 \approx 3\frac{B^2}{h_0^2},\qquad(11.72)$$

which agrees with the lubrication calculation in the next section.

11.4.2 Long-Wavelength Model

In order to address the case of large-amplitude waving, the calculation for the swimming speed of the sheet may be carried out in a different mathematical limit. As proposed by Katz (1974) for sinusoidal waves, we may consider the long-wavelength (or lubrication) limit in which $h_0 \ll k^{-1}$ but for which the waving amplitude B does not necessarily have to be small compared to h_0.

Setup

Following Chan et al. (2005), we consider the setup illustrated in Fig. 11.8 but now allow the wave, $y_s(x, t)$, to have an arbitrary periodic profile that is not necessarily sinusoidal. The wavelength is $2\pi/k$, the frequency is ω and the amplitude B is subject to the geometrical constraint $B < h_0$ so that the wave never actually touches the wall.

In the classical lubrication setup, the dimensional Stokes equations take their long-wavelength form (Leal, 2007),

$$\frac{\partial p}{\partial x} = \mu\frac{\partial^2 u_x}{\partial y^2},\quad \frac{\partial p}{\partial y} = 0,\qquad(11.73a)$$

where u_x is the x component of flow velocity between the swimmer and the wall and p the dynamic pressure in the fluid. As in Section 7.1 we solve the swimming problem in the wave frame, so that the boundary condition on the surface is

$$u_x(x, h_0) = U - V.\qquad(11.74)$$

Using notation from Section 7.1, the wavelengths measured in the lab frame and along the sheet are approximately equal in the lubrication limit, $\Lambda \approx \lambda$, so that $Q \approx -V$. Since $\mathbf{t} \approx \mathbf{e}_x$, the boundary condition on the sheet, Eq. (7.3), becomes in the waving frame

$$u_x(x, y_s) = -V.\qquad(11.75)$$

Lubrication Solution

We first integrate Eq. (11.73) analytically for u_x using the boundary conditions in Eqs. (11.74)–(11.75) and obtain the velocity as

$$u_x = \frac{1}{2\mu}\frac{\partial p}{\partial x}(y - y_s)(y - h_0) - V + U\frac{y - y_s}{h_0 - y_s}, \qquad (11.76)$$

where we have used the fact that the pressure in the fluid is not a function of y. We next consider the flow rate Q in the gap between the swimmer and the wall and apply mass conservation to get

$$Q \equiv \int_{y_s}^{h_0} u_x \, dy, \qquad \frac{\partial Q}{\partial x} = 0. \qquad (11.77)$$

The flow rate is evaluated directly by integrating Eq. (11.76) and, using the shorthand notation $h(x) \equiv h_0 - y_s$ to denote the gap between swimmer and wall, it is found to be

$$Q = -\frac{h^3}{12\mu}\frac{\partial p}{\partial x} - Vh + \frac{1}{2}Uh. \qquad (11.78)$$

The value of Q is then obtained by enforcing that no net pressure jump is created over one wavelength (i.e. by requiring periodicity of the pressure),

$$\int_0^{2\pi/k} \frac{\partial p}{\partial x} \, dx = 0, \qquad (11.79)$$

which, using Eq. (11.78), leads to the first equality linking Q, U and V,

$$Q = \left(\frac{1}{2}U - V\right)\frac{I_2}{I_3}, \qquad (11.80)$$

where we have defined the integrals I_n computed over one wavelength of the wave

$$I_n \equiv \int_0^{2\pi/k} \left(\frac{1}{h(x)}\right)^n dx. \qquad (11.81)$$

To determine the value of the swimming speed of the sheet, U, we invoke the constraint of force-free swimming along the x direction. Since the stress field in a Stokes flow is divergence-free, Eq. (2.13), a zero net force constraint on the swimmer is equivalent to zero force on the surface, which is easier to evaluate since the wall is straight and thus only shear stresses contribute. Requiring zero force on one period of the wall is then written as

$$\int_0^{2\pi/k} \mu\frac{\partial u_x}{\partial y}\Big|_{(x,h_0)} dx = 0, \qquad (11.82)$$

which, using Eq. (11.76), leads to

$$\int_0^{2\pi/k} \left(\frac{1}{2}\frac{\partial p}{\partial x}h + \frac{U}{h} \right) dx = 0. \tag{11.83}$$

After using the equality in Eq. (11.78), the integral constraint obtained in Eq. (11.83) leads to the second equality linking Q, U and V,

$$3QI_2 = (2U - 3V)I_1. \tag{11.84}$$

Combining the relationships derived in Eqs. (11.80) and (11.84) leads to the value of the swimming speed, U, as a function of the wave speed, V, as

$$\frac{U}{V} = \frac{6(1-A)}{4-3A}, \quad A = \frac{I_2^2}{I_1 I_3}. \tag{11.85}$$

Discussion

We first note that all dependence on the geometry of the wave in Eq. (11.85) is embedded in a single dimensionless shape parameter, A. This general result allows us to address a variety of problems in viscous propulsion near surfaces, for example gastropod locomotion (Chan et al., 2005). In the context of microorganism locomotion, it can be used to recover the results for the sinusoidally varying waving sheet derived by Katz (1974). In that case, we pick $y_s = B \sin x$, so that the relevant integrals I_n ($1 \leq n \leq 3$) may be obtained analytically as

$$I_1 = \frac{2\pi}{h_0} \frac{1}{[1-(B/h_0)^2]^{1/2}}, \tag{11.86a}$$

$$I_2 = \frac{2\pi}{h_0^2} \frac{1}{[1-(B/h_0)^2]^{3/2}}, \tag{11.86b}$$

$$I_3 = \frac{\pi}{h_0^3} \frac{[2+(B/h_0)^2]}{[1-(B/h_0)^2]^{5/2}}. \tag{11.86c}$$

Therefore the dimensionless ratio of integrals A appearing in Eq. (11.85) is given by

$$A = \frac{2}{2+(B/h_0)^2}, \tag{11.87}$$

and the swimming speed of the sheet is obtained as

$$\frac{U}{V} = \frac{3}{2+(h_0/B)^2}. \tag{11.88}$$

Recall that this result was obtained in the long-wavelength limit and only requires that $B < h_0$. Clearly the result in Eq. (11.88) shows that the swimming speed increases when the distance to the wall decreases, and similar to the small-amplitude

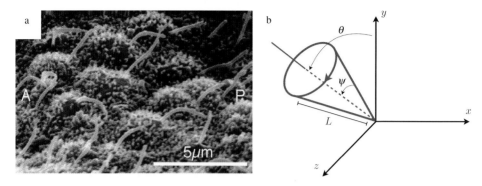

Figure 11.9 Nodal cilia. (a) Scanning electron micrograph of nodal cilia in a mouse embryo tilted in the anterior (A) - posterior (P) direction (Nonaka et al., 2005, reproduced under Creative Commons license). (b) Kinematics of a nodal cilium modelled as a straight rod of length L sweeping a cone of semi-angle ψ in the CW direction when viewed from above the (x, z) surface. The axis of the cone is tilted by an angle θ from the y direction in the (y, z) plane and the net flow is induced along x.

result we see that surfaces increase propulsion speeds systematically. The maximum value of U is obtained when h_0 approaches B, at which point U is approximately equal to the wave speed (which is thus an upper bound). Note that in the limit of small-amplitude waving, $B/h_0 \ll 1$, an expansion of Eq. (11.88) leads to the approximate value

$$\frac{U}{V} \approx 3\frac{B^2}{h_0^2},\tag{11.89}$$

which agrees with Eq. (11.72) derived using the small-amplitude expansion.

11.5 Wall-Bound Cilia

The final aspect of motion near walls we address is the role of surfaces on the ability of cilia to generate forces and flow. We illustrated in Chapter 3 the deformation kinematics of individual cilia in the context of a continuum model of metachronal waves (see Fig. 3.2). Here we zoom in to the level of an individual cilium and quantify the flow rate induced by its motion. We distinguish between two types of deformation, namely the rotation of rigid nodal cilia and the two-stroke motion of flexible cilia.

11.5.1 Nodal Cilia

During the early development stages of vertebrate embryos (about a week after fertilisation), flows created by cilia are responsible for the right-left symmetry-breaking of their internal body plan (Nonaka et al., 1998; Shiratori and Hamada, 2006). Termed nodal cilia, they are small (2–3 μm), located inside a small structure called the ventral node and often widely separated from each other (Fig. 11.9(a)). Unlike the propulsive cilia seen in Chapter 3 and studied below, nodal cilia move rigidly along approximately conical surfaces, which are tilted in the anterior-posterior direction, and as a result create a net flow along the right-left axis. Mathematical modelling has been used to compute this flow and characterise its impact on embryo development, and we present here a model to predict the flow rate of an individually rotating cilium (Smith et al., 2011).

Flow Rate Induced by a Point Force

In order to characterise the net flow induced by an individual cilium, we first need to calculate the instantaneous flow rate induced by a point force above a surface. Consider a stokeslet of magnitude $F_\parallel$ oriented in the x direction and located at distance h above a no-slip surface. The image flow for the parallel stokeslet is given in Eq. (11.8) and the total flow is the sum in Eq. (11.1) of the original point force and all its images. The flow rate induced in the x direction by the point force is given by the integral of the x component of the velocity,

$$Q = \int_{-\infty}^{+\infty} \int_{0}^{\infty} u_x(x, y, z) \mathrm{d}y \, \mathrm{d}z, \qquad (11.90)$$

while the net flows in the other two directions are zero by symmetry.

By mass conservation, the value of Q does not depend on the location x where the integral in Eq. (11.90) is evaluated, and thus it can be computed in the far field, $|x| \gg h$. In that limit, the total flow (stokeslet plus all its images) becomes an xy stresslet, the sum of two force dipoles. This is apparent visually from the images in Fig. 11.3(b) or can be deduced from the spatial decay of the terms in Eq. (11.8). The stresslet is composed of two dipoles, namely: (i) a y dipole coming from the two opposite x point forces (along x above the surface and $-x$ below the surface) and (ii) a x dipole of y point forces, which is part of the image system in Eq. (11.8). Together these dipoles combine to a symmetric stresslet with a $\sim 1/r^2$ spatial decay. Since the source dipole in Eq. (11.8) decays faster spatially, as $1/r^3$, its contribution to the flow rate in Eq. (11.90) can be neglected compared to that of the stresslet. This stresslet is also used in Chapter 12 when we model hydrodynamic interactions between cilia and their synchronisation.

Far from the location of the point force, adding the x components of Eq. (11.2) and Eq. (11.19a) leads to the far-field value of the velocity,

$$u_{\|,x}(\mathbf{r}; h) \approx \frac{3F_{\|}h}{2\pi\mu} \frac{x^2 y}{r^5}, \tag{11.91}$$

and the flow rate can then be calculated exactly as

$$Q = \frac{3F_{\|}h}{2\pi\mu} \int_{-\infty}^{+\infty} \int_0^\infty \frac{x^2 y}{(x^2 + y^2 + z^2)^{5/2}} \, dy \, dz = \frac{F_{\|}h}{\pi\mu}. \tag{11.92}$$

The flow rate scales linearly with the torque applied by the force on the fluid by the point on the surface immediately below it, $F_{\|}h$, a scaling that can also be obtained using dimensional analysis and writing $Q = \mathcal{F}(F_{\|}, h, \mu)$. Note that a rotlet above a surface also leads to a finite flow rate (see Exercises).

Geometrical Model of Nodal Cilium

A model proposed by Smith et al. (2008) allows us to calculate the net flow induced by the rotation of a nodal cilium analytically. The cilium is assumed to be a straight rigid rod of length L sweeping a cone of semi-angle ψ with frequency ω in the positive sense in the (x, z) plane (i.e. CW when viewed from above the surface; see notation in Fig. 11.9(b)).

In the case where the axis of the cone is normal to the surface (i.e. along the y axis), material points along the cilium are located at $\mathbf{x}'(s, t)$, where s is the arclength, follow circular trajectories of radius $s \sin\psi$ and remain at height $s \cos\psi$ above the surface. Thus we can write their kinematics as

$$\begin{pmatrix} x'(s, t) \\ y'(s, t) \\ z'(s, t) \end{pmatrix} = \begin{pmatrix} s \sin\psi \cos(\omega t) \\ s \cos\psi \\ s \sin\psi \sin(\omega t) \end{pmatrix}. \tag{11.93}$$

In order to induce a net flow rate and model biological cilia, the conical surface needs to be tilted away from the surface normal (see Fig. 11.9(a)). We denote the tilt angle of the cone axis from the y direction in the (y, z) plane by θ. The motion of material points on the cilium, $\mathbf{x}(s, t)$, can be deduced in that case from that in Eq. (11.93) by applying a rotation around the x axis,

$$\begin{pmatrix} x(s, t) \\ y(s, t) \\ z(s, t) \end{pmatrix} = \begin{pmatrix} 1 & 0 & 0 \\ 0 & \cos\theta & -\sin\theta \\ 0 & \sin\theta & \cos\theta \end{pmatrix} \begin{pmatrix} x'(s, t) \\ y'(s, t) \\ z'(s, t) \end{pmatrix}, \tag{11.94}$$

so that we obtain the cilium kinematics

$$\begin{pmatrix} x(s, t) \\ y(s, t) \\ z(s, t) \end{pmatrix} = \begin{pmatrix} s \sin\psi \cos(\omega t) \\ s \cos\psi \cos\theta - s \sin\psi \sin\theta \sin(\omega t) \\ s \cos\psi \sin\theta + s \sin\psi \cos\theta \sin(\omega t) \end{pmatrix}. \tag{11.95}$$

By symmetry, the flow created by the motion in Eq. (11.95) along the z direction during one half of the period is cancelled out during the other half, so that the net flow is induced along x only (there can be no flow in y due to the presence of the surface). This flow rate can be computed using Eq. (11.92). If we denote the force acting on the fluid along the x direction per unit length of the cilium by $f_x(s,t)$, the total flow rate is given by

$$Q = \int_0^L \frac{f_x(s,t)y(s,t)}{\pi\mu}\,ds. \tag{11.96}$$

To compute the value of f_x we use resistive-force theory, Eq. (5.7), and recall that the force on the fluid is equal and opposite to the force on the cilium. During the cone-sweeping motion, the tangent to the cilium and its instantaneous velocity are orthogonal to each other, so that $f_x = c_\perp u_x$, and therefore the flow rate in Eq. (11.96) becomes

$$Q = \frac{c_\perp}{\pi\mu} \int_0^L y(s,t)\frac{\partial x}{\partial t}(s,t)\,ds. \tag{11.97}$$

The integral in Eq. (11.97) can be evaluated easily by substituting the kinematics from Eq. (11.95), and we get a time-varying flow rate of the form

$$Q = \langle Q \rangle + Q' \sin(\omega t), \tag{11.98}$$

with

$$\langle Q \rangle = \frac{\omega c_\perp L^3}{6\pi\mu} \sin^2\psi \sin\theta, \quad Q' = -\frac{\omega c_\perp L^3}{3\pi\mu} \sin\psi \cos\psi \cos\theta. \tag{11.99}$$

Note that the results in Eq. (11.99) are consistent with dimensional analysis since, if we had written a flow rate of the form $Q = \mathcal{F}(L, \omega, \theta, \psi, \mu, c_\perp)$, we would have predicted a result scaling as $Q = \omega L^3 f(\theta, \psi, c_\perp/\mu)$.

Interpretation and Optimisation

The mean flow induced by the nodal cilium occurs along the x direction since $\langle Q \rangle > 0$, which agrees with intuition. Indeed, we see from Eq. (11.92) that the flow rate increases with the distance to the wall, and therefore the net flow is oriented in the direction where the force parallel to the surface is further away from wall, which as seen in Fig. 11.9 is the positive x direction. Physically, the motion of a nodal cilium can be interpreted as having effective and recovery strokes, similar to the propulsive cilia from Chapter 3, with the effective stroke in the upper part of the cone (further away from the surface) and the recovery stroke in the bottom half. Note also that both the mean and unsteady parts of the flow rate in Eq. (11.99) are independent of the viscosity of the fluid (since $c_\perp$ scales with μ) but depend only on the geometry of the cone (through the length L and angles ψ and θ). The cubic increase in $\langle Q \rangle$ with L indicates a strong sensitivity to cilium length.

What is the motion of the cilium leading to the largest net flow created? Since the cilium is always located above the surface, it is apparent from Fig. 11.9 that the angles ψ and θ are subject to the geometrical constraint $\psi + \theta < 90°$. The mean flow in Eq. (11.99), $\langle Q \rangle$, increases with both ψ and θ so the maximum flow rate is obtained when both angles can reach their maximal values, and therefore in the limit $\psi + \theta \approx 90°$ (but note that, in this limit, the cilium touches the wall during the recovery stroke and thus the model would be expected to break down locally). In that case the two angles are related through $\sin\theta = \cos\psi$ so that the mean flow rate scales as $\langle Q \rangle \sim \sin^2\psi \cos\psi$, which is maximised when $\tan\psi = \sqrt{2}$. The conical motion of the cilium leading to the largest flow occurs therefore for $\psi \approx 54.7°$ and thus $\theta \approx 90° - 54.7° = 35.3°$.

The predictions of this simple optimisation procedure agree well with experiments. An investigation on the kinematics of nodal cilia for three vertebrate species (mouse, rabbit and medaka fish) reported conical kinematics with $\theta = 40° \pm 10°$ and $\psi + \theta \approx 90°$ (Okada et al., 2005). Similarly, experiments on mouse cilia showed a cone tilt in the anterior–posterior direction with average $\theta \approx 26.6°$, while the optimality of the condition $\psi + \theta = 90°$ was verified using macroscale experiments (Nonaka et al., 2005).

11.5.2 Two-Stroke Motion of Flexible Cilia

Unlike nodal cilia, propulsive cilia such as those illustrated in Fig. 3.2 for the organisms *Opalina* and *Paramecium* are flexible and deform periodically in an asymmetric way. The flagella of individual somatic cells of the green alga *Volvox carteri* display similar deformation kinematics, as shown in Fig. 11.10(a). Such asymmetric kinematics are referred to as a 'two-stroke' motion. During the effective stroke, the cilium extends into the fluid to create a large force while during the recovery stroke it moves closer to the surface and decreases its drag, averaging over time to a net force induced in the direction of the effective force.

Spherical Model

The first model, popular in addressing interactions in cilia arrays (see Chapter 12), replaces the geometry of the whole cilium by an effective sphere. As carried out quantitatively by Brumley et al. (2014) for *Volvox carteri*, we can model the action of the flagellum on the fluid (Fig. 11.10(a)) by that of its net instantaneous force acting at its centre of mass (Fig. 11.10(b)). Although this is admittedly a simplistic model, it provides an intuitive approach to quantifying the role played by surfaces on the relationship between the trajectories of the cilia and the flows they create.

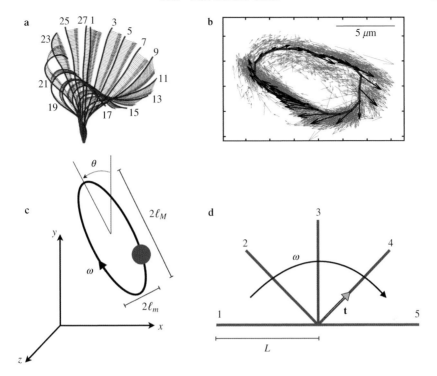

Figure 11.10 Two-stroke cilium. (a) Similar to the propulsive cilia illustrated in Fig. 3.2, flagella of the somatic cells of *Volvox carteri* deform in a two-stroke motion; arrows show the instantaneous velocity distribution along the flagellum (indexed by frame number, frame rate 1000 fps). (b) Force exerted on the fluid by the beating flagellum plotted using arrows placed at its centre of mass; light red shows the net force for each frame while black displays the force averaged over 1000 frames (Brumley et al., 2014). (c) Spherical model of the situation in (b) where the centre of mass of the cilium is represented by a sphere of radius a moving along an elliptical trajectory (minor axis ℓ_m, major axis ℓ_M) tilted by angle θ in the (x, y) plane. (d) Two-stroke model with an effective stroke where the cilium of length L and tangent vector $\mathbf{t}$ extends into the fluid and sweeps the fluid at rate ω while during the recovery stroke it slides along the surface.

Consider a sphere of radius a located at position $(x(t), y(t), 0)$ above a no-slip surface. Assuming that the sphere is sufficiently far from the wall, the x component of the force induced by the sphere on the fluid, F_x, is well approximated by the infinite-fluid Stokes's law

$$F_x \approx 6\pi\mu a \frac{\mathrm{d}x}{\mathrm{d}t}.$$
(11.100)

The flow rate induced instantaneously is given by Eq. (11.92) as

$$Q = \frac{F_x y}{\pi\mu},$$
(11.101)

and therefore the time-averaged flow rate during a periodic motion of the sphere is obtained as

$$\langle Q \rangle = 6a \left\langle y \frac{dx}{dt} \right\rangle. \qquad (11.102)$$

This result, similar to Eq. (11.97), indicates that the net flow rate does not depend on the viscosity of the fluid, but only on the geometry of the trajectory of the sphere in the fluid.

Motivated by the *Volvox* experiments shown in Fig. 11.10(b), let us focus on the situation where the sphere moves along elliptical trajectories. Consider first the case where the sphere traces out an ellipse of minor axis ℓ_m along the x direction and major axis ℓ_M along y with frequency ω. The position of sphere (x', y') about the centre of the ellipse is given by

$$\begin{pmatrix} x'(s,t) \\ y'(s,t) \end{pmatrix} = \begin{pmatrix} \ell_m \cos(\omega t) \\ -\ell_M \sin(\omega t) \end{pmatrix}, \qquad (11.103)$$

and the case $\omega < 0$ represents CCW motion while for $\omega > 0$ the sphere travels along the ellipse in the CW direction.

To obtain trajectories along a tilted ellipse in the (x, y) plane, as relevant for the experimental results, we apply a rotation matrix of angle θ about the z axis (see notation in Fig. 11.10(c)), which leads to coordinates of material points about the centre given by

$$\begin{pmatrix} x(s,t) \\ y(s,t) \end{pmatrix} = \begin{pmatrix} \cos\theta & -\sin\theta \\ \sin\theta & \cos\theta \end{pmatrix} \begin{pmatrix} x'(s,t) \\ y'(s,t) \end{pmatrix}. \qquad (11.104)$$

For a sphere located at a mean height y_0 above the surface, the trajectories of the sphere from Eq. (11.104) become

$$\begin{pmatrix} x(s,t) \\ y(s,t) \end{pmatrix} = \begin{pmatrix} \ell_m \cos\theta \cos(\omega t) + \ell_M \sin\theta \sin(\omega t) \\ y_0 + \ell_m \sin\theta \cos(\omega t) - \ell_M \cos\theta \sin(\omega t) \end{pmatrix}. \qquad (11.105)$$

Using the trajectories from Eq. (11.105) in Eq. (11.102) finally leads to the flow rate given by

$$\langle Q \rangle = 3a\omega\ell_m\ell_M. \qquad (11.106)$$

The net flow induced by the trajectory of the sphere occurs along the $+x$ direction if $\omega > 0$, i.e. when the sphere traces out the ellipse in a CW sense. This is in agreement with physical intuition when viewing the sphere motion as that of the centre of mass of the cilium (Fig. 11.10(a) and (b)). The case $\omega > 0$ corresponds to the situation where the effective stroke, away from the wall, is directed along $+x$ and the recovery stroke along $-x$, and therefore we expect the net flow to also be along $+x$. Interestingly, we note that in this model the net flow rate in Eq. (11.106) is independent of the value of the angle θ by which the ellipse is tilted.

Two-Stroke Model

A second model may be proposed where the two strokes are treated separately. In the effective stroke, the cilium of length L is assumed to be straight and to sweep the fluid in the (x, y) plane at frequency ω from $-x$ to $+x$ (Fig. 11.10(d)). The flow rate created during this effective stroke is denoted $\langle Q_{\text{eff}} \rangle$. To compute its value, we apply the formula derived above for filaments, Eq. (11.96), and use resistive-force theory to evaluate the force density along the straight rotating cilium as $\mathbf{f}(s, t) = c_\perp \mathbf{u}(s, t)$, where the velocity of a point at arclength s along the cilium of unit tangent $\mathbf{t}$ is given by $\mathbf{u}(s, t) = -\omega \mathbf{e}_z \times (s\mathbf{t})$. The x component of the force density is therefore obtained as

$$f_x(t) = c_\perp s\omega \cos \theta(t), \tag{11.107}$$

where $\theta(t)$ is the instantaneous angle between the cilium and the y direction, ranging from $-\pi$ to π during the effective stroke. Since a point at arclength s along the cilium is at height $y = s \cos \theta$ above the surface, evaluating the integral in Eq. (11.96) leads to the instantaneous flow rate

$$Q_{\text{eff}} = \frac{\omega c_\perp L^3}{3\pi\mu} \cos^2 \theta, \tag{11.108}$$

and the average value during the effective stroke

$$\langle Q_{\text{eff}} \rangle = \frac{\omega c_\perp L^3}{3\pi\mu} \langle \cos^2 \theta \rangle = \frac{\omega c_\perp L^3}{6\pi\mu}. \tag{11.109}$$

During the recovery stroke, the cilium is assumed to bend back to its initial location by remaining along the surface at $y = 0$. According to Eq. (11.92), no net flow is induced during the recovery stroke and therefore $\langle Q_{\text{rec}} \rangle = 0$. The total flow, given as the sum of effective and recovery strokes, $\langle Q \rangle = \langle Q_{\text{eff}} \rangle + \langle Q_{\text{rec}} \rangle$, is thus finally given by Eq. (11.109). Note that the scaling of the flow rate with L and ω, and its independence of the fluid viscosity, is identical to that obtained in Eq. (11.99) for nodal cilia and could have also been obtained from dimensional analysis.

Further Reading

The hydrodynamic attraction of pusher cells and their reorientation by walls were analysed in this chapter for no-slip surfaces. Both phenomena are in fact general results valid for a wide range of surface properties and rheology (Lopez and Lauga,

2014). A detailed look at the time-dependent trajectories of swimming cells near flat surfaces uncovers a range of possible dynamical behaviour, including complex scattering (Drescher et al., 2011; Spagnolie and Lauga, 2012; Kantsler et al., 2013; Lushi et al., 2017), and the importance of near-field interactions between cells and surfaces (Bianchi et al., 2017).

The circular swimming of bacteria near boundaries was modelled using a far-field approach, but similar results are obtained using full numerical computations (Ramia et al., 1993; Giacché et al., 2010) or models where the cells are assumed to remain close to the surface (Lauga et al., 2006; Di Leonardo et al., 2011). The radius of curvature of the cell trajectories is highly sensitive to the distance between the cell and the wall, which can lead to an amplification of Brownian fluctuations (Li et al., 2008). The direction of rotation along the circles depends also on the physicochemical characteristics and the rheological nature of the surface (Lemelle et al., 2010; Morse et al., 2013; Lemelle et al., 2013; Lopez and Lauga, 2014). The combination of attraction by surfaces and circular motion leads to accumulation of swimming bacteria along the edges of microfluidic devices (DiLuzio et al., 2005).

The interactions of cells with boundaries was limited in this chapter to the idealised scenario of infinite flat walls, and new physical behaviours arise in the presence of geometrically complex surfaces. Isolated obstacles lead to the capture and scattering of swimmers (Takagi et al., 2014; Spagnolie et al., 2015), while arrays of obstacles can both trap and disperse swimmers depending on the obstacle volume fraction (Chamolly et al., 2017). Strong confinement can modify the trajectories of swimmers from steady and unidirectional to periodic and three-dimensional (Zhu et al., 2013).

Complex geometries can also be exploited to sculpt the dynamics of swimmers and generate interesting applications. Funnels can be used to induce a spatial accumulation of bacteria (Galajda et al., 2007) and planar ratchet-like structures allow swimming cells to transport colloidal particles in predefined areas (Koumakis et al., 2013). Physical traps in ratchet geometries have also been devised to harness the mechanical propulsive work done by flagellated bacteria (Di Leonardo et al., 2010) before the dynamics of their flagellar filaments allow them to escape (Cisneros et al., 2006; Kühn et al., 2017).

The study of the asymmetric dynamics of two-stroke cilia has a long scientific history (Gray, 1928; Blake and Sleigh, 1974) and recent efforts have harnessed computational methods to gain quantitative insight on dynamics and energy expenditure (Gueron and Liron, 1993). For example, the total energy expended by a cilium in its power stroke is about five times that during the recovery stroke (Gueron and Levit-Gurevich, 1999). The two-stroke motion was shown computationally to be the consequence of an energetic optimisation whereby the cilium creates the maximum flow for a fixed amount of internal energy expenditure (Eloy and Lauga, 2012).

Exercises

1. A cell swims a distance h from a flat free surface and is tilted at an angle θ from it (so that $\theta = 0$ means the cell is parallel to the surface; see Fig. 11.5). The flow created by the cell is well approximated by an axisymmetric force dipole of magnitude $\mathcal{P}$. Show that all boundary conditions on the free surface (no shear and no penetration) are satisfied by the superposition of the original dipole and a mirror-image dipole located on the other side of the surface at a distance h away. Deduce the value of the perturbation flow induced by the free surface on the cell and compare it with the no-slip result in Eq. (11.14).

2. The orientation of the cell in Exercise 1 in an external flow may be described using Jeffery's equation with a shape factor $B \geqslant 0$. Calculate the hydrodynamic reorientation of the cell induced by the presence of the free surface, as a function of θ, and compare it with the no-slip result in Eq. (11.31).

3. A cell swims at a distance h from a flat no-slip surface and is tilted at an angle θ from it (see notation in Fig. 11.5). Compute the value of the wall-induced flow velocity on the cell, thereby extending the parallel result in Eq. (11.14) to all possible orientations.

4. Calculate the flow rate, Q, induced by a rotlet of strength R parallel to a no-slip surface at a distance h above it and show that it is independent of h. Could this be expected from dimensional analysis?

5. A two-dimensional sheet undergoes longitudinal waving deformation of dimensionless amplitude a with material points (x_s, y_s) on the sheet given by $x_s = \epsilon a \sin(x - t)$ and $y_s = 0$. The sheet is located at a distance d above a rigid surface. Calculate the swimming speed of the sheet at leading order in ϵ and show that it is independent of the distance from the surface.

12

Hydrodynamic Synchronisation

It has long been observed that an ensemble of flagella, such as those of two nearby swimming spermatozoa or on a single biflagellated alga, can synchronise their periodic beating. Similarly, dense ciliary arrays deform in a coordinated fashion and display metachronal waves. Among the possible mechanisms leading to phase-locking of these biological oscillators, a long-standing hypothesis is that hydrodynamic interactions may provide a systematic route towards synchronisation. While many biological cases involve ingredients other than hydrodynamics, in this chapter we focus on the role played by fluid mechanics and highlight how interactions through the viscous fluid may lead to synchronised beating consistent with experiments. We start by considering the case where flagella, or cilia, are anchored on a surface or on an organism. We use a minimal model of spheres undergoing cyclic motion above a surface and interacting hydrodynamically in the far field. We show that in-phase synchronisation can be achieved if the spheres move along compliant paths or if the forcing responsible for their motion is phase-dependent. The resulting dynamics leads to the classical Adler equation and captures experimental observations. We then address the synchronisation of free-swimming cells such as spermatozoa. Using a two-dimensional model we show that the additional swimming degree of freedom may lead to passive synchronisation in a manner that depends only on the geometry, but might not minimise energy dissipation. In contrast, active synchronisation always leads to in-phase swimming, as observed in experiments.

12.1 Synchronisation of Anchored Flagella and Cilia

In the first part of this chapter, we consider the situation in which interacting flagella or cilia deform periodically around a fixed mean position. This is the relevant scenario when these appendages are anchored and cannot move relative to one another. They could be attached to a surface, e.g. cilia in a variety of settings (Brennen and Winet, 1977; Niedermayer et al., 2008), or could be individual flagella

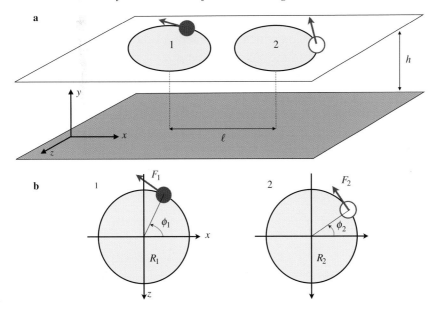

Figure 12.1 Minimal model to study the synchronisation of two anchored flagella or cilia. Two spheres of radius a (1 and 2) move along circular orbits. (a) The orbits are located at a distance h above a no-slip surface and are separated by a mean distance ℓ along the x direction, while y is the direction normal to the surface. (b) Each sphere is driven by a force F_i tangent to the circular orbit of radius R_i and its location is parametrised by the angle ϕ_i (with $i = 1, 2$).

anchored on the body of an organism, such as the algae *Chlamydomonas* (Goldstein et al., 2009) and *Volvox* (Brumley et al., 2012). Using the analytical models of Niedermayer et al. (2008) and Uchida and Golestanian (2011), we show that hydrodynamic interactions lead generically to phase-locking when coupled with a mechanism for the intrinsic phase of these oscillators to change.

12.1.1 Minimal Model: Spheres on Cyclic Paths

Following Niedermayer et al. (2008), we consider a model in which interacting flagella or cilia are replaced by identical spheres (radius a) interacting hydrodynamically. As motivated in Section 11.5.2, each sphere undergoes periodic motion on an approximately circular orbit. For simplicity we assume here that each orbit lies in a plane parallel to and at a distance h above a no-slip surface. The case in which the orbit plane is tilted compared to the no-slip surface can be solved numerically and leads to similar results (Brumley et al., 2012).

The model is illustrated in Fig. 12.1. The circular orbits are separated by a mean distance ℓ along the x direction, while the y axis is directed along the normal to the no-slip wall. For each sphere i, we assume that the orbit takes the form of an instantaneous circle of radius R_i on which the location of the sphere is measured using the angle ϕ_i. The motion of the sphere on the orbit is driven by an instantaneous force of magnitude F_i directed along the vector instantaneously tangent to circle, $\mathbf{e}_{\phi_i}$ (i.e. $\mathbf{F}_i = F_i \mathbf{e}_{\phi_i}$). For the specific model below we consider the case where each orbit is tethered elastically to a reference configuration of radius R_0, and the same setup is used in Section 12.1.6 in the absence of elastic compliance.

We start by characterising the dynamics of an isolated sphere, say sphere 1. In the Cartesian frame (x, y, z) shown in Fig. 12.1, the radial and azimuthal vectors along the circular trajectory remain parallel to the no-slip surface as

$$\mathbf{e}_{r_1} = \begin{pmatrix} \cos \phi_1 \\ 0 \\ -\sin \phi_1 \end{pmatrix}, \quad \mathbf{e}_{\phi_1} = \begin{pmatrix} -\sin \phi_1 \\ 0 \\ -\cos \phi_1 \end{pmatrix}. \tag{12.1}$$

The preferred intrinsic waving motion of each flagellum or cilium is modelled as follows. In the absence of any inertial effects, the viscous drag on the sphere in the radial direction balances an elastic restoring force that acts to bring the orbit back to a preferred circle of radius R_0. Denoting by ζ the viscous resistance coefficient for the sphere ($\zeta \approx 6\pi\mu a$ if the sphere is far from the wall, $a \ll h$, with μ the dynamic viscosity of the fluid) and by λ the strength of the elastic restoring force (with dimension of force per length), the radial dynamics takes the form

$$\zeta \dot{R}_1 = -\lambda(R_1 - R_0), \tag{12.2}$$

where we use dots to denote time derivatives. The relevant timescale for the sphere to adjust radially to the preferred orbit is

$$\tau = \frac{\zeta}{\lambda}. \tag{12.3}$$

This is $O(10^{-4} \text{ s})$ in the case of cilia (Niedermayer et al., 2008). Notably, since cilia and eukaryotic flagella oscillate with typical frequencies, ω, on the order of tens of hertz, we are in the limit $\tau\omega \ll 1$.

In the tangential direction, the viscous drag on the sphere balances the driving force and we obtain the angular dynamics

$$\zeta R_1 \dot{\phi}_1 = F_1. \tag{12.4}$$

For any initial condition, after a short time on the order of τ, the solution to Eqs. (12.2) and (12.4) is given by $R_1 = R_0$ (orbit on the preferred circle) with the sphere rotating with frequency $\omega_1 = F_1/(\zeta R_0)$.

12.1.2 Hydrodynamic Interactions near a Surface

When two (or more) spheres move along their orbits, their motion creates flows that lead to hydrodynamic interactions. Specifically, since the circular trajectories result from externally applied tangential forces, each sphere acts on the fluid as a stokeslet. We assume that the separation between the spheres is always larger than their distance to the surface, $\ell \gg h$, so that they interact hydrodynamically only in the far field.

While the calculation can be carried out in the absence of a surface (see Exercises), we focus here on the case where the spheres move near a no-slip wall, as relevant for example to the case of beating cilia. In that limit, and as seen in Section 11.5, the flow induced by a point force above a no-slip surface is that of a stresslet. Specifically, using the same notation as Section 11.5, a force $\mathbf{F} = F_{\|}\mathbf{e}_x$ parallel to the surface leads to the flow in Eq. (11.91), which is associated with the general stresslet field from Eq. (9.12) with $\mathbf{S} = -2hF_{\|}(\mathbf{e}_x\mathbf{e}_y + \mathbf{e}_y\mathbf{e}_x)$. The flow velocity created by the sphere in the plane $y = h$ parallel to the surface and containing the circular orbit is given by

$$\mathbf{u}(\mathbf{r}) = \frac{3h^2}{2\pi\mu}\frac{(\mathbf{F}\cdot\mathbf{r})\mathbf{r}}{r^5}, \tag{12.5}$$

where $\mathbf{r} = (x,0,z)$ is the vector pointing from the centre of the sphere to the point of evaluation of the flow. In the case of two spheres as in Fig. 12.1, we denote by $\mathbf{u}_{2\to1}$ the flow induced by the motion of sphere 2 at the location of sphere 1 and by $\mathbf{r}_{2\to1}$ the vector joining them, i.e. $\mathbf{r}_{2\to1} = \mathbf{r}_1 - \mathbf{r}_2$. In that case, Eq. (12.5) implies that

$$\mathbf{u}_{2\to1} = \mathbf{u}(\mathbf{r}_{2\to1}) = \frac{3h^2}{2\pi\mu}\frac{(\mathbf{F}_2\cdot\mathbf{r}_{2\to1})\mathbf{r}_{2\to1}}{|\mathbf{r}_{2\to1}|^5}, \tag{12.6}$$

while the result for $\mathbf{u}_{1\to2}$ is similar under a $1 \leftrightarrow 2$ symmetry.

12.1.3 Synchronisation via Flexibility

Dynamics

Focusing now on the case of two interacting spheres, the radial and tangential balances of forces on sphere 1, which were given by Eqs. (12.2)–(12.4) when the sphere was isolated, become

$$\zeta(\dot{R}_1 - \mathbf{u}_{2\to1}\cdot\mathbf{e}_{r_1}) = \lambda(R_0 - R_1), \tag{12.7a}$$

$$\zeta(R_1\dot{\phi}_1 - \mathbf{u}_{2\to1}\cdot\mathbf{e}_{\phi_1}) = F_1 = \zeta R_0\omega_1, \tag{12.7b}$$

where we have used the fact that the drag on the sphere is proportional to its velocity relative to that of the background fluid (see Chapter 10).

Hydrodynamic Interactions

The velocity induced by sphere 2 at the location of sphere 1, given by Eq. (12.5), is known explicitly since the force created by each sphere arises primarily from its tangential motion and therefore $\mathbf{F}_2 \approx \zeta R_2 \dot{\phi}_2 \mathbf{e}_{\phi_2}$. Indeed, any perturbation from this value arises from forces due to the flow created by the other sphere, which decay to zero in the limit $\ell \gg h$. At leading order in the distance between the spheres, the force entering Eq. (12.6) therefore only comes from the external tangential motion. In that case, Eq. (12.5) becomes

$$\mathbf{u}_{2\to1} \approx 9ah^2 R_2 \dot{\phi}_2 \frac{(\mathbf{e}_{\phi_2} \cdot \mathbf{r}_{2\to1})\mathbf{r}_{2\to1}}{|\mathbf{r}_{2\to1}|^5}. \tag{12.8}$$

In the far field, $\ell \gg R_i$, we have $\mathbf{r}_{2\to1} \approx -\ell \mathbf{e}_x$ and thus $\mathbf{e}_{\phi_2} \cdot \mathbf{r}_{2\to1} \approx \ell \sin \phi_2$ so that

$$\mathbf{u}_{2\to1} \approx -\frac{9ah^2}{\ell^3} R_2 \dot{\phi}_2 \sin \phi_2 \, \mathbf{e}_x. \tag{12.9}$$

Radial Dynamics

Using the result from Eq. (12.9), the radial force balance, Eq. (12.7a), becomes, using the vector components in Eq. (12.1),

$$\dot{R}_1 = -\frac{9ah^2}{\ell^3} R_2 \dot{\phi}_2 \cos \phi_1 \sin \phi_2 + \frac{(R_0 - R_1)}{\tau}, \tag{12.10}$$

where the timescale τ is defined in Eq. (12.3). If we further assume that the elasto-hydrodynamic relaxation is fast compared to the radial dynamics induced by hydrodynamic interactions, i.e. $\tau \ll R_1/\dot{R}_1$ (an assumption that we verify later), we obtain a quasi-steady equation for the radius as

$$R_1 = R_0 - \beta R_2 \tau \dot{\phi}_2 \cos \phi_1 \sin \phi_2, \tag{12.11}$$

where we have introduced the dimensionless number β as

$$\beta \equiv \frac{9ah^2}{\ell^3}. \tag{12.12}$$

Note that since $a \ll h$ and $h \ll \ell$, β is small. We see later that its value plays a critical role in the synchronisation dynamics. The spatial decay, $\beta \sim 1/\ell^3$, results from hydrodynamic interactions mediated by walls, and there is a slower decay, $\beta \sim 1/\ell$, in the absence of a surface (see Exercises).

Angular Dynamics

Next we use Eqs. (12.1) and (12.9) to transform the force balance along the tangential direction, Eq. (12.7b), into

$$\dot{\phi}_1 = \frac{R_0}{R_1} \omega_1 + \beta \frac{R_2}{R_1} \dot{\phi}_2 \sin \phi_1 \sin \phi_2. \tag{12.13}$$

We may then invert Eq. (12.11) using a Taylor expansion of R_1 in β to obtain

$$\frac{R_0}{R_1} \approx 1 + \beta \frac{R_2}{R_0} \tau \dot{\phi}_2 \cos \phi_1 \sin \phi_2. \tag{12.14}$$

Furthermore, in the absence of hydrodynamic interactions we have $R_2 = R_0$ and $\dot{\phi}_2 = \omega_2$, so that at leading order in β the ratio of radii is

$$\frac{R_0}{R_1} \approx 1 + \beta \tau \omega_2 \cos \phi_1 \sin \phi_2. \tag{12.15}$$

Combining this result with the angular dynamics, Eq. (12.13), and exploiting the leading-order results $R_2/R_1 = 1$ and $\dot{\phi}_2 = \omega_2$ lead to the angular dynamics for sphere 1 as

$$\dot{\phi}_1 = \omega_1 + \beta \tau \omega_1 \omega_2 \cos \phi_1 \sin \phi_2 + \beta \omega_2 \sin \phi_1 \sin \phi_2, \tag{12.16}$$

plus terms that decay faster with ℓ (i.e. higher order in β). By symmetry, sphere 2 obeys similar dynamic equations and thus

$$\dot{\phi}_2 = \omega_2 + \beta \tau \omega_1 \omega_2 \cos \phi_2 \sin \phi_1 + \beta \omega_1 \sin \phi_1 \sin \phi_2. \tag{12.17}$$

Synchronisation

The system of equations, Eqs. (12.16)–(12.17), may be solved exactly in the case where the forcings on the two spheres are identical, so that the rotation frequencies are equal, i.e. $\omega_1 = \omega_2$. In that case, denoting the phase difference by $\Delta \equiv \phi_1 - \phi_2$ and subtracting Eq. (12.17) from Eq. (12.16) lead to

$$\dot{\Delta} = \beta \tau \omega^2 (\cos \phi_1 \sin \phi_2 - \cos \phi_2 \sin \phi_1) = -\beta \tau \omega^2 \sin \Delta. \tag{12.18}$$

We therefore see that the phase dynamics is governed by the Adler equation already seen in Chapter 10 (swimmers with biased locomotion) and Chapter 11 (upstream swimming near surfaces). Since Eq. (12.18) is separable, it can be integrated. Starting from an initial condition Δ_0, its exact solution is given by

$$\Delta(t) = 2 \arctan \left[\tan \left(\frac{\Delta_0}{2} \right) e^{-\beta \tau \omega^2 t} \right]. \tag{12.19}$$

Regardless of the initial phase difference, the coupled system of oscillators synchronises to an in-phase configuration ($\Delta = 0$) at large times. For this synchronisation to occur, two separate physical ingredients are required. First, the two spheres need to interact hydrodynamically, since otherwise we have $\beta = 0$ and thus no phase change, $\dot{\Delta} = 0$. Second, elastic compliance of the cyclic paths is also required. Indeed, in the limit where the spheres are attached rigidly to their circular orbits, we obtain $\lambda = \infty$, so $\tau = 0$ and again $\dot{\Delta} = 0$. Note that while we have focused here on the case with the wall, similar physics occurs in the absence of the nearby surface where synchronisation to the in-phase configuration is also attained (see Exercises).

As is evident from the solution in Eq. (12.19), the relevant timescale for synchronisation of the two oscillators, T_ϕ, is given by

$$T_\phi = \frac{1}{\beta \tau \omega^2} \sim \frac{1}{\tau \omega^2} \frac{\ell^3}{ah^2}, \tag{12.20}$$

and phase-locking is attained faster if the spheres are closer to each other or if they are larger. Further, we see that the value of T_ϕ increases when h decreases, which is expected since the surface screens hydrodynamic interactions.

Comparing this timescale to the intrinsic frequency of each orbit, and using the fact that both β and $\tau \omega$ are small parameters, we have

$$\omega T_\phi = \frac{1}{\beta \tau \omega} \gg 1, \tag{12.21}$$

and thus the synchronisation process occurs over many rotations of the spheres on their orbits.

Finally, we note that to derive our results we assumed that $\tau \ll R_i/\dot{R}_i$ so that the radius of the orbit could evolve in a quasi-steady fashion. This is equivalent to assuming that $\tau \ll T_\phi$. The ratio of these two timescales is given by

$$\frac{\tau}{T_\phi} = \beta(\tau \omega)^2 \ll 1, \tag{12.22}$$

and therefore the assumption is verified *a posteriori*.

Mean Phase

A final point of interest is the fact that the mean frequency of the oscillators increases progressively as they become synchronised. To see this, we define $\varphi \equiv \phi_1 + \phi_2$ and add Eq. (12.17) to Eq. (12.16) to obtain

$$\dot{\varphi} = 2\omega + \beta \tau \omega^2 \sin \varphi + \beta \omega (\cos \Delta - \cos \varphi). \tag{12.23}$$

We may then average Eq. (12.23) over one period of the total phase, φ. This results in trigonometric functions of φ averaging out to approximately zero, and we obtain

$$\langle \dot{\varphi} \rangle \approx 2\omega + \beta \omega \cos \Delta, \tag{12.24}$$

where $\langle \cdots \rangle$ denotes the phase-averaged rate of change. The result in Eq. (12.24) shows that when Δ decreases, the mean phase increases. When the two oscillators are fully synchronised they both rotate with mean frequency

$$\frac{1}{2} \langle \dot{\varphi} \rangle \approx \omega \left(1 + \frac{\beta}{2} \right), \tag{12.25}$$

which is larger than their original frequency, ω.

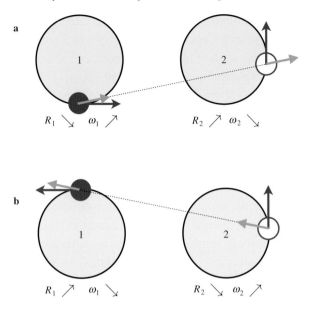

Figure 12.2 Physics of hydrodynamic synchronisation via flexibility. (a) When sphere 1 is lagging behind sphere 2, the flows induced by the rotation of the spheres (grey arrows) lead to a decrease in the radius of the orbit of sphere 1 but an increase in that of sphere 2, which are accompanied by an increase in the rotation frequency of sphere 1 and a decrease in that of sphere 2, and thus a tendency for the two spheres to reduce their phase difference. (b) When sphere 1 is ahead of sphere 2 the opposite happens, and sphere 1 rotates more slowly (increase in orbit size) while sphere 2 rotates faster (decrease), here again leading to rotation in synchrony.

12.1.4 Physical Mechanism of Synchronisation

Why does the combination of hydrodynamic interactions and elastic compliance of the orbits lead to synchronisation of the spheres? The fundamental physical mechanism behind the tendency to phase-lock can be illustrated graphically, as shown in Fig. 12.2.

Consider the case where sphere 1 is lagging behind sphere 2. This is the situation shown in Fig. 12.2(a), where we use dark arrows to represent the local tangential forcing on the sphere and grey arrows to show the direction of the flow velocities experienced by each sphere due to the motion of the other. The flow experienced by sphere 1 due to the motion of sphere 2 acts to decrease the radius of its orbit, R_1. Since $F_1 \sim R_1\omega_1$, this induces an increase in the rotation frequency of sphere 1 (note that this is not due to the conservation of angular momentum since there is no inertia). Similarly, the flow from sphere 1 acts to increase the radius R_2 of the orbit of sphere 2, which leads to a decrease in the frequency ω_2. Although sphere 1 lags behind sphere 2, the combination of hydrodynamic interactions with compliance of their orbits leads to sphere 1 speeding up and sphere 2 slowing down, and as a consequence to a decrease in their phase difference.

By symmetry, when sphere 2 is lagging behind sphere 1, the rotation speed of sphere 2 increases while sphere 1 slows down, resulting again in a decrease in the phase difference (Fig. 12.2(b)). When coupled with flexibility of the oscillator, hydrodynamic interactions therefore lead to synchronisation.

12.1.5 Comparison with Experiments: *Chlamydomonas reinhardtii* **Flagella**

The main prediction of the model is the equation satisfied by the phase difference between the oscillators, namely the Adler equation (Eq. (12.18)), whose solution leads systematically to in-phase synchronisation (Eq. (12.19)). Detailed experimental work has shown that this model captures the phase dynamics of the flagella on the unicellular biflagellate alga *Chlamydomonas reinhardtii* remarkably well (Goldstein et al., 2009). The two flagella of the organism are observed to beat in synchrony for long periods of time interrupted by short phase 'slip' events, where the flagella lose synchrony abruptly and then recover it progressively in $O(10)$ beats.

This dynamics is illustrated in Fig. 12.3. Small observation windows are set up for each flagellum (Fig. 12.3(a)) and used to plot the time variation of the phase, $\Delta(t)$, between the oscillators (Figs. 12.3(b) and (c)); note that in-phase beating are cases where $\Delta = 2\pi n$, where n is an integer. In Fig. 12.3(c) we plot the time dynamics of the phase (relative to its mean) during the short events where the phases lose synchronisation and recover it slowly; thin lines capture individual events, while the thick dashed line shows the averages of the measurements, and the thick solid line represents the analytical solution of the model, Eq. (12.19). Clearly, the Adler equation is able to capture experimental observations quantitatively.

The case of flagella on distinct *Chlamydomonas* cells was addressed by subsequent experimental work (Brumley et al., 2014). Hydrodynamic interactions were shown to be at the origin of the synchronisation in this situation by varying the distance ℓ between the cells and showing that the coupling strength scaled as $1/\ell$, in agreement with the interactions of two free-space stokeslet flows (see Exercises).

12.1.6 Synchronisation via Phase-Dependent Forcing

While some amount of flexibility has long been thought of as a requirement for synchronisation of flagella and cilia, a different physical mechanism was put forward that does not rely on the presence of compliance in the system. Instead, as shown theoretically by Uchida and Golestanian (2011) and confirmed experimentally by

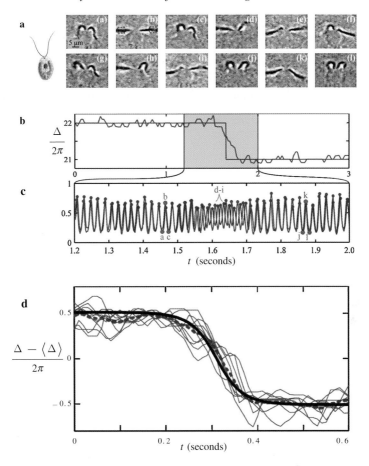

Figure 12.3 Hydrodynamic synchronisation of the biflagellate green alga *Chlamydomonas reinhardtii* (reprinted figure with permission from Goldstein et al. (2009), Copyright 2009 by the American Physical Society). The presence of the two flagella on the cell (a, left) is captured in small observation windows (a, right) from which a phase difference, Δ, is measured. The two flagella alternate between periods of in-phase beating and periods of phase slips where synchrony is lost and recovered over about 10 beats (b and c). The phase dynamics during phase slips (d; individual dynamics plotted as thin lines with the dashed line showing their average) is well captured by the solution to the Adler equation (Eq. (12.19), thick line). Drawing in (a) courtesy of Jacques Lauga.

Kotar et al. (2013), phase-locking can also be attained for coupled oscillators moving along rigid orbits and driven by phase-dependent forces.

Model

This result can be illustrated with the same model as in the previous section. We ignore the radial dynamics for each sphere (i.e. Eq. (12.7a)) and instead assume that sphere i moves along a circle of fixed radius R_i. The angular force balance for sphere 1, Eq. (12.7b), becomes

$$\zeta(R_1\dot{\phi}_1 - \mathbf{u}_{2\to1} \cdot \mathbf{e}_{\phi_1}) = F_1(\phi_1),\tag{12.26}$$

where the new physical ingredient is that the azimuthal driving force is assumed to be a known function of the phase of the oscillator, denoted by $\mathbf{F}_i(\phi_i) = F_i(\phi_i)\mathbf{e}_{\phi_i}$ for each sphere.

The velocity induced by hydrodynamic interactions, $\mathbf{u}_{2\to1}$, is still given by Eq. (12.6). As above we consider the far-field limit where $\mathbf{r}_{2\to1} \approx -\ell\mathbf{e}_x$ so that $(\mathbf{F}_2 \cdot \mathbf{r}_{2\to1}) \approx \ell F_2(\phi_2)\sin\phi_2$ and thus

$$\mathbf{u}_{2\to1} = \mathbf{u}(\mathbf{r}_{2\to1}) \approx -\frac{3h^2}{2\pi\mu\ell^3}F_2(\phi_2)\sin\phi_2\,\mathbf{e}_x.\tag{12.27}$$

Given the coordinates of the tangent vector in Eq. (12.1), we obtain

$$\mathbf{u}_{2\to1}\cdot\mathbf{e}_{\phi_2} = \frac{3h^2}{2\pi\mu\ell^3}F_2(\phi_2)\sin\phi_1\sin\phi_2,\tag{12.28}$$

so that the tangential force balance, Eq. (12.26), becomes

$$\dot{\phi}_1 = \frac{F_1(\phi_1)}{\zeta R_1} + \frac{3h^2}{2\pi\mu\ell^3 R_1}F_2(\phi_2)\sin\phi_1\sin\phi_2.\tag{12.29}$$

The phase dynamics for sphere 2 may be deduced by symmetry, leading to

$$\dot{\phi}_2 = \frac{F_2(\phi_2)}{\zeta R_2} + \frac{3h^2}{2\pi\mu\ell^3 R_2}F_1(\phi_1)\sin\phi_1\sin\phi_2.\tag{12.30}$$

Synchronisation

To proceed, we next assume that the two spheres are identical so that $R_i = R_0$ and $F_i(\phi_i) = F(\phi_i)$ for some prescribed smooth function F. The dynamics of the spheres, Eqs. (12.29)–(12.30), then obeys the coupled differential equations

$$\dot{\phi}_1 = \frac{1}{\zeta R_0}\left(F(\phi_1) + \beta F(\phi_2)\sin\phi_1\sin\phi_2\right),\tag{12.31a}$$

$$\dot{\phi}_2 = \frac{1}{\zeta R_0}\left(F(\phi_2) + \beta F(\phi_1)\sin\phi_1\sin\phi_2\right),\tag{12.31b}$$

with β defined in Eq. (12.12).

The system in Eq. (12.31) admits as a solution the phase-locked state $\phi_1(t) = \phi_2(t) = \phi(t)$, where

$$\dot{\phi} = \frac{F(\phi)}{\zeta R_0}\left(1 + \beta\sin^2\phi\right).\tag{12.32}$$

To assess the tendency of the system to phase-lock, we carry out a perturbation around this state, and write $\phi_2(t) = \phi(t)$ and $\phi_1(t) = \phi(t) + \Delta(t)$, with $|\Delta| \ll 1$. Subtracting Eq. (12.31b) from Eq. (12.31a) and linearising around $\Delta = 0$ lead to

$$\dot{\Delta} = \frac{F'(\phi)}{\zeta R_0} \left(1 - \beta \sin^2 \phi\right) \Delta. \tag{12.33}$$

The solution for Δ of this linear differential equation grows, or decays, exponentially depending on the sign of the growth rate, $\hat{\sigma} \equiv \dot{\Delta}/\Delta$. Denoting by T_0 the period of oscillation of the base state, defined as

$$T_0 = \int_0^{2\pi} \frac{d\phi}{\dot{\phi}}, \tag{12.34}$$

the value of the growth rate averaged over one period of oscillation, $\langle \hat{\sigma} \rangle$, is evaluated as

$$\langle \hat{\sigma} \rangle = \frac{1}{T_0} \int_0^{T_0} \frac{\dot{\Delta}}{\Delta} \, dt = \frac{1}{T_0} \int_0^{2\pi} \frac{\dot{\Delta}}{\Delta} \frac{d\phi}{\dot{\phi}}, \tag{12.35}$$

which using Eq. (12.32) is obtained explicitly as

$$\langle \hat{\sigma} \rangle = \frac{1}{T_0} \int_0^{2\pi} \frac{F'(\phi) \left(1 - \beta \sin^2 \phi\right)}{F(\phi) \left(1 + \beta \sin^2 \phi\right)} \, d\phi. \tag{12.36}$$

In the absence of hydrodynamic interactions, $\beta = 0$, the integrand in Eq. (12.36) is an exact derivative of a periodic function. Therefore, $\langle \hat{\sigma} \rangle = 0$ and the phase shift does not change on average. With hydrodynamic interactions, we can take advantage of the fact that β is a small parameter to expand Eq. (12.36) in powers of β and obtain

$$\langle \hat{\sigma} \rangle = \frac{\beta}{T_0} \int_0^{2\pi} \frac{F'(\phi)}{F(\phi)} \cos 2\phi \, d\phi + O(\beta^2). \tag{12.37}$$

The sign of the average growth rate is therefore controlled by the Fourier decomposition of $F'(\phi)/F(\phi)$. For example, considering a general force profile of the form

$$F = F_0 \left[1 + \sum_{n \geq 1} A_n \sin(n\phi + \delta_n) \right], \tag{12.38}$$

then at leading order in the A_n coefficients, the averaged growth rate is given by (Uchida and Golestanian, 2011)

$$\langle \hat{\sigma} \rangle = \frac{2\pi \beta A_2}{T_0} \cos \delta_2. \tag{12.39}$$

Synchronisation to the in-phase configuration occurs when $A_2 \cos \delta_2 < 0$, so that $\langle \hat{\sigma} \rangle < 0$, e.g. for a force of the form $F(\phi) = F_0 \left[1 - \alpha^2 \sin(2\phi)\right]$ with $\alpha \ll 1$.

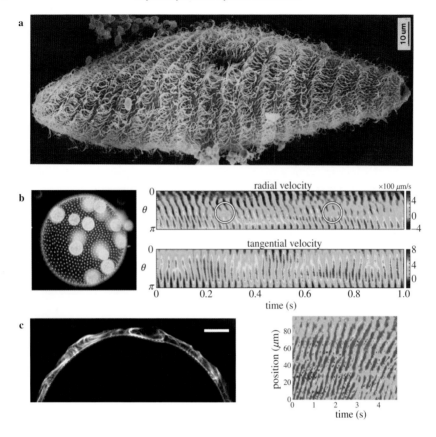

Figure 12.4 Metachronal waves in biological and synthetic systems. (a) Ciliary waves on the surface of the protozoon *Paramecium* with wavelengths on the order of a few microns (Tamm (1972) reproduced with permission). (b) Flagella from different somatic cells on the outer surface of *Volvox carteri* induce radial and tangential velocities with wave-like coordination (adapted from Brumley et al. (2015); cell picture courtesy of Raymond Goldstein). (c) A mixture of micro-tubules and molecular motors can self-organise into active filaments beating periodically with metachronal coordination (scale bar: 40 μm; Sanchez et al. (2011) reproduced with permission).

More generally, it is clear from Eq. (12.37) that a wide range of force profiles can be found that are associated with a tendency to phase-lock, i.e. $\langle \hat{\sigma} \rangle < 0$, and therefore phase-dependent forcing appears to offer a generic driving mechanism for synchronisation in rigid systems.

12.1.7 Metachronal Waves

Our approach to synchronisation has so far only concerned pairs of coupled oscilla-tors. In some instances relevant to biology, phase-locking occurs for many flagella interacting collectively. The most studied example is that of ciliary arrays where large numbers of cilia deform as metachronal waves. In that case, the kinematics of each cilium are identical to those of its neighbours but shifted in time by a small phase difference, resulting in collective waves of deformation. The two main types of metachronal waves were introduced in Section 3.6 and discussed in the context of the waving sheet model.

Classical experimental work on metachronal waves focused on the protozoon *Paramecium*, as illustrated in Fig. 12.4(a) (Tamm, 1972), and on cilia in the lung mucosa of animals (Brennen and Winet, 1977). Examples of wave-like deformation quantified using modern tracking techniques include flagella on the surface of multicellular *Volvox carteri* cells (Fig. 12.4(b), Brumley et al., 2015) and non-biological systems composed of microtubules and molecular motors that self-organise into active, periodically beating filaments (Fig. 12.4(c), Sanchez et al., 2011).

N-Sphere Model

A model for the phase-locking of an oscillator array can be proposed by extending the two-sphere configuration of Section 12.1.3 to a one-dimensional line of $N > 2$ interacting spheres. The angular equation obtained in the case of two spheres, Eq. (12.16), generalises naturally to N spheres as

$$\dot{\phi}_i = \omega_i + \beta \sum_{j \neq i} \omega_j \sin \phi_i \sin \phi_j + \beta \tau \sum_{j \neq i} \omega_i \omega_j \cos \phi_i \sin \phi_j. \tag{12.40}$$

This sequence of coupled equations can be rewritten by introducing the sum and difference in phase for each pair of oscillators as

$$\dot{\phi}_i = \omega_i + \frac{\beta}{2} \sum_{j \neq i} \omega_j [\cos(\phi_i - \phi_j) - \cos(\phi_i + \phi_j)]$$

$$+ \frac{\beta \tau}{2} \sum_{j \neq i} \omega_i \omega_j [\sin(\phi_i + \phi_j) - \sin(\phi_i - \phi_j)]. \tag{12.41}$$

Assuming that the physics with N coupled oscillators is similar to the case of two spheres, we anticipate that the separation of timescales seen in Section 12.1.3 is applicable here too. The sums of phases, $\phi_i + \phi_j$ in Eq. (12.41), are thus expected to increase on the fast timescale as $(\omega_i + \omega_j)t$, whereas the phase differences, $\phi_i - \phi_j$, evolve at a much smaller rate. If we average Eq. (12.41) over the fast timescale, the trigonometric functions of the sums of phases approximately average to zero and we are left with the slowly varying evolution equations

$$\dot{\phi}_i \approx \omega_i + \frac{\beta}{2} \sum_{j \neq i} \omega_j \cos(\phi_i - \phi_j) - \frac{\beta \tau}{2} \sum_{j \neq i} \omega_i \omega_j \sin(\phi_i - \phi_j). \tag{12.42}$$

Waves

The conditions for the existence of metachronal waves can be derived if we assume for simplicity that all oscillators are identical, so that $\omega_i = \omega$. In that case, Eq. (12.42) becomes

$$\dot{\phi}_i = \omega + \frac{\beta\omega}{2}\sum_{j\neq i}\cos(\phi_i - \phi_j) - \frac{\beta\tau\omega^2}{2}\sum_{j\neq i}\sin(\phi_i - \phi_j). \qquad (12.43)$$

Defining the phase difference between two neighbours as $\Delta_i \equiv \phi_{i+1} - \phi_i$ and assuming nearest-neighbour interactions only, Eq. (12.43) transforms into

$$\dot{\phi}_i = \omega + \frac{\beta\omega}{2}(\cos\Delta_i + \cos\Delta_{i-1}) + \frac{\beta\tau\omega^2}{2}(\sin\Delta_i - \sin\Delta_{i-1}). \qquad (12.44)$$

Metachronal waves are solutions with a constant phase shift between neighbours, $\Delta_i = \Delta$. Examining Eq. (12.44), we see that the sine terms disappear in this case, while the cosine terms are equal. Provided that they are compatible with the boundary conditions, metachronal waves therefore exist if each oscillator rotates at the same rate $\dot{\phi}_i = \bar{\omega}$ given by Eq. (12.44) as

$$\bar{\omega} = \omega(1 + \beta\cos\Delta). \qquad (12.45)$$

Note that the factor of $1/2$ from Eq. (12.25) has disappeared because each oscillator now has two neighbours.

To understand which metachronal waves are expected to be observable experimentally, one has to investigate the stability of the phase-locked state in Eq. (12.45). This can be done by subtracting Eq. (12.45) for oscillators i from $i + 1$, leading to the sequence of coupled nonlinear differential equations

$$\dot{\Delta}_i = \frac{\beta\omega}{2}(\cos\Delta_{i+1} - \cos\Delta_{i-1}) + \frac{\beta\tau\omega^2}{2}(\sin\Delta_{i+1} + \sin\Delta_{i-1} - 2\sin\Delta_i). \qquad (12.46)$$

The stability characteristics of Eq. (12.46) can be tackled in a discrete fashion, as in Niedermayer et al. (2008), allowing us to uncover which of the metachronal wave solutions are unstable. Here we propose an alternative analysis in the continuum limit.

Continuum Approximation

Instead of a distribution of phase differences living on a discrete lattice, we now describe the metachronal waves as a continuous phase field, $\Delta(x)$. This approximation is a suitable model provided we focus on spatial variations of the phase difference occurring on length scales much larger than the typical distance between each oscillator, ℓ.

In that continuum approach, the difference terms in Eq. (12.46) can be interpreted as centred finite-difference approximations of spatial derivatives,

$$\cos \Delta_{i+1} - \cos \Delta_{i-1} \approx 2\ell \frac{\partial(\cos \Delta)}{\partial x}, \tag{12.47a}$$

$$\sin \Delta_{i+1} + \sin \Delta_{i-1} - 2 \sin \Delta_i \approx \ell^2 \frac{\partial^2(\sin \Delta)}{\partial x^2}, \tag{12.47b}$$

so that in the continuum limit Eq. (12.46) becomes the partial differential equation

$$\frac{\partial \Delta}{\partial t} = \beta \omega \ell \frac{\partial(\cos \Delta)}{\partial x} + \frac{\beta \tau \omega^2 \ell^2}{2} \frac{\partial^2(\sin \Delta)}{\partial x^2}. \tag{12.48}$$

The stability of metachronal wave solutions to Eq. (12.48), $\Delta(x) = \Delta_0$, can be investigated by perturbing them as

$$\Delta(x, t) = \Delta_0 + \delta(x, t), \tag{12.49}$$

with $|\delta| \ll 1$. A Taylor expansion of Eq. (12.48) leads to an evolution equation for the perturbation as

$$\left(\frac{\partial}{\partial t} + \beta \omega \ell \sin \Delta_0 \frac{\partial}{\partial x} \right) \delta = \frac{\beta \tau \omega^2 \ell^2 \cos \Delta_0}{2} \frac{\partial^2 \delta}{\partial x^2}. \tag{12.50}$$

In the case where $\cos \Delta_0 < 0$, this advection–diffusion equation has a negative diffusion constant, which classically results in solutions growing in time. This can be seen, for example, by Fourier-transforming the perturbation δ and examining the fate of each Fourier mode separately. Substituting a mode of the form $\delta(x, t) = \delta_0(t)e^{ikx}$ into Eq. (12.50) leads to the time evolution for each mode as

$$\left(\frac{\partial}{\partial t} + ik\beta \omega \ell \sin \Delta_0 \right) \delta_0 = -\frac{k^2 \beta \tau \omega^2 \ell^2 \cos \Delta_0}{2} \delta_0. \tag{12.51}$$

When $\cos \Delta_0 < 0$, the solution of Eq. (12.51) for δ_0 is exponentially growing in time, and the associated metachronal mode is unstable. In contrast, if $\cos \Delta_0 > 0$, any Fourier mode of the perturbation decays exponentially in time, and the metachronal wave is stable. In that case, Eq. (12.45) shows that $\bar{\omega} > \omega$ and oscillators in stable metachronal waves rotate with frequencies faster than individual orbits.

Beyond a stability analysis, the appearance of metachronal waves may be investigated for this spherical model by solving Eq. (12.43) numerically. For oscillator arrays with periodic boundary conditions, metachronal waves are indeed observed to form spontaneously if the orbits are sufficiently flexible (Niedermayer et al., 2008).

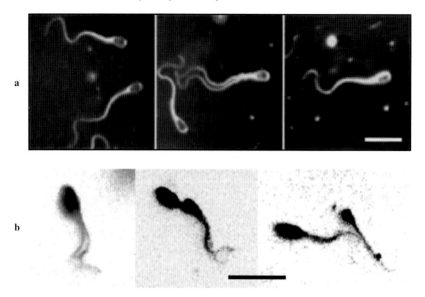

Figure 12.5 Synchronisation of the flagella on two independent spermatozoa. (a) Bull spermatozoa swimming towards each other end up in a co-swimming state with synchronised flagella (Woolley et al. (2009) reproduced with permission). (b) Co-swimming human spermatozoa with phase-locked flagella separate eventually from each other (reprinted figure with permission from Yang et al. (2008), Copyright 2008 by the American Physical Society). Scale bars: 25 μm.

12.2 Synchronisation of Swimming Cells

In the previous section we characterised phase-locking for oscillators with time-periodic motion around a fixed position, motivated by the motion of flagella and cilia anchored on a surface or on an organism. A somewhat different form of synchronisation is observed for the flagella of spermatozoa. As reported by Gray (1928), individual spermatozoa, each propelled by a single active flagellum, often swim in close proximity to each other and as a result phase-lock their flagella for extended periods of time. This is illustrated in the case of bull spermatozoa in Fig. 12.5(a) (Woolley et al., 2009) and human spermatozoa in Fig. 12.5(b) (Yang et al., 2008). This tendency of co-swimming cells to phase-lock was also observed in computational studies (Fauci, 1990; Fauci and McDonald, 1995; Yang et al., 2008).

Compared to the case of anchored flagella and cilia, swimming cells possess an additional degree of freedom, namely their relative position. As a result, it would be possible for phase-locking to be induced passively purely through the difference in swimming velocities between the cells. Using a two-dimensional model, we show below that indeed two different scenarios enable the flagella of free-swimming

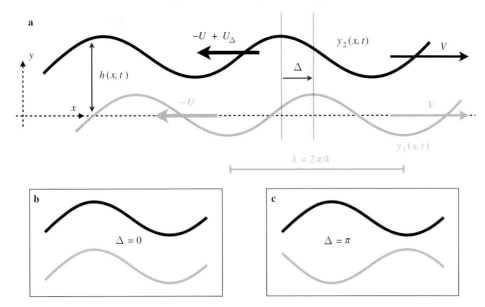

Figure 12.6   Synchronisation of two waving swimmers. (a) Two waving sheets
located instantaneously at $y_1(x,t)$ and $y_2(x,t)$ and separated by a distance $h(x,t)$
deform by waves propagating with speed V along the x direction. The phase difference between the swimmers is denoted by Δ. Sheet 1 swims with velocity $-U\mathbf{e}_x$
while sheet 2 has a relative velocity $U_\Delta\mathbf{e}_x$ above that of swimmer 1. (b) $\Delta = 0$ is
the in-phase swimming configuration. (c) When $\Delta = \pi$ the two sheets swim with
opposite phases.

cells to synchronise, namely: (i) a passive mechanism applicable to cells with fixed
waveforms, which relies entirely on geometry and might result in cells stuck in a
local maximum of energy expenditure; and (ii) an active mechanism for cells able
to modulate their waveform as a function of their environment, which always leads
to in-phase swimming and the energetic minimum.

12.2.1 Two-Dimensional Model

In order to tackle the synchronisation of co-swimming cells, we revisit the waving sheet model of Chapters 3 and 11 and modify it to account for the presence
of a second swimmer. The model is illustrated in Fig. 12.6(a). Two waving sheets
swim parallel to each other while separated by a fixed mean distance h_0. In our approach below we first assume the waveforms to be rigid and then consider the role
of flexibility. The instantaneous waveform of the bottom sheet is denoted $y_1(x,t)$
while the top sheet is at $y_2(x,t)$, and they both deform as transverse travelling
waves propagating along the x direction with wavelength $\lambda = 2\pi/k$, frequency ω

and wave speed $V = \omega/k$. We denote the instantaneous distance between the sheets by $h(x,t) \equiv y_2 - y_1$.

The phase shift between the two swimmers is measured by the instantaneous signed phase difference $\Delta(t)$, defined such that a peak of swimmer 1 is trailing a peak of swimmer 2 in the direction of swimming by $(\Delta + 2\pi n)/k$, where n is an integer (Fig. 12.6). The situation with $\Delta = 0$ is in-phase swimming (Fig. 12.6(b)) while opposite phase corresponds to $\Delta = \pi$ (Fig. 12.6(c)).

As seen in Chapters 3 and 11, locomotion occurs in the direction opposite to the direction of wave propagation. We therefore assume that the instantaneous locomotion velocity of swimmer 1 is $-U\mathbf{e}_x$ while that of swimmer 2 is $(-U + U_\Delta)\mathbf{e}_x$. The speed U_Δ is thus the relative speed between swimmer 2 and swimmer 1 and when $U_\Delta > 0$ the phase difference between the swimmers decreases. More precisely, the rate of change of the phase difference is related to the relative velocity between the swimmers by

$$\dot{\Delta} = -kU_\Delta, \tag{12.52}$$

which, in essence, is a differential equation for the phase difference. The formula in Eq. (12.52) can either be intuited geometrically or derived in the following simple fashion. Consider swimmers whose shapes are travelling waves of the form $y \sim \cos(kx - \omega t)$. The location of a peak along swimmer 1 is, without loss of generality, $x_1(t) = \omega t/k$ while on the phase-shifted swimmer 2 it is $x_2(t) = [\omega t - \Delta(t)]/k$. The relative speed of the two swimmers is then obtained as $U_\Delta = \dot{x}_2 - \dot{x}_1 = -\dot{\Delta}/k$, which is Eq. (12.52).

12.2.2 Taylor's Energy Argument

In his original paper introducing the waving sheet model, Taylor proposed a fluid mechanical argument for the tendency of co-moving swimmers to prefer an in-phase configuration for their flagella (Taylor, 1951). Specifically, he computed the rate of viscous dissipation in the fluid between the two sheets (equal to the rate of work of the swimmers), and showed that it was always minimal for in-phase swimming. We now reproduce this result.

Setup

Consider a waving sheet undergoing sinusoidally varying transverse deformations of amplitude B. As with the other waving sheet calculations from Chapters 3 and 11, we non-dimensionalise lengths as k^{-1} using the wavenumber, time as ω^{-1} using the frequency, and pressure as $\mu\omega$, where μ is the dynamic viscosity of the fluid. We keep the same notation for convenience, including U and U_Δ, but they are understood implicitly to be dimensionless in this section. The dimensionless mean distance between the sheets is denoted $d = kh_0$. Using the same notation as in Chapter 3, the material points on the two swimmers are then located at

$$y_1 = \epsilon b \sin(x - t), \tag{12.53a}$$

$$y_2 = d + \epsilon b \sin(x - t + \Delta). \tag{12.53b}$$

We solve for the flow between the two swimmers using a streamfunction, ψ, defined as

$$u_x = \frac{\partial \psi}{\partial y}, \quad u_y = -\frac{\partial \psi}{\partial x}, \tag{12.54}$$

and the solution to the biharmonic equation

$$\nabla^4 \psi = 0. \tag{12.55}$$

Moving into the frame translating with velocity $-U\mathbf{e}_x$, the boundary conditions for the flow on the sheet are given by

$$\left. \frac{\partial \psi}{\partial y} \right|_{(x,y_1)} = 0, \tag{12.56a}$$

$$\left. \frac{\partial \psi}{\partial x} \right|_{(x,y_1)} = \epsilon b \cos(x - t), \tag{12.56b}$$

$$\left. \frac{\partial \psi}{\partial y} \right|_{(x,y_2)} = U_\Delta, \tag{12.56c}$$

$$\left. \frac{\partial \psi}{\partial x} \right|_{(x,y_2)} = \epsilon b \cos(x - t + \Delta). \tag{12.56d}$$

Flow at Order ϵ

Following Taylor's argument, we calculate the leading-order rate of work of the swimmers on the fluid, $\dot{W}$, which is equal to the rate of viscous dissipation in the fluid and given by Eq. (2.18). We use a regular perturbation expansion in the small parameter ϵ as in Chapters 3 and 11, i.e.

$$\psi = \epsilon \psi_1 + \epsilon^2 \psi_2 + \cdots, \tag{12.57a}$$

$$U = \epsilon U_1 + \epsilon^2 U_2 + \cdots. \tag{12.57b}$$

Since the rate of work is a quantity varying quadratically with the flow, we only need to compute the streamfunction at order one, ψ_1, to obtain the leading-order value of the rate of work (occurring at order two), $\dot{W}_2$, as a function of b and Δ.

Using complex notation, the boundary conditions in Eq. (12.56) at order ϵ become

$$\left.\frac{\partial \psi_1}{\partial y}\right|_{(x,0)} = 0, \tag{12.58a}$$

$$\left.\frac{\partial \psi_1}{\partial x}\right|_{(x,0)} = be^{i(x-t)}, \tag{12.58b}$$

$$\left.\frac{\partial \psi_1}{\partial y}\right|_{(x,d)} = U_{\Delta,1}, \tag{12.58c}$$

$$\left.\frac{\partial \psi_1}{\partial x}\right|_{(x,d)} = be^{i\Delta}e^{i(x-t)}. \tag{12.58d}$$

The general unit-speed 2π-periodic solution to Eq. (12.55) is given in Eq. (11.62), and reproduced here for convenience as

$$\psi = \psi_0 + Ky + Ly^2 + My^3 \tag{12.59}$$

$$+ \mathcal{R}\left\{\sum_{n\geqslant 1}\left[(G_n + yH_n)\cosh\left(n(y-d)\right)\right]e^{in(x-t)}\right\}$$

$$+ \mathcal{R}\left\{\sum_{n\geqslant 1}\left[(I_n + yJ_n)\sinh\left(n(y-d)\right)\right]e^{in(x-t)}\right\}.$$

Similarly to Chapter 11 we have $L = 0$ and $M = 0$ and the value of ψ_0 is irrelevant. Furthermore, given the form of the boundary conditions in Eq. (12.58), only the $n = 1$ mode in Eq. (11.62) is nonzero and we may write the solution as

$$\psi_1 = K_1 y + \left[(G_1 + yH_1)\cosh(y-d) + (I_1 + yJ_1)\sinh(y-d)\right]e^{i(x-t)}, \tag{12.60}$$

with five constants to be determined, namely K_1, G_1, H_1, I_1, J_1. Comparing Eqs. (12.58a) and (12.58c), we see that the constant K_1 can be consistent with both equations only if both $U_{\Delta,1} = 0$ and $K_1 = 0$. The fact that no relative swimming can occur at order ϵ was expected by symmetry since any swimming has to be even in ϵ.

Leading-Order Rate of Work

The four boundary conditions can in theory be used to determine the values of the four constants G_1, H_1, I_1, J_1. However, in order to compute the rate of work of the swimmers we do not need to calculate all of them. Indeed, the equation for the rate of work at leading order for each sheet is given Eq. (3.34) to within a $\pm$ sign necessary to account for the correct direction of the normal on each sheet, i.e.

$$\dot{W}_2 = -\int_0^{2\pi} u_{1,x}\sigma_{1,xy}\Big|_{y=0}\,\mathrm{d}x - \int_0^{2\pi} u_{1,y}\sigma_{1,yy}\Big|_{y=0}\,\mathrm{d}x$$
$$+ \int_0^{2\pi} u_{1,x}\sigma_{1,xy}\Big|_{y=d}\,\mathrm{d}x + \int_0^{2\pi} u_{1,y}\sigma_{1,yy}\Big|_{y=d}\,\mathrm{d}x. \qquad (12.61)$$

Since the x component of the velocity is zero at order ϵ, only the second integral in each line of Eq. (12.61) contributes and we obtain the total rate of work per period as

$$\dot{W}_2 = -\int_0^{2\pi} u_{1,y}\sigma_{1,yy}\Big|_{y=0}\,\mathrm{d}x + \int_0^{2\pi} u_{1,y}\sigma_{1,yy}\Big|_{y=d}\,\mathrm{d}x. \qquad (12.62)$$

Furthermore, the yy component of the normal rate of strain is given by

$$e_{1,yy} = \frac{\partial u_{1,y}}{\partial y} = \frac{\partial^2 \psi_1}{\partial x \partial y}, \qquad (12.63)$$

which, given Eqs. (12.58a) and (12.58c), is identically zero on both swimming sheets at $O(\epsilon)$. The only contribution to normal stresses on the swimmers is therefore the pressure, $\sigma_{1,yy} = -p_1$, and Eq. (12.62) becomes

$$\dot{W}_2 = \int_0^{2\pi} u_{1,y}p_1\Big|_{y=0}\,\mathrm{d}x - \int_0^{2\pi} u_{1,y}p_1\Big|_{y=d}\,\mathrm{d}x. \qquad (12.64)$$

By symmetry, an equal amount of energy rate is contributed by both terms in Eq. (12.64). This can be seen by performing reflectional symmetries in x and y, followed by kinematic reversibility, leaving the value of each integral unchanged but a swapping of the roles of swimmers 1 and 2. Both integrals in Eq. (12.64) are therefore equal and we only need to compute one of them (say, on swimmer 2), so the total rate of work is given by

$$\dot{W}_2 = -2\int_0^{2\pi} u_{1,y}p_1\Big|_{y=d}\,\mathrm{d}x. \qquad (12.65)$$

The non-dimensional pressure obeys the Stokes equations

$$\frac{\partial p_1}{\partial x} = \frac{\partial}{\partial y}\nabla^2\psi_1, \qquad (12.66)$$

and, using Eq. (12.60), is given by

$$p_1 = -2i[H_1\cosh(y-d) + J_1\sinh(y-d)]e^{i(x-t)}. \qquad (12.67)$$

Since we evaluate the value of the pressure at $y = d$, and since the velocity component $u_{1,y} = -\partial\psi_1/\partial x$ in Eq. (12.65) is known explicitly from Eq. (12.58d), the only constant required in order to evaluate Eq. (12.65) is H_1, and we obtain

$$\dot{W}_2 = 4\pi b[\mathcal{I}(H_1)\cos\Delta - \mathcal{R}(H_1)\sin\Delta], \qquad (12.68)$$

where $\mathcal{I}$ and $\mathcal{R}$ denote taking the imaginary and real parts, respectively.

Enforcing the four boundary conditions in Eq. (12.58) leads to the system

$$i(G_1\cosh d - I_1\sinh d) = b, \qquad (12.69a)$$

$$-(G_1 + J_1)\sinh d + (H_1 + I_1)\cosh d = 0, \qquad (12.69b)$$

$$i(G_1 + dH_1) = be^{i\Delta}, \qquad (12.69c)$$

$$H_1 + I_1 + dJ_1 = 0. \qquad (12.69d)$$

From this we can solve for H_1 as

$$H_1 = ibe^{i\Delta}\left(\frac{d + \cosh d\sinh d}{\sinh^2 d - d^2}\right) - ib\left(\frac{\sinh d + d\cosh d}{\sinh^2 d - d^2}\right), \qquad (12.70)$$

and therefore Eq. (12.68) leads to the leading-order rate of work as

$$\dot{W}_2 = 4\pi b^2\left[\left(\frac{d + \cosh d\sinh d}{\sinh^2 d - d^2}\right) - \cos\Delta\left(\frac{\sinh d + d\cosh d}{\sinh^2 d - d^2}\right)\right]. \qquad (12.71)$$

Discussion

The result in Eq. (12.71) is always positive, as expected. The minimum value of $\dot{W}_2(\Delta)$ is obtained when $\cos\Delta = 1$ (in-phase swimming, $\Delta = 0$) and is associated with the rate of work

$$\dot{W}_2(0) = 4\pi b^2\frac{\cosh d - 1}{\sinh d + d}. \qquad (12.72)$$

In contrast, the maximum energetic cost is obtained when $\cos\Delta = -1$ (i.e. in the opposite-phase configuration, $\Delta = \pi$), for which

$$\dot{W}_2(\pi) = 4\pi b^2\frac{\cosh d + 1}{\sinh d - d}. \qquad (12.73)$$

Between these two extrema, the rate of work of the swimmers increases monotonically with Δ, which is Taylor's original result.

The ratio between the in-phase and opposite-phase energetic costs is given by

$$\frac{\dot{W}_2(\pi)}{\dot{W}_2(0)} = \frac{(\cosh d + 1)(\sinh d + d)}{(\cosh d - 1)(\sinh d - d)}, \qquad (12.74)$$

and its value can become very large when d is small. For example, for $d = 1$, $\dot{W}_2(\pi)/\dot{W}_2(0) \approx 5.5$ while for $d = 1/2$, $\dot{W}_2(\pi)/\dot{W}_2(0) \approx 800$. In the limit where the swimmers are close to each other, $d \ll 1$, using the Taylor expansions $\cosh d \approx 1 + d^2/2$ and $\sinh d \approx d + d^3/6$, we obtain the leading-order value of Eq. (12.71) in $1/d$ as

$$\dot{W}_2 \approx \frac{24\pi b^2}{d^3} (1 - \cos \Delta), \tag{12.75}$$

which is recovered in the lubrication limit below (Eq. (12.114)).

The calculation in this section shows that two waving sheets always minimise their energy expenditure for swimming if they can remain in the in-phase configuration. While we assumed a perfectly sinusoidal waveform, a more general derivation using arbitrary Fourier series shows that the result remains true for any periodic waveform (Elfring and Lauga, 2011a), which is also recovered in the long-wavelength limit in Section 12.2.5.

12.2.3 Synchronisation Requires Geometrical Symmetry Breaking

While Taylor's argument shows that swimmers would save energy by waving in phase, it does not indicate whether the distribution of forces in the fluid would lead to this synchronisation dynamically. As swimmers self-propel at low Reynolds number, they are subject to the symmetry and reversibility constraints of the Stokes equations. One important consequence in the context of synchronisation is that swimmers need broken symmetries in their waveforms in order to have any hope of synchronising their flagella.

To explain this result, it is best to consider a pair of swimmers with (infinite) symmetric waveforms and show that hydrodynamic interactions alone cannot lead to their synchronisation (Elfring and Lauga, 2009). Consider the setup in Fig. 12.7(a) with two identical flagella parallel to each other and deforming as travelling waves (indicated by dashed arrows). The swimmers are phase-shifted and the bottom swimmer (grey solid line) is displaced to the right of the top swimmer (black solid line) by the phase difference Δ along the x axis. Without loss of generality, we assume that their relative speed, U_Δ, acts to decrease the value of Δ and synchronise the flagella.

To show that synchronisation is impossible we assume that the waveforms of the flagella are symmetric under two mirror-image symmetries, namely a front-back symmetry by the y axis and a right-left symmetry by the x axis. These symmetries are obeyed by sinewaves, which we employ for illustration in Fig. 12.7. With this assumption we start by using a mirror-image symmetry by the y axis (denoted S_y) to transform the original setup with swimmers propelling to the left (Fig. 12.7(a)) into one where the swimmers propel to the right, their waves propagate to the left and they still synchronise (Fig. 12.7(b)). Note that it is now the top swimmer (black line) that is shifted from the bottom one (grey) by Δ along x.

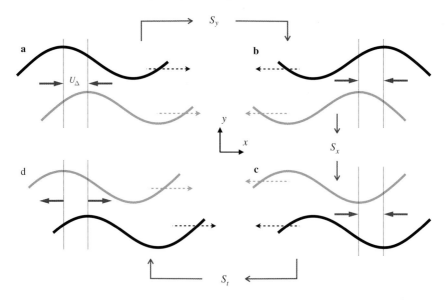

Figure 12.7 Swimmers with symmetric flagellar waveforms cannot synchronise in Stokes flows. (a) Two identical and parallel-waving flagella deform as travelling waves propagating to the right (dashed arrows). Their waveforms are identical under both front-back and right-left symmetries and are assumed to synchronise hydrodynamically. (b) A mirror-image symmetry by the y axis, S_y, leaves the waveforms unchanged but changes the sign of the phase shift between the swimmers and the sign of the wave propagation. (c) A mirror-image symmetry by the x axis, S_x, leaves also the waveforms unchanged and restores the sign of the phase shift between the swimmers from (a). (d) A time-reversal symmetry returns the direction of the wave propagation to that from (a) but changes the sign of the relative speed between the swimmers, U_Δ. The configurations of (a) and (d) are identical except for the sign of U_Δ and therefore $U_\Delta = 0$.

We then use a second mirror-image symmetry by the x axis (denoted S_x) to transform the configuration in Fig. 12.7(b) into one where the swimmer on top and the one at the bottom have switched places (Fig. 12.7(c)). The bottom swimmer (black line) is now displaced by Δ along the x axis compared to the top swimmer (grey) and they still synchronise their flagella.

We exploit finally the time-reversal symmetry property of Stokes flows (see Chapter 2) and reverse instantaneously the direction of propagation of the waves in Fig. 12.7(c). This leads to the configuration in Fig. 12.7(d), where the geometry is identical to the one in Fig. 12.7(c) except that the waves now propagate to the right. As a result of kinematic reversibility, all velocities from Fig. 12.7(c) are reversed in Fig. 12.7(d) and in particular the value of the relative speed between the swimmers, U_Δ, now acts to increase the phase difference, Δ.

Comparing the new setup in Fig. 12.7(d) with the original configuration in Fig. 12.7(a) we see that geometry, relative positions and waving motion of the swimmers are identical except that the signs of the relative speeds U_Δ in Fig. 12.7(a) and Fig. 12.7(d) are opposite. Note that we would have reached the same conclusion if we had assumed in Fig. 12.7(a) that U_Δ acted to increase the phase difference, in which case the configuration in Fig. 12.7(d) would now have led to synchronisation.

Since Stokes flows have unique solutions (see Chapter 2), the flow problems in Fig. 12.7(a) and Fig. 12.7(d) have to be identical. The relative speed between the swimmers must therefore satisfy $U_\Delta = -U_\Delta$ and thus $U_\Delta = 0$. The forces from Stokes flows can therefore not lead to the synchronisation of waving swimmers whose waveforms obey two mirror-image symmetries. For example, pure sinewaves cannot synchronise.

The three symmetry arguments used to rule out synchronisation suggest three ways in which breaking symmetries may result in hydrodynamic synchronisation. First, swimmers with front-back asymmetric flagellar waveforms are not identical under the S_y symmetry (Fig. 12.7(a) to (b)). In that case, synchronisation is possible, may occur passively and, as shown below in Section 12.2.5, the location of the final phase difference depends only on the wave geometry. Alternatively, swimmers may modulate the amplitude of their waveforms, which forbids the use of the S_x symmetry (Fig. 12.7(b) to (c)). This is the mechanism used by active flexible swimmers to synchronise in phase and is detailed in Section 12.2.6. Finally, the use of the time-reversal symmetry S_t is forbidden in a non-Newtonian fluid, and in that case symmetric shapes can synchronise hydrodynamically (see Chapter 15, Questions 9 and 10).

12.2.4 Long-Wavelength Synchronisation: Flow and Forces

In order to characterise the dynamics of synchronisation for waveforms with broken symmetries, we solve for the fluid dynamics between swimmers with arbitrary waveforms. Though this can be achieved numerically in general (Elfring and Lauga, 2011b), we can make analytical progress by assuming that the two swimmers in Fig. 12.6 deform in the long-wavelength limit, $hk \ll 1$. In that case, the typical length scale in the fluid in between the cells over which velocity gradients are expected to occur (h) is much smaller than the relevant length scale in the outer domain (k^{-1}). In this long-wavelength limit, we may then ignore the flow problem outside the swimmers and focus on solving for the fluid dynamics between the waving sheets. We first present below the general derivation for arbitrary periodic waveforms and then address separately the cases of passive and active synchronisation.

Setup

Similarly to the waving filament calculation from Chapter 7 and using the same notation, we solve the problem by first moving into the swimming frame of sheet 1 and then into its wave frame. In order to be consistent with the calculation in Section 11.4.2 for swimming near walls, we keep the derivation dimensional. In the long-wavelength limit, the Stokes equations become the lubrication equations (Leal, 2007)

$$\frac{\partial p}{\partial x} = \mu \frac{\partial^2 u_x}{\partial y^2}, \quad \frac{\partial p}{\partial y} = 0. \tag{12.76}$$

In the limit where $hk \ll 1$, material points in the wave frame of swimmer 1 move with velocity $-Q\mathbf{t} \approx -V\mathbf{e}_x$ (i.e. Eq. (11.75)) while material points on swimmer 2 have velocity $-Q\mathbf{t} + U_\Delta \mathbf{e}_x \approx (-V + U_\Delta)\mathbf{e}_x$. The boundary conditions for Eq. (12.76) are therefore

$$u_x(y_1) = -V, \tag{12.77a}$$

$$u_x(y_2) = -V + U_\Delta. \tag{12.77b}$$

Flow and Mass Conservation

The solution for the velocity is easily found as

$$u_x = \frac{1}{2\mu} \frac{\partial p}{\partial x}(y - y_1)(y - y_2) + U_\Delta \frac{y - y_1}{h} - V, \tag{12.78}$$

with $h = y_2 - y_1$. There are two parameters to solve for in Eq. (12.78), namely the pressure gradient, $\partial p / \partial x$, and the relative speed, U_Δ, which are obtained using mass conservation and the overall force balance.

We first compute the flow rate between the swimmers, Q, as

$$Q \equiv \int_{y_1}^{y_2} u_x \, dy = -\frac{h^3}{12\mu} \frac{\partial p}{\partial x} - Vh + \frac{h}{2} U_\Delta. \tag{12.79}$$

Mass conservation is enforced by integrating the divergence of the velocity field as

$$\int_{y_1}^{y_2} \left(\frac{\partial u_x}{\partial x} + \frac{\partial u_y}{\partial y} \right) dy = 0, \tag{12.80}$$

leading to the balance

$$\frac{\partial Q}{\partial x} = u_x(y_2) \frac{\partial y_2}{\partial x} - u_x(y_1) \frac{\partial y_1}{\partial x} - u_y(y_2) + u_y(y_1), \tag{12.81}$$

where we recall that u_x and u_y are velocity components along x and y.

In order to simplify Eq. (12.81), we make use of the general kinematics for a deforming boundary, stating that the vertical velocity is equal to the material derivative of the boundary position,

$$u_y(y_s) = \left(\frac{\partial y}{\partial t} + u_x \frac{\partial y}{\partial x} \right) \Big|_{y=y_s}, \tag{12.82}$$

where y_s is the time-varying location of one of the waving sheets, $y_s = y_1$ or y_2. For swimmer 1, there is no change in shape (since we are in its wave frame) and thus

$$u_y(y_1) = u_x(y_1) \frac{\partial y_1}{\partial x}. \tag{12.83}$$

In contrast, the shape of swimmer 2 changes with time and Eq. (12.82) applies fully. However, it can be simplified by noting that the shape would not be changing if we moved in the frame translating with speed U_Δ. Thus

$$\left(\frac{\partial}{\partial t} + U_\Delta \frac{\partial}{\partial x} \right) y_2 = 0. \tag{12.84}$$

Combining Eq. (12.82) with Eq. (12.84) leads to

$$u_y(y_2) = [u_x(y_2) - U_\Delta] \frac{\partial y_2}{\partial x}. \tag{12.85}$$

Together with Eq. (12.83), the mass conservation equation, Eq. (12.81), becomes

$$\frac{\partial Q}{\partial x} = U_\Delta \frac{\partial y_2}{\partial x}. \tag{12.86}$$

Substituting the value of the flow rate, Eq. (12.79), leads to the first relationship between the pressure gradient and the relative speed of the swimmers,

$$-\frac{1}{12\mu} \frac{\partial}{\partial x} \left(h^3 \frac{\partial p}{\partial x} \right) = \frac{\partial}{\partial x} \left[Vh + U_\Delta \left(\frac{y_1 + y_2}{2} \right) \right]. \tag{12.87}$$

This is associated with the constraint of periodicity of the pressure, i.e. no net pressure gradient generated over one wavelength $\lambda = 2\pi/k$ of the swimmers.

Force Balance

The second equation comes from the overall force balance on the swimmers. Recalling that the stress tensor in a Stokes flow is divergence-free, an integration of Eq. (2.13) in the fluid between the swimmers and over one wavelength shows that the forces on the swimmers are equal and opposite. Enforcing one of the swimmers to be force-free (e.g. swimmer 2) therefore leads automatically to force-free motion for the other one (swimmer 1).

By symmetry, the only relevant component of the hydrodynamic force on the swimmers is along x. The force per unit area acting along x is given by

$\mathbf{e}_x \cdot \boldsymbol{\sigma} \cdot \mathbf{n} = \sigma_{xx} n_x + \sigma_{xy} n_y$. In the long-wavelength limit, the normal vector to swimmer 2 directed into the fluid is

$$\mathbf{n}_2 = \left(\frac{\partial y_2}{\partial x}, -1 \right). \tag{12.88}$$

Using the usual long-wavelength scalings where normal viscous stresses can be neglected (Leal, 2007) leads to hydrodynamic surface forces

$$\mathbf{e}_x \cdot \boldsymbol{\sigma} \cdot \mathbf{n}|_{y_2} \approx -p \frac{\partial y_2}{\partial x} - \mu \frac{\partial u_x}{\partial y}\Big|_{y_2}, \tag{12.89}$$

plus higher-order terms (recall that the value of the pressure does not depend on y).

The total hydrodynamic force over one wavelength of swimmer 2, $\mathcal{F}_x$, is then obtained by integrating Eq. (12.89) as

$$\mathcal{F}_x = \int_0^\lambda \left(-p \frac{\partial y_2}{\partial x} - \mu \frac{\partial u_x}{\partial y}\Big|_{y_2} \right) dx = \int_0^\lambda \left(y_2 \frac{\partial p}{\partial x} - \mu \frac{\partial u_x}{\partial y}\Big|_{y_2} \right) dx, \tag{12.90}$$

where the second integral is a consequence of integration by parts and periodicity of the pressure. The value of the shear stress is obtained by differentiating Eq. (12.78) as

$$\mu \frac{\partial u_x}{\partial y}\Big|_{y_2} = \frac{h}{2} \frac{\partial p}{\partial x} + \mu U_\Delta \frac{1}{h}, \tag{12.91}$$

so that the force in Eq. (12.90) becomes

$$\mathcal{F}_x = \int_0^\lambda \left[\frac{(y_1 + y_2)}{2} \frac{\partial p}{\partial x} - \mu U_\Delta \frac{1}{h} \right] dx. \tag{12.92}$$

Enforcing force-free swimming, $\mathcal{F}_x = 0$, leads therefore to the second relationship between the relative speed between the swimmers and the pressure gradient as

$$U_\Delta \int_0^\lambda \frac{1}{h} \, dx = \int_0^\lambda \frac{(y_1 + y_2)}{2\mu} \frac{\partial p}{\partial x} \, dx. \tag{12.93}$$

While it is possible to combine Eqs. (12.87) and (12.93) to obtain a general formula for the value U_Δ, the result is convoluted and not very enlightening. More intuitive physical insight can be gained by focusing on two special cases. In Section 12.2.5 we first assume that the waveforms are rigid and show that synchronisation may occur passively, while in Section 12.2.6 we consider the active synchronisation of flexible swimmers.

12.2.5 Passive Synchronisation of Rigid Swimmers

We first consider here the situation where the flagellar waveforms are fixed. This is the relevant situation for rigid swimmers whose shapes are not modified by the fluid stresses. In that case the swimmers can synchronise passively but it is a slow effect that occurs at a rate scaling as the fourth power of the waving amplitude.

Hydrodynamic Force

To characterise passive synchronisation, we assume that there is no relative speed, $U_\Delta = 0$, and compute the hydrodynamic force acting on each of the swimmers. Following the argument in the previous section, the total force acting on both swimmers is zero, and therefore the force acting on swimmer 1 is equal and opposite to that acting on swimmer 2.

The value of the pressure gradient is obtained by integrating Eq. (12.87). Assuming no net pressure over a wavelength, as in Eq. (11.79), leads to

$$\frac{\partial p}{\partial x} = 12\mu V \left(\frac{I_2}{I_3 h^3} - \frac{1}{h^2} \right), \tag{12.94}$$

where the integrals of the form I_n are defined in Eq. (11.81). Using Eq. (12.94), we then obtain the hydrodynamic force on swimmer 2 from Eq. (12.92) as

$$\mathcal{F}_x = 6\mu V \int_0^\lambda (y_1 + y_2) \left(\frac{I_2}{I_3 h^3} - \frac{1}{h^2} \right) dx. \tag{12.95}$$

Fixed Points

The formula in Eq. (12.95) allows us to characterise the synchronisation dynamics. If $\mathcal{F}_x = 0$, then the phase difference remains constant and is a fixed point of the dynamics. Two fixed points are easily identified from Eq. (12.95): (i) for in-phase swimming ($\Delta = 0$), $h = h_0$ is constant, and $\partial p / \partial x = 0$ so $\mathcal{F}_x = 0$; (ii) if instead the swimmers are in opposite phase ($\Delta = \pi$), then $y_1 + y_2$ is a constant and thus, due to the periodicity of the pressure, we also have $\mathcal{F}_x = 0$. These two fixed points are the only ones in the long-wavelength limit (Elfring and Lauga, 2011a). For any other phase difference, we have $\mathcal{F}_x \neq 0$ and the phase difference changes in time as $\dot{\Delta} \sim -U_\Delta \sim -\mathcal{F}_x$. The system of two swimmers then evolves dynamically to the one of the two fixed points that is stable.

Small-Amplitude Limit

In order to characterise the stability of the fixed points analytically, we first evaluate the hydrodynamic force, Eq. (12.95), in the limit of small-amplitude waving. We write

$$h = h_0 \left[1 + \alpha(x) \right], \tag{12.96}$$

where $|\alpha(x)| \ll 1$ and h_0 is the mean value of h so that $\int_0^\lambda \alpha \, dx = 0$. We then proceed to Taylor-expand the expressions in powers of α all the way to $O(\alpha^3)$. Using the expansion

$$\frac{1}{h^n} = \frac{1}{h_0^n} \left[1 - n\alpha + \frac{n(n+1)}{2}\alpha^2 - \frac{n(n+1)(n+2)}{6}\alpha^3 + O(\alpha^4) \right], \qquad (12.97)$$

we obtain, up to $O(\alpha^3)$,

$$\frac{I_2}{I_3 h^3} - \frac{1}{h^2} = \frac{1}{h_0^2}\left[-\alpha + 3\alpha^2 - 6\alpha^3 + (3\alpha - 1)\frac{3}{\lambda}\int_0^\lambda \alpha^2 \, dx + \frac{6}{\lambda}\int_0^\lambda \alpha^3 \, dx \right]. \qquad (12.98)$$

We next write $y_1 \equiv \eta_1$ and $y_2 \equiv h_0 + \eta_2$, where the functions η_i have zero mean, $\int_0^\lambda \eta_i \, dx = 0$ $(i = 1, 2)$; α is thus defined as $\alpha = (\eta_2 - \eta_1)/h_0$. We further assume that the shapes are identical and that they are right-left symmetric in the frame of the swimmer, i.e. $\eta_i(x + \lambda/2) = -\eta_i(x)$. This assumption applies to all microorganisms that swim along a straight line on average (such as the majority of spermatozoa). In contrast, the asymmetry between the front and the back of the organisms is left arbitrary.

With this notation and these assumptions, we can use the result of Eq. (12.98) in Eq. (12.95) to see that the expression for the force involves integral terms of the form

$$\mathcal{K}_n = \int_0^\lambda (\eta_1 + \eta_2)\alpha^n \, dx, \qquad n = 0, 1, 2, 3. \qquad (12.99)$$

The first three of these integrals are zero: (i) $\mathcal{K}_0 = 0$ because the functions η_i have zero mean; (ii) $\mathcal{K}_1 \sim \int(\eta_2^2 - \eta_1^2) = 0$ because the waveforms are assumed to be identical; (iii) $\mathcal{K}_2 = 0$ because of the right-left symmetry property of the waveforms. The first nonzero integral is therefore $\mathcal{K}_3$, which leads to the expression for the hydrodynamic force at leading order in the wave amplitudes as

$$\mathcal{F}_x = -\frac{36\mu V}{h_0^2} \int_0^\lambda (\eta_1 + \eta_2)\alpha^3 \, dx = -\frac{36\mu V}{h_0^5} \int_0^\lambda (\eta_1 + \eta_2)(\eta_2 - \eta_1)^3 \, dx. \qquad (12.100)$$

To simplify the discussion, we non-dimensionalise the waveforms and write $k\eta_1(x) = bg(kx)$, where b is the dimensionless wave amplitude (notation chosen to be consistent with the waving-sheet results in Chapter 3) and g is the dimensionless shape of the waveform. The shape for swimmer 2 is then given by $k\eta_2 = bg(kx + \Delta)$. Using the same notation, where x is now dimensionless, the right-left symmetry of g is written as $g(x + \pi) = -g(x)$. The dimensional force in Eq. (12.100) becomes, over one dimensionless wavelength,

$$\mathcal{F}_x = -\frac{36\mu V b^4}{(kh_0)^5} \int_0^{2\pi} [g(x) + g(x + \Delta)] [g(x + \Delta) - g(x)]^3 \, dx. \qquad (12.101)$$

Clearly $\Delta = 0, \pi$ are associated with zero force. Note that the scaling $\mathcal{F}_x \sim b^4$ with the fourth power of the wave amplitude is consistent with asymptotic calculations carried out outside the long-wavelength limit (Elfring and Lauga, 2011a).

Stability of Fixed Points

The stability of the fixed point at $\Delta = 0$ is examined by writing $\Delta = 0 + \delta$ and using a Taylor expansion of Eq. (12.101) for small values of δ. The hydrodynamic force near the fixed point, $\mathcal{F}_{x,0}$, is obtained at leading order as

$$\mathcal{F}_{x,0} = -\frac{72\mu V b^4}{(kh_0)^5} \delta^3 \int_0^{2\pi} g(x) \left[g'(x)\right]^3 \, dx. \qquad (12.102)$$

The stability near the fixed point at $\Delta = \pi$ involves a bit more algebra. Writing $\Delta = \pi + \delta$, a Taylor expansion of the integrand in Eq. (12.101) for small values of δ leads to

$$[g(x) - g(x + \Delta)]^3 = 8g^3 + 12\delta g^2 g' + 6\delta^2 (g^2 g'' + gg'^2) + O(\delta^3), \qquad (12.103)$$

so that

$$[g(x) + g(x + \Delta)] [g(x + \Delta) - g(x)]^3 = \delta(8g^3 g') \qquad (12.104)$$
$$+ \delta^2 (4g^3 g'' + 12g^2 g'^2)$$
$$+ \delta^3 \left(\frac{4}{3} g^3 g''' + 12g^2 g' g'' + 6gg'^3\right),$$

plus terms of order δ^4. Integrating Eq. (12.104) over a wavelength and cancelling out all the terms that are exact derivatives and thus average out to zero, we obtain the force near π, $\mathcal{F}_{x,\pi}$, at leading order as

$$\mathcal{F}_{x,\pi} = \frac{72\mu V b^4}{(kh_0)^5} \delta^3 \int_0^{2\pi} g(x) \left[g'(x)\right]^3 \, dx. \qquad (12.105)$$

Passive Synchronisation Is Set by Geometry

Remarkably, the hydrodynamic forces near the two fixed points, Eqs. (12.102) and (12.105), are equal and opposite. Depending on the sign of the shape integral $\int gg'^3$, unless it is exactly zero, one of the two forces is therefore necessarily of the form $\mathcal{F}_x \sim \delta^3$, leading to the phase dynamics near its associated fixed point of the form $\dot{\delta} \sim -\delta^3$, indicating a stable fixed point. Given the opposite signs of Eqs. (12.102) and (12.105), the other fixed point in this case has a hydrodynamic force scaling as $\mathcal{F}_x \sim -\delta^3$, leading to the phase dynamics $\dot{\delta} \sim \delta^3$, which is unstable.

We therefore obtain the remarkable result that, in the case where the waving swimmers are identical and their waveforms are fixed, they always synchronise to a stable fixed point whose value ($\Delta = 0$ or π) is set purely by the geometry of the waveform (Elfring and Lauga, 2009, 2011b).

Note that in the particular case where the waveform is front-back symmetric, the dimensionless function g satisfies

$$g(\pi - x) = g(x), \quad 0 \leqslant x \leqslant \pi, \tag{12.106}$$

leading to

$$\int_0^{2\pi} g(x) \left[g'(x)\right]^3 \, dx = 0, \tag{12.107}$$

and therefore no hydrodynamic force. In agreement with Section 12.2.3, front-back symmetric waveforms do not synchronise.

Rate of Work and Synchronisation

We demonstrated that, depending on the wave geometry, the swimmers always synchronise passively in phase or in opposite phase. How do these dynamical results compare to Taylor's argument that the in-phase configuration leads to the minimum rate of work of the swimmers?

In the long-wavelength limit, the rate of energy dissipation per unit length in the z direction in the fluid between the swimmers is given by

$$\dot{W} = \int_0^\lambda \int_{y_1}^{y_2} \mu \left(\frac{\partial u_x}{\partial y}\right)^2 \, dy \, dx. \tag{12.108}$$

The shear rate is computed from Eq. (12.78) as

$$\frac{\partial u_x}{\partial y} = \frac{1}{2\mu} \frac{\partial p}{\partial x} (2y - y_1 - y_2), \tag{12.109}$$

and with the pressure gradient in Eq. (12.94) we obtain

$$\dot{W} = 12\mu V^2 \int_0^\lambda h^3 \left(\frac{I_2}{I_3 h^3} - \frac{1}{h^2}\right)^2 \, dx. \tag{12.110}$$

At small amplitude we have the expansion

$$h^3 \left(\frac{I_2}{I_3 h^3} - \frac{1}{h^2}\right)^2 = \frac{1}{h_0} \left[\alpha^2 + O(\alpha^3)\right], \tag{12.111}$$

and therefore we obtain, at leading order,

$$\dot{W} = \frac{12\mu V^2}{h_0} \int_0^\lambda \alpha^2 \, dx = \frac{12\mu V^2}{h_0^3} \int_0^\lambda (\eta_2 - \eta_1)^2 \, dx. \tag{12.112}$$

This can be rewritten after non-dimensionalising x and the waveforms as

$$\dot{W} = \frac{12\mu b^2 V^2}{(kh_0)^3} \int_0^{2\pi} [g(x + \Delta) - g(x)]^2 \, dx. \tag{12.113}$$

For a sinewave, this integral can be evaluated exactly, leading to

$$\dot{W} = \frac{24\pi\mu b^2 V^2}{(kh_0)^3} (1 - \cos \Delta), \tag{12.114}$$

which is the dimensional version of Taylor's result, Eq. (12.75).

In the case of an arbitrary waveform, we use Taylor expansions to perturb the power in Eq. (12.113) near the fixed points. Near the in-phase configuration, $\Delta = 0 + \delta$ with δ small, we obtain, at leading order in δ,

$$\dot{W}\big|_0 = \frac{12\mu b^2 V^2}{(kh_0)^3} \delta^2 \int_0^{2\pi} [g'(x)]^2 \, dx, \tag{12.115}$$

which is a minimum for all waveforms.

In the case where the phase difference is near the opposite-phase configuration, $\Delta = \pi + \delta$, we make use of the expansion

$$[g(x) - g(x + \Delta)]^2 = 4g^2 + 4\delta gg' + \delta^2(g'^2 + 2gg'') + O(\delta^3), \tag{12.116}$$

and integrate to obtain the local rate of dissipation for small values of δ up to order δ^2 as

$$\dot{W}\big|_\pi = \frac{12\mu b^2 V^2}{(kh_0)^3} \left(4 \int_0^{2\pi} [g(x)]^2 \, dx - \delta^2 \int_0^{2\pi} [g'(x)]^2 \, dx \right). \tag{12.117}$$

This is clearly a local maximum of energy dissipation for any waveform g.

In agreement with Taylor's argument, the rate of work of swimmers with arbitrary waveforms is always minimum at the in-phase point, and maximum at the opposite phase. However, this may be at odds with the dynamics. In case where $\Delta = 0$ is the stable fixed point, the dynamics of passive synchronisation brings the swimmers to the energetically favourable configuration so that dynamic and energy considerations agree. However, in the opposite case where it is $\Delta = \pi$ that is stable, the two swimmers are pushed dynamically into the least favourable energetic configuration, leading to a situation where dynamic and energy arguments are at odds with one another.

12.2.6 Active Synchronisation of Flexible Swimmers

In the previous section we assumed that the swimmers actuated their flagella with identical, rigid waveforms and we obtained passive synchronisation at a rate scaling with the fourth power of the wave amplitude. Since we know from Chapter 7 that eukaryotic flagella are active flexible filaments, we consider now the case where the waving swimmers are modelled as internally forced flexible sheets and solve for the dynamic balance governing their waveforms (Elfring and Lauga, 2011b). In that case, we obtain active synchronisation at a rate proportional to the second power of the wave amplitude.

Forces and Synchronisation Speed

In order to proceed, we anticipate that it is sufficient to determine the shape of each swimmer in the linear limit (which will be verified later). Assuming that the relative speed, U_Δ, scales at least as the square of the waving amplitude, linearising the equation for the pressure gradient, Eq. (12.87), leads to

$$\frac{\partial p}{\partial x} = \frac{12\mu V}{h_0^3}(\eta_1 - \eta_2), \tag{12.118}$$

with no integration constant since the pressure is periodic. Substituting in the equation for the dimensional hydrodynamic force, Eq. (12.92), leads to

$$\mathcal{F}_x = \frac{6\mu V}{h_0^3} \int_0^\lambda \left(\eta_1^2 - \eta_2^2\right) dx - \mu U_\Delta \frac{2\pi}{kh_0}, \tag{12.119}$$

which is equivalent to keeping the $O(\alpha)$ term in Eq. (12.98). Enforcing force-free motion, $\mathcal{F}_x = 0$, allows us finally to determine the value of the relative speed between the two swimmers, U_Δ, as

$$U_\Delta = \frac{3Vk}{\pi h_0^2} \int_0^\lambda \left(\eta_1^2 - \eta_2^2\right) dx. \tag{12.120}$$

Note that the scaling with the wave amplitude, $U_\Delta \sim \eta^2$, justifies *a posteriori* the linearisation of the pressure gradient in Eq. (12.118). If the two waveforms are identical, then $U_\Delta = 0$. As we see below, the hydrodynamic interactions between the swimmers lead to different waving amplitudes, resulting in $U_\Delta \neq 0$ and a time evolution of the phase.

Active Waveforms

In order to solve for the waveforms of the swimmers, η_i, we revisit the balance of forces seen in Chapter 7 for active flagella. For three-dimensional filaments, the shape was a solution to the linearised active elasto-hydrodynamic equation in Eq. (7.53), namely

$$B\frac{\partial^4 y}{\partial x^4} + c_\perp \frac{\partial y}{\partial t} = f_a(x,t), \tag{12.121}$$

which quantified the balance between elastic forces (first term on the left-hand side), hydrodynamic drag (second term) and active, internal force density (right-hand side).

The same physical balance applies to the case of two waving swimmers. Let us denote by σ the hydrodynamic stress tensor in the fluid between the sheets and by $\mathbf{n}_i$ the unit normal on swimmer i directed into the fluid. The hydrodynamic term in Eq. (12.121) is minus the y component of the viscous force due to the flow, which in two dimensions is equal to $-\mathbf{e}_y \cdot \sigma \cdot \mathbf{n}_i$. The force balance for the two waving sheets along the y direction in Eq. (12.121) becomes therefore

$$B\frac{\partial^4 \eta_1}{\partial x^4} - \mathbf{e}_y \cdot \sigma \cdot \mathbf{n}_1|_{y_1} = f_1(x,t), \tag{12.122a}$$

$$B\frac{\partial^4 \eta_2}{\partial x^4} - \mathbf{e}_y \cdot \sigma \cdot \mathbf{n}_2|_{y_2} = f_2(x,t), \tag{12.122b}$$

where we have used f_i to denote the active two-dimensional force density acting on sheet i (with dimension of force per unit area) and where the two-dimensional bending modulus B now has dimension of force times length. In the long-wavelength limit, the unit normals are given by $\mathbf{n}_1 = \mathbf{e}_y$ and $\mathbf{n}_2 = -\mathbf{e}_y$, and $\mathbf{e}_y \cdot \sigma \cdot \mathbf{e}_y \approx -p$, so that the system in Eq. (12.122) becomes

$$B\frac{\partial^4 \eta_1}{\partial x^4} + p = f_1(x,t), \tag{12.123a}$$

$$B\frac{\partial^4 \eta_2}{\partial x^4} - p = f_2(x,t). \tag{12.123b}$$

The equation we have for the pressure involves its gradient, Eq. (12.118), so we take one more spatial derivative of Eq. (12.123), and use the result for $\partial p/\partial x$ to obtain

$$B\frac{\partial^5 \eta_1}{\partial x^5} + \frac{12\mu V}{h_0^3}(\eta_1 - \eta_2) = \frac{\partial f_1}{\partial x}, \tag{12.124a}$$

$$B\frac{\partial^5 \eta_2}{\partial x^5} - \frac{12\mu V}{h_0^3}(\eta_1 - \eta_2) = \frac{\partial f_2}{\partial x}. \tag{12.124b}$$

Adding and subtracting these equations, and using the notation $\eta_+ \equiv \eta_1 + \eta_2$ and $\eta_- \equiv \eta_2 - \eta_1$, we then get

$$B\frac{\partial^5 \eta_+}{\partial x^5} = \frac{\partial(f_1 + f_2)}{\partial x}, \tag{12.125a}$$

$$B\frac{\partial^5 \eta_-}{\partial x^5} + \frac{24\mu V}{h_0^3}\eta_- = \frac{\partial(f_2 - f_1)}{\partial x}. \tag{12.125b}$$

We assume that the internal forces acting on each swimmer are identical but are shifted by the phase difference Δ. Taking them to be pure sinewaves for simplicity we may then write

$$f_1 = f_0 \cos(kx - \omega t) = f_0 \mathcal{R}\left\{ e^{i(kx - \omega t)} \right\}, \tag{12.126a}$$

$$f_2 = f_0 \cos(kx - \omega t + \Delta) = f_0 \mathcal{R}\left\{ e^{i\Delta} e^{i(kx - \omega t)} \right\}, \tag{12.126b}$$

where f_0 is a real, positive amplitude. Using complex notation and writing $\eta_\pm = \mathcal{R}\left\{ \gamma_\pm e^{i(kx - \omega t)} \right\}$ we then obtain

$$ik^5 B \gamma_+ = ik f_0 \left(e^{i\Delta} + 1 \right), \tag{12.127a}$$

$$\left(ik^5 B + \frac{24\mu V}{h_0^3} \right) \gamma_- = ik f_0 \left(e^{i\Delta} - 1 \right), \tag{12.127b}$$

whose solutions are

$$\gamma_+ = \frac{f_0}{Bk^4} \left(e^{i\Delta} + 1 \right), \tag{12.128a}$$

$$\gamma_- = \frac{ikh_0^3 f_0}{24\mu V + ik^5 Bh_0^3} \left(e^{i\Delta} - 1 \right). \tag{12.128b}$$

Active Synchronisation

With the shapes of the two waving sheets solved for, we may calculate their relative swimming speed. We first rewrite the formula for U_Δ, Eq. (12.120), as

$$U_\Delta = -\frac{3Vk}{\pi h_0^2} \int_0^\lambda \eta_+ \eta_- \, dx, \tag{12.129}$$

which can be evaluated using complex notation as

$$U_\Delta = -\frac{3V}{h_0^2} \mathcal{R}\left\{ \gamma_+^* \gamma_- \right\}, \tag{12.130}$$

where $*$ denotes complex conjugate. Given the solution in Eq. (12.128), the relative speed between the swimmers becomes

$$U_\Delta = U_0 \sin \Delta, \quad U_0 = \frac{144 \mu V^2 f_0^2 h_0}{Bk^3 \left(B^2 h_0^6 k^{10} + 576 \mu^2 V^2 \right)}. \tag{12.131}$$

Using Eq. (12.52), this leads to a differential equation for the phase difference Δ as

$$\dot{\Delta} = -kU_\Delta = -kU_0 \sin \Delta. \tag{12.132}$$

Once again we obtain the Adler equation, with two fixed points. Since $kU_0 > 0$, the in-phase fixed point, $\Delta = 0$, is always stable while the opposite-phase configuration, $\Delta = \pi$, is unstable. Following the analytical solution given in Eq. (12.19), we see that for all initial conditions the phase difference decays exponentially to zero. Active synchronisation thus leads systematically to swimmers locked in the in-phase configuration, and therefore in their minimum of energy dissipation.

Note that the dependence of the typical synchronisation speed, U_0, on the parameters of the swimmers can be recast using dimensionless numbers as

$$\frac{U_0}{V} \sim \left(\frac{f_0}{Bk^3}\right)^2 \frac{kh_0 \text{Sp}^3}{(kh_0)^6 + \text{Sp}^6}, \tag{12.133}$$

where we ignored all numerical prefactors in Eq. (12.131), used that $V = \omega/k$ and introduced the two-dimensional sperm number, analogous to the three-dimensional version from Chapter 7, as

$$\text{Sp}^3 = \frac{\mu\omega}{Bk^3}. \tag{12.134}$$

The synchronisation dynamics thus depends on the value of Sp, on the dimensionless separation distance between the swimmers (kh_0) and on the dimensionless magnitude of the active force ($f_0/(Bk^3)$), with the expected quadratic dependence on the force. No synchronisation occurs in the rigid limit ($\text{Sp} = 0$) and U_0 vanishes when the swimmers are far from each other, $kh_0 \gg 1$.

Physical Mechanism of Synchronisation

The physical mechanism behind the active synchronisation can be understood by realising that the difference between the swimming speeds of the two sheets arises from the modulation of their waving amplitudes resulting from hydrodynamic interactions. This can be seen by inspecting the solution in Eq. (12.128). When the two swimmers are exactly in phase ($\Delta = 0$, resulting in $\gamma_- = 0$) or in opposite phase ($\Delta = \pi$, leading to $\gamma_+ = 0$) the two waving amplitudes are identical, and the swimming speeds are equal (Fig. 12.8(a)). For any other value of phase, the waving amplitudes of the two swimmers differ, so the quadratic integrals in Eq. (12.120) do not cancel out and their relative phase difference evolves in time.

The fact that the in-phase configuration is the stable phase-locked state can be interpreted physically by computing the waving amplitude of each swimmer for a small deviation from the stable point $\Delta = 0$. Using Eq. (12.128) and assuming $|\Delta|$ to be small, we obtain the wave amplitudes as

$$|\eta_1|^2 = \left(\frac{f_0}{Bk^4}\right)^2 (1 + \Delta\Gamma), \tag{12.135a}$$

$$|\eta_2|^2 = \left(\frac{f_0}{Bk^4}\right)^2 (1 - \Delta\Gamma), \tag{12.135b}$$

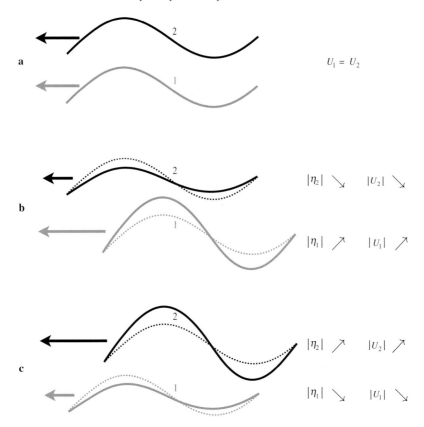

Figure 12.8 Physical mechanism for in-phase synchronisation of active swimmers. (a) When in phase, the two swimmers have identical waving amplitudes
and swim at the same speed. (b) If the two swimmers from (a) are perturbed so
that $\Delta > 0$, the amplitude of wave 2 decreases so it swims more slowly and that
of wave 1 increases so it swims faster. Both effects tend to decrease the value of
Δ, and return the two swimmers to the in-phase configuration. (c) If instead the
perturbation of the phase difference is such that $\Delta < 0$, the changes in amplitude
are reversed so that wave 2 increases in amplitude and swims faster while wave
1 decreases its amplitude and slows down, and the magnitude of the phase difference decreases. In both cases, hydrodynamic forces bring the two swimmers back
to the stable in-phase configuration.

where the dimensionless quantity Γ is given by

$$\Gamma = \frac{\Lambda}{1 + \Lambda^2}, \quad \Lambda = 24 \frac{\mu V}{k^5 B h_0^3} = 24 \left(\frac{\mathrm{Sp}}{k h_0} \right)^3. \tag{12.136}$$

These results allow us to interpret the tendency to synchronise in phase. If swimmer 2 happens to swim in front of swimmer 1 ($\Delta > 0$, Fig. 12.8(b)), swimmer 1
experiences an increase in its waving amplitude and therefore swims faster, while

swimmer 2 sees its amplitude decreasing and thus swims more slowly. Both of these allow swimmer 1 to catch up with swimmer 2 and to decrease the phase difference Δ. Similarly, if swimmer 2 is behind swimmer 1 ($\Delta < 0$, Fig. 12.8(c)), the amplitude of swimmer 1 decreases so it slows down while that of swimmer 2 increases so it can swim faster and catch up with swimmer 1; here again, the magnitude of the phase difference decreases. To achieve active synchronisation, the compliance of the waveforms is therefore crucial, as it allows variations in the waving amplitudes and a modulation of the speed of each swimmer. This physical mechanism is similar to the one discussed in Section 12.1.4 for the synchronisation of anchored flagella on compliant paths.

Further Reading

Taylor's energy argument for the ubiquity of in-phase beating was derived in this chapter in two dimensions. The result remains correct for three-dimensional flagella periodically waving in identical planes and the minimum of dissipation always occurs for in-phase motion (Mettot and Lauga, 2011). When instead the flagella beat in different planes, the most favourable energetic situation may switch to the opposite-phase configuration, which can be rationalised using a model of two periodically oscillating spheres (see Exercises).

The mechanics of metachronal waves of cilia has received much attention in the literature. A mean-field approach was proposed by Blake (1972) in which the flow velocity induced by each cilium is set by the flow rate from Eq. (11.92), with a force density obtained from resistive-force theory using the velocity of a cilium relative to the fluid, leading to an integro-differential equation to solve for the mean flow. An active, elastic-filament model was put forward by Guirao and Joanny (2007), who showed theoretically that aligned ciliary arrays could spontaneously break symmetries and generate net flows. More detailed three-dimensional models have been derived for the active mechanical deformation of cilia and their hydrodynamic coupling, such as elastic filaments driven by load-dependent internal forcing (Gueron et al., 1997) or by a geometric clutch switching between effective and recovery strokes (Elgeti and Gompper, 2013). Numerical simulations show that not only do metachronal waves appear spontaneously, but they are also energetically advantageous and lead to transport of the fluid with minimum energetic cost (Gueron and Levit-Gurevich, 1999; Osterman and Vilfan, 2011).

Although this chapter was focused on the role played by fluid dynamics, several non-hydrodynamic synchronisation mechanisms have also been proposed for cells and their appendages, with ongoing debates regarding their relative importance in experiments. For example, the internal mechanical coupling of the basal bodies inside *Chlamydomonas* cells plays a role in the synchronisation of their two flagella (Quaranta et al., 2015; Wan and Goldstein, 2016). For the same organism, the coupling between the flagella via the time-dependent motion of the cell to which they are anchored can lead to synchronisation without the need for hydrodynamic interactions (Bennett and Golestanian, 2013; Geyer et al., 2013)

A topic not addressed in this book but where synchronisation plays an important role is the interaction of bacterial flagellar filaments for peritrichous bacteria. During swimming the flagellar filaments are bundled behind the cell body and rotate in synchrony. Cells can modify their swimming directions by changing the rotation directions of some of their filaments, leading to unbundling of the filament assembly, and quick reorientation of the cell (Turner et al., 2000). The repeated assembly ('bundling') and disassembly ('unbundling') of the flagellar bundles require synchronised rotation (Macnab, 1977) and have been at the centre of many studies. Both experiments (Kim et al., 2003) and numerical simulations (Flores et al., 2005; Reigh et al., 2012, 2013) demonstrate that bundling and unbundling are physical processes driven by hydrodynamic interactions, which can be facilitated by the coupling to the cell body (Powers, 2002) and captured by long-wavelength models (Man et al., 2017). Without some form of flexibility, either from the hook or the flagellar filaments, no synchronisation is possible between rotating helical filaments (Kim and Powers, 2004; Reichert and Stark, 2005).

Beyond the biological realm, artificial systems have also been proposed to study synchronisation driven by hydrodynamic interactions. Spherical colloids, made to oscillate between two potential wells according to a 'clutch' mechanism where potentials are switched on or off according to the location of the colloids, perfectly synchronise in opposite phase solely due to hydrodynamic interactions (Kotar et al., 2010; see Exercises). Similarly, asymmetric rotors rotated by identical external torques display in-phase synchronisation induced by hydrodynamic interactions (Di Leonardo et al., 2012). Related work includes the generation of waves in driven colloids (rowers) moving along straight lines and interacting hydrodynamically (Lagomarsino et al., 2002, 2003; Wollin and Stark, 2011), while rotated paddles have been used to model interactions between rotated flagellar filaments (Qian et al., 2009).

Exercises

1. The synchronisation of two interacting spheres on compliant circular paths was derived in Section 12.1.1 in the presence of a rigid surface. Show that, in an unbounded fluid, the same setup also leads to Adler's equation for the phase difference Δ,

$$\dot{\Delta} = -\beta \tau \omega^2 \sin \Delta,$$

with a coupling strength, β, decaying as $1/\ell$ where ℓ is the distance between the spheres.

2. The dynamics of two interacting spheres rotating with different intrinsic frequencies, ω_1 and ω_2, on compliant circular paths was obtained as

$$\dot{\phi}_1 = \omega_1 + \beta\tau\omega_1\omega_2\cos\phi_1\sin\phi_2 + \beta\omega_2\sin\phi_1\sin\phi_2,$$
$$\dot{\phi}_2 = \omega_2 + \beta\tau\omega_1\omega_2\cos\phi_2\sin\phi_1 + \beta\omega_1\sin\phi_1\sin\phi_2.$$

Derive the dynamics of the phase difference at leading order in $\beta\tau\omega_1$, $\beta\tau\omega_2$ and $\omega_1 - \omega_2$. Show that a stable phase-locked state where $\phi_1(t) - \phi_2(t)$ is constant exists, provided the difference between the two intrinsic frequencies is sufficiently small.

3. The synchronisation of two interacting spheres using the phase-dependent forcing mechanism was addressed in Section 12.1.6, with the time-averaged growth rate of the phase difference, $\langle\hat{\sigma}\rangle$, obtained as

$$\langle\hat{\sigma}\rangle = \frac{\beta}{T_0}\int_0^{2\pi}\frac{F'(\phi)}{F(\phi)}\cos 2\phi\, d\phi,$$

at first order in β. In the case where the spheres are driven by the phase-dependent force $F/F_0 = 1 + \sum A_n\sin(n\phi + \delta_n)$ with $\cos\delta_2 = 0$, compute the leading-order value of $\langle\hat{\sigma}\rangle$ in powers of A_n.

4. Two identical spheres of radius a oscillate periodically along straight paths tilted by angles β_1 and β_2 relative to the x direction. The instantaneous velocities of the spheres are given by $\mathbf{U}_1 = U_1(t)(\cos\beta_1\mathbf{e}_x + \sin\beta_1\mathbf{e}_y)$ and $\mathbf{U}_2 = U_2(t)(\cos\beta_2\mathbf{e}_x + \sin\beta_2\mathbf{e}_y)$, with speeds U_1 and U_2 oscillating periodically as $U_1(t) = U_0\cos(\omega t + \phi)$ and $U_2(t) = U_0\cos(\omega t - \phi)$. The centres of the two paths are separated by $d\mathbf{e}_x$ with $d \gg a$. Calculate the rate of work of the moving spheres on the fluid at first order in $a/d \ll 1$. Show that the phase difference, 2ϕ, leading to minimum overall energy expenditure is either π or 0 depending on the relative orientation of the two planes of beating. Show that, when the beating planes are perpendicular to each other, the total dissipation is independent of the phase and that, for a given orientation β_1, there is an optimal orientation β_2 that leads to minimum energy dissipation.

5. Taylor's energy argument for the synchronisation of two waving sheets was reproduced in Section 12.2.2 in the case of transverse waving motion. Carry out the small-amplitude calculation in the case of longitudinal waving motion, i.e. when the material points on the two sheets are described by $x_1 = \epsilon a\cos(x-t)$ and $x_2 = d + \epsilon a\cos(x - t + \Delta)$. Show that in this case the rate of work of the

sheets on the fluid is maximum in the case where they are in phase and minimum in the opposite-phase configuration.

6. Carry out the calculation from Exercise 5 for the rate of work of two sheets undergoing solely longitudinal waving motion in the long-wavelength (lubrication) limit. Show that your results agree with those of Exercise 5.

7. The study by Kotar et al. (2010) shows experimentally and theoretically that colloidal beads driven by a geometrical clutch mechanism and interacting hydrodynamically synchronise to an anti-phase configuration. Consider two spherical particles, each driven in oscillations along a line by periodically switching the position of a harmonic potential according to the position of the particle. For the first particle, the harmonic potential oscillates between being centred at $x = -A$ and at $x = A$, and similarly for the second particle at a distance $d > 2A$ away. In both cases the switch is activated when the sphere approaches to within a fixed distance from the minimum of the potential. Use symmetry arguments to explain why both the in-phase and opposite-phase configurations are fixed points of the dynamics. By using a sketch and arguments similar to those illustrated in Fig. 12.2, explain why hydrodynamic interactions amplify a small perturbation to the in-phase case while they decrease one in the opposite-phase configuration, thereby explaining heuristically the stability of the opposite-phase fixed point.

13

Diffusion and Noisy Swimming

After having examined the deterministic motility of swimming cells, we now turn to their interactions with a fluctuating environment. We consider the motion of microorganisms subject to thermal noise, a situation relevant to the locomotion of small bacteria. This allows us to introduce two modelling approaches, namely a discrete framework (along with ensemble averaging) and a continuum probabilistic framework, both of which we adapt for the modelling of collective dynamics in Chapter 14. We first review Brownian motion in translation and rotation for a passive particle, introduce all the relevant timescales for its dynamics, show how the statistical properties of its trajectory can be captured with both discrete and continuum frameworks, and apply these concepts to the diffusion of cells. By adding a swimming velocity to the particle, we next show how thermal noise affects the motion of swimming microorganisms and in turn how the noisy run-and-tumble motion of bacteria can be described as an effective diffusive process.

13.1 Brownian Motion

In this first section, we focus on the impact of thermal noise on passive particles. Thermal noise leads to diffusion in both translation and rotation, which we review in the classical case of a spherical particle. Beyond thermal noise, biological cells are of course subject to much non-thermal, active noise. For example, bacteria experience fluctuations of the motor rotation due to changes in the flux of protons. Similarly, the internal flagellar forcing for swimming eukaryotes fluctuates with ATP concentration and the activity of molecular motors. The modelling of such non-thermal fluctuations is beyond the scope of this book, but the results below may be extended to such cases by introducing the concept of effective temperature.

Brownian motion can be tackled mathematically via two modelling approaches. First we introduce a discrete model where particles are subject to noisy forces (Langevin framework), giving rise to trajectories as individual realisations, which can then be averaged to obtain statistical properties. We discover in particular that the dynamics of an individual particle is characterised by four different timescales, and that fluid mechanics plays an important role since viscous friction governs ultimately the ability to diffuse (Stokes–Einstein relationship). We next turn to the alternative, a continuum probabilistic model (Fokker–Planck and Smoluchowski frameworks) allowing us to compute directly the probability distribution function in position and orientation of the particle. Although these are different approaches, they lead to identical statistical descriptions of trajectories, and can both be adapted to model collective locomotion (see Chapter 14). Throughout this section, the derivations we present are standard and we refer to classical work for further insight (Chandrasekhar, 1943; Batchelor, 1976; Ermak and McCammon, 1978; Brady and Bossis, 1988; Mazo, 2002; Romanczuk et al., 2012).

13.1.1 Discrete Framework: Langevin Equations

We consider the diffusion of a passive rigid sphere of radius a. In the discrete Langevin approach the mean motion of the particles is treated deterministically and is subject to Newton's laws, while the fluctuations are tackled stochastically using random forces with known statistical properties. If we denote the position of the particle by $\mathbf{X}(t)$, Newton's second law applied to the particle of mass m is

$$m\frac{\mathrm{d}^2\mathbf{X}}{\mathrm{d}t^2} = \mathbf{F}_H + \mathbf{F}_B, \tag{13.1}$$

where $\mathbf{F}_H$ is the deterministic hydrodynamic friction force and $\mathbf{F}_B$ the stochastic Brownian force arising from molecular collisions. This thermal force fluctuates with a very high molecular frequency. Its correlation timescale, t_1, is set by the speed of sound and the size of molecules, and for water is approximately $t_1 = O(10^{-13}\,\mathrm{s})$. On timescales much larger than t_1, it is therefore appropriate to describe $\mathbf{F}_B$ as a zero-mean stationary process:

$$\langle \mathbf{F}_B \rangle = \mathbf{0}, \tag{13.2}$$

where we use brackets to denote ensemble averages over many realisations. In Eq. (13.1), the hydrodynamic force is given by Stokes drag, $\mathbf{F}_H = -\mathbf{U}/M$, where M is the mobility of the sphere, $M = (6\pi\mu a)^{-1}$. Since the velocity is the time rate of change of the position, we may rewrite Eq. (13.1) as

$$m\left(\frac{\mathrm{d}^2\mathbf{X}}{\mathrm{d}t^2} + \frac{1}{mM}\frac{\mathrm{d}\mathbf{X}}{\mathrm{d}t}\right) = \mathbf{F}_B. \tag{13.3}$$

This makes apparent a second timescale, $t_2 = Mm$, which is the characteristic time for the particle to adapt inertially to any change in the forcing (i.e. the timescale for relaxation of the particle momentum). For a density-matched sphere of radius 1 μm in water this is $t_2 = m/(6\pi\mu a) = 2\rho a^2/(9\mu) = O(10^{-7}$ s), which is orders of magnitude larger than t_1.

Since the noise force has zero mean, taking the ensemble average of Eq. (13.3) for a sphere with zero initial position and velocity leads to

$$\langle \mathbf{X} \rangle = \mathbf{0}. \tag{13.4}$$

Although the particle is moving, it is equally likely to go in any direction and therefore it has zero mean position. The second moment of the particle position is next obtained by dotting Eq. (13.3) by $\mathbf{X}$ to get

$$m \left(\mathbf{X} \cdot \frac{d^2\mathbf{X}}{dt^2} + \frac{1}{mM} \mathbf{X} \cdot \frac{d\mathbf{X}}{dt} \right) = \mathbf{F}_B \cdot \mathbf{X}. \tag{13.5}$$

We then recognise that

$$\mathbf{X} \cdot \frac{d^2\mathbf{X}}{dt^2} = \frac{d}{dt} \left(\mathbf{X} \cdot \frac{d\mathbf{X}}{dt} \right) - \mathbf{U} \cdot \mathbf{U}, \tag{13.6}$$

so that Eq. (13.5) simplifies to

$$\frac{1}{2} m \left[\left(\frac{1}{mM} + \frac{d}{dt} \right) \frac{d(\mathbf{X} \cdot \mathbf{X})}{dt} \right] = \mathbf{F}_B \cdot \mathbf{X} + m\mathbf{U} \cdot \mathbf{U}. \tag{13.7}$$

We next evaluate the ensemble averages of both sides of Eq. (13.7). Since the fluctuating force and the position of the particle vary on very different timescales, we follow Batchelor (1976) and neglect $\langle \mathbf{F}_B \cdot \mathbf{X} \rangle$ compared to the other terms in Eq. (13.7) (see Exercises for a justification). Assuming that the quantity $\langle \mathbf{U} \cdot \mathbf{U} \rangle$ does not depend on time, multiplying Eq. (13.7) by the integrating factor $e^{t/(Mm)}$, we obtain an exact derivative,

$$\frac{d}{dt} \left(e^{t/(Mm)} \frac{d\langle \mathbf{X} \cdot \mathbf{X} \rangle}{dt} \right) = 2\langle \mathbf{U} \cdot \mathbf{U} \rangle e^{t/(Mm)}, \tag{13.8}$$

which integrates to

$$\frac{d\langle \mathbf{X} \cdot \mathbf{X} \rangle}{dt} = 2Mm\langle \mathbf{U} \cdot \mathbf{U} \rangle \left(1 - e^{-t/(Mm)} \right). \tag{13.9}$$

Although the fluctuating force does not appear explicitly in Eq. (13.9), it is present indirectly through $\langle \mathbf{U} \cdot \mathbf{U} \rangle$, whose value may be found by invoking the principle of equipartition of energy from statistical physics (Landau and Lifshitz, 1980). At thermal equilibrium all degrees of freedom have average energy $\frac{1}{2}k_BT$, where k_B is Boltzmann's constant ($k_B \approx 1.38 \times 10^{-23}$ J/K) and T the absolute temperature.

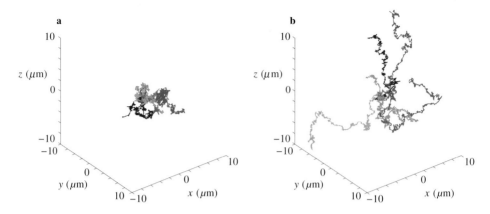

Figure 13.1 Passive vs. active diffusion. (a) Brownian motion of a passive spherical particle of radius 1 μm in water at $T = 300$ K during a time interval of 30 s; five different trajectories are shown. (b) Same setup but now the particle is active and swims with constant speed $U = 1$ μm/s.

Due to its three degrees of freedom in translation, the mean kinetic energy of the particle is therefore given by

$$\frac{1}{2}m\langle \mathbf{U} \cdot \mathbf{U}\rangle = \frac{3}{2}k_BT. \qquad (13.10)$$

At long times, $t \gg Mm$, Eq. (13.9) leads thus to

$$\frac{d\langle \mathbf{X} \cdot \mathbf{X}\rangle}{dt} = 6Mk_BT, \qquad (13.11)$$

and the mean square position of the particle varies as

$$\langle \mathbf{X} \cdot \mathbf{X}\rangle = 6Mk_BTt. \qquad (13.12)$$

A scaling for the mean square displacement $r^2 \sim t$ is characteristic of a diffusive behaviour. The result in Eq. (13.12) leads to the appearance of a third timescale, $t_3 = a^2/(Mk_BT)$, the characteristic time for the particle to be transported diffusively by thermal fluctuations on a length scale approximately equal to its own size. For a 1 μm sphere in water at room temperature ($T = 300$ K as in all examples below), we have $Mk_BT \approx 2 \times 10^{-13}$ m^2/s leading to the timescale $t_3 = O(5$ s$)$. Five different trajectories illustrating the stochastic nature of the particle motion are shown for a 1 μm sphere in Fig. 13.1(a) over a time interval of 30 s.

13.1.2 Brownian Dynamics Limit

As a simplification to the Langevin approach, the so-called Brownian dynamics limit describes the motion of the particle on timescales larger than t_2, thus allowing us to neglect inertial adjustments in Eq. (13.3). This limit is appropriate for a large particle whose momentum reaches equilibrium before the position of the particle has changed appreciably. Using our notation, this means that it is a valid approach provided that $t_3 \gg t_2$, a limit satisfied by micron-sized particles and cells. If we assume that the relaxation of the momentum is instantaneous, Eq. (13.3) simplifies to

$$\frac{d\mathbf{X}}{dt} = M\mathbf{F}_B. \tag{13.13}$$

Since $t_3 \gg t_1$, the zero-mean Brownian force, $\mathbf{F}_B$, can be modelled as delta-correlated in time, and an important result gives us access to the magnitude of this correlation (i.e. the rms value of the random force). Writing

$$\langle F_i^B(t) F_j^B(t') \rangle = \frac{\gamma}{3} \delta_{ij} \delta(t - t'), \tag{13.14}$$

we proceed to find the value of γ consistent with the results obtained using the Langevin framework. For a particle initially at the origin, we first integrate Eq. (13.13) as

$$\mathbf{X}(t) = M \int_0^t \mathbf{F}_B(t') \, dt', \tag{13.15}$$

so that

$$\langle \mathbf{X} \cdot \mathbf{X} \rangle = M^2 \int_0^t \int_0^t \langle \mathbf{F}_B(t') \cdot \mathbf{F}_B(t'') \rangle \, dt'' \, dt'. \tag{13.16}$$

Using Eq. (13.14) we obtain

$$\langle \mathbf{X} \cdot \mathbf{X} \rangle = M^2 \int_0^t \int_0^t \gamma \delta(t' - t'') \, dt'' \, dt' = M^2 \int_0^t \gamma \, dt' = \gamma M^2 t. \tag{13.17}$$

For the result of Eq. (13.17) to be consistent with Eq. (13.11) we require

$$\gamma M^2 = 6Mk_B T, \tag{13.18}$$

and thus the magnitude of the correlation is given by

$$\gamma = \frac{6k_B T}{M}. \tag{13.19}$$

The result in Eq. (13.19) is called the fluctuation–dissipation theorem, linking the rms magnitude of the fluctuating force, dissipation in the fluid (through the mobility M) and the magnitude of the thermal energy. While we have derived the result above in the case of a sphere, a generalised tensorial version exists for more complex shapes (Brady and Bossis, 1988).

13.1.3 Probabilistic Framework

Instead of describing individual realisations of the particle trajectory and using ensemble averaging to obtain their statistical properties, an alternative approach consists in computing directly the probability distribution function for the position of the particle.

Fokker–Planck equation

For the purpose of illustration, let us consider the simple situation of a spherical particle diffusing along the x direction only. In that case, the one-dimensional version of the Brownian dynamics in Eq. (13.13) can be written using notation from stochastic calculus as

$$dx = \alpha dW, \quad \alpha = (2Mk_BT)^{1/2}, \tag{13.20}$$

where the noise W is a continuous stochastic function called a Wiener process, whose infinitesimal increment obeys

$$\langle dW \rangle = 0, \quad (dW)^2 = dt. \tag{13.21}$$

Importantly, the second property in Eq. (13.21) is true without taking ensemble averages, a result from stochastic calculus known as Itô's lemma (Jacobs, 2010).

To describe the stochastic dynamics of the particle in a probabilistic fashion, we introduce the probability distribution function (pdf) $\Psi(x,t)$ to find the particle at position x at time t. As a probability distribution function it has the properties that it is non-negative everywhere and that its total integral is normalised to 1 (Graham, 2018). We next define an arbitrary test function, $f[x(t)]$, whose ensemble average, denoted by $\langle \ldots \rangle$, is computed as an integral over all possible states of x weighted by its pdf Ψ, i.e.

$$\langle f \rangle(t) = \int \Psi(x,t) f(x) \, dx. \tag{13.22}$$

Since f is a function of x, we use a Taylor expansion and the chain rule combined with Eq. (13.20) to write its increment, df, as

$$df = f[x + dx] - f[x] = f'[x]\alpha dW + \frac{1}{2}f''[x]\alpha^2 dW^2 + O(dW^3). \tag{13.23}$$

Using Eq. (13.21) we may replace $(dW)^2$ by dt and obtain

$$df = f'[x]\alpha dW + \frac{1}{2}f''[x]\alpha^2 dt + O(dWdt). \tag{13.24}$$

Taking averages allows us to drop the terms involving dW and we arrive at

$$d\langle f \rangle = \frac{1}{2}\langle f''[x] \rangle \alpha^2 dt + O(dt^2), \tag{13.25}$$

which leads in the limit $dt \to 0$ to

$$\frac{d\langle f \rangle}{dt} = \frac{\alpha^2}{2}\langle f'' \rangle. \tag{13.26}$$

In order to derive the equation satisfied by the probability distribution function, we rewrite the ensemble averages in Eq. (13.26) as integrals involving Ψ. Using the definition from Eq. (13.22), the equality in Eq. (13.26) becomes, after two integrations by parts,

$$\int \frac{\partial \Psi}{\partial t} f(x)\, dx = \frac{\alpha^2}{2} \int \frac{\partial^2 \Psi}{\partial x^2} f(x)\, dx. \tag{13.27}$$

Since the test function f in Eq. (13.27) is arbitrary, we obtain necessarily

$$\frac{\partial \Psi}{\partial t} = \frac{\alpha^2}{2}\frac{\partial^2 \Psi}{\partial x^2}, \tag{13.28}$$

which can be rewritten as

$$\frac{\partial \Psi}{\partial t} = D\frac{\partial^2 \Psi}{\partial x^2}, \quad D = Mk_B T. \tag{13.29}$$

The equation for Ψ is termed the Fokker–Planck equation and, in this particular example, it is here a simple diffusion equation. Advection can also be included in Eq. (13.20) using a similar formalism; see Exercises.

Smoluchowski Conservation Equation

We obtained the Fokker–Planck equation, Eq. (13.29), using a stochastic approach. It can be recast in the framework of statistical physics by rewriting it as a conservation equation,

$$\frac{\partial \Psi}{\partial t} + \frac{\partial}{\partial x}(u\Psi) = 0, \tag{13.30}$$

where the speed $u = MF$ is proportional to a force F written formally as

$$F = -\frac{\partial}{\partial x}(k_B T \ln \Psi). \tag{13.31}$$

The force F has therefore the interpretation of a thermodynamic force, obtained as the gradient of a free energy, $\mathcal{U} \equiv k_B T \ln \Psi$, akin to a chemical potential in thermodynamics.

Written this way, the equation for Ψ is often known as the Smoluchowski equation, and it is straightforward to generalise it to three dimensions. The conservation equation becomes in that case

$$\frac{\partial \Psi}{\partial t} + \nabla \cdot (\mathbf{u}\Psi) = 0, \quad \mathbf{u} = M\mathbf{F}, \tag{13.32}$$

where $\mathbf{u}(\mathbf{r}, t)$ is the velocity of the particle at position $\mathbf{r}$ under the thermodynamic force, $\mathbf{F}$, given by the gradient of the free energy, $\mathcal{U}$, as

$$\mathbf{F} = -\nabla \mathcal{U}, \quad \mathcal{U} = k_B T \ln \Psi. \tag{13.33}$$

Substituting Eq. (13.33) into Eq. (13.32) leads to the three-dimensional diffusion equation

$$\frac{\partial \Psi}{\partial t} = D \nabla^2 \Psi, \tag{13.34}$$

with a diffusivity D given by the Stokes–Einstein relationship,

$$D = M k_B T, \tag{13.35}$$

a classical result that can be generalised for rigid particles of arbitrary shape as a tensorial identity (Brady and Bossis, 1988).

Diffusive Behaviour

The result in Eq. (13.35) may be recovered alternatively by examining the statistical properties of the particle trajectory. For a particle initially at $\mathbf{X} = \mathbf{0}$, the initial condition to Eq. (13.34) is a delta function, so the solution at $t \geq 0$ is given by the classical Gaussian profile

$$\Psi(\mathbf{r}, t) = \frac{1}{(4\pi D t)^{3/2}} \exp\left(-\frac{r^2}{4Dt}\right), \quad r = |\mathbf{r}|. \tag{13.36}$$

Since the solution in Eq. (13.36) is spherically symmetric, the mean position is zero, $\langle \mathbf{X} \rangle = \mathbf{0}$. The mean square value, $\langle \mathbf{X} \cdot \mathbf{X} \rangle = \langle r^2 \rangle$, may be obtained using spherical coordinates by substituting the result of Eq. (13.36) as

$$\langle \mathbf{X} \cdot \mathbf{X} \rangle = \iiint r^2 \Psi(\mathbf{r}, t) \, dV = 4\pi \int_0^\infty \frac{r^4}{(4\pi D t)^{3/2}} \exp\left(-\frac{r^2}{4Dt}\right) dr. \tag{13.37}$$

Using the change of variable $u = r/(4Dt)^{1/2}$, this leads to diffusion as

$$\langle \mathbf{X} \cdot \mathbf{X} \rangle = 6Dt. \tag{13.38}$$

Comparing Eq. (13.12) with Eq. (13.38) we deduce $D = M k_B T$, which is the Stokes–Einstein relationship from Eq. (13.35).

13.1.4 Rotational Diffusion

The previous sections focused on the random translational motion of the particle due to fluctuating thermal forces. Molecular collisions in the fluid also lead to fluctuating thermal torques and therefore the orientation of the particle is subject to a similar stochastic process.

Diffusion Equation on a Sphere

A probabilistic approach analogous to the one introduced in Section 13.1.3 can be introduced in the case of rotational diffusion. The probability distribution function, denoted Ψ, now measures the distribution of orientation of the particle, and the mobility is equal to the viscous rotational mobility of the spherical particle, $M_R = (8\pi\mu a^3)^{-1}$ (see Eq. (9.21)). The Stokes–Einstein relationship still applies and is written in this case as

$$D_R = M_R k_B T, \tag{13.39}$$

where the rotational diffusivity, D_R, now has dimensions of a rate (strictly speaking, rad^2/s; see below).

In the context of locomotion, an important consequence of rotational diffusion is the loss of any preferred orientation, and in particular the swimming direction of a microorganism (see Section 13.2.2). Consider a spherical particle with a body-fixed unit vector $\mathbf{p}$; this is taken later as the vector along which the cell self-propels. Describing the two degrees of freedom for the orientation of $\mathbf{p}$ using spherical coordinates with $0 \leqslant \theta \leqslant \pi$ the polar angle and $0 \leqslant \phi < 2\pi$ the azimuthal angle, the angular probability distribution function $\Psi(\theta, \phi, t)$ satisfies the diffusion equation analogous to Eq. (13.34) but now on the unit sphere,

$$\frac{\partial \Psi}{\partial t} = D_R \nabla_S^2 \Psi, \quad \nabla_S^2 \cdots \equiv \frac{1}{\sin\theta} \frac{\partial}{\partial\theta} \left(\sin\theta \frac{\partial}{\partial\theta} \cdots \right) + \frac{1}{\sin^2\theta} \frac{\partial^2}{\partial\phi^2} (\cdots). \tag{13.40}$$

The general solution to Eq. (13.40) may be expressed using spherical harmonics (Yosida, 1949; Riley et al., 1999). At long times the probability distribution function is uniform and equal to $\Psi = 1/(4\pi)$ (since the integral of Ψ on the unit sphere is 1). At short times, the diffusive scaling in Eq. (13.38) may be obtained by computing $\langle \theta^2 \rangle(t)$ using the following general identity.

Angular Averages

Consider a smooth function $g(\theta)$ of the polar angle. Its expected value is given by

$$\langle g(\theta) \rangle(t) = \iint g(\theta) \Psi(\theta, \phi, t) \sin\theta \, d\theta \, d\phi, \tag{13.41}$$

and its rate of change is

$$\frac{d}{dt} \langle g(\theta) \rangle(t) = \iint g(\theta) \frac{\partial \Psi}{\partial t} \sin\theta \, d\theta \, d\phi. \tag{13.42}$$

Starting at $t = 0$ with the particle oriented along $\theta = 0$, the probability distribution function is expected by symmetry to be independent of the azimuthal angle ϕ, and using Eq. (13.40) we obtain

$$\frac{d}{dt}\langle g(\theta)\rangle(t) = D_R \iint g(\theta)\frac{\partial}{\partial\theta}\left(\sin\theta\frac{\partial\Psi}{\partial\theta}\right) d\theta\, d\phi. \qquad (13.43)$$

Rewriting the integrand as

$$g(\theta)\frac{\partial}{\partial\theta}\left(\sin\theta\frac{\partial\Psi}{\partial\theta}\right) = \frac{\partial}{\partial\theta}\left(g\sin\theta\frac{\partial\Psi}{\partial\theta} - \Psi\sin\theta\frac{\partial g}{\partial\theta}\right) + \frac{\partial}{\partial\theta}\left(\sin\theta\frac{\partial g}{\partial\theta}\right)\Psi,$$
$$(13.44)$$

the exact derivatives integrate to zero since $\sin\theta = 0$ at $\theta = 0, \pi$, leading to

$$\frac{d}{dt}\langle g(\theta)\rangle(t) = D_R \iint \frac{\partial}{\partial\theta}\left(\sin\theta\frac{\partial g}{\partial\theta}\right)\Psi\, d\theta\, d\phi. \qquad (13.45)$$

Mean Square Angle

Applying the result of Eq. (13.45) to the function $g(\theta) = \theta^2$ leads to

$$\frac{d}{dt}\langle\theta^2\rangle(t) = 2D_R \iint (1 + \theta\cot\theta)\Psi\sin\theta\, d\theta\, d\phi = 2D_R(1 + \langle\theta\cot\theta\rangle). \qquad (13.46)$$

At early times, the polar angle is small and thus $\theta\cot\theta \approx 1$, so we get

$$\langle\theta^2\rangle \approx 4D_R t. \qquad (13.47)$$

We therefore obtain a diffusive scaling for the polar angle with a factor of 4 instead of the factor of 6 from Eq. (13.38) because the orientation of the unit vector $\mathbf{p}$ is characterised by two (instead of three) degrees of freedom. Note that the equality in Eq. (13.47) makes it clear why the dimensions of D_R are rad^2/s.

Inspecting Eq. (13.47) reveals a fourth timescale, $t_4 = D_R^{-1}$, as the relevant timescale for an order-one reorientation of the particle due to thermal fluctuations. For a 1 μm sphere in water at room temperature, we obtain $D_R^{-1} = O(5\text{ s})$. Comparing the two timescales t_3 and t_4 for the spherical particle, we find

$$\frac{t_3}{t_4} = O\left(\frac{a^2/D}{D_R^{-1}}\right) = \frac{a^2 M_R k_B T}{M k_B T} = \frac{a^2 M_R}{M} = O(1), \qquad (13.48)$$

so a rigid sphere always diffuses a distance comparable to its own size in the same time that it reorients appreciably. The result is very different for an elongated particle, as we see in Section 13.2.1 for the diffusion of bacteria.

Exponential Loss of Orientation

A final consequence of rotational diffusion useful in the context of locomotion is the exponential loss of orientation. This is best seen by computing the time variation of the mean cosine of the polar angle, $\langle\cos\theta\rangle(t) = \langle\mathbf{p}(t)\cdot\mathbf{p}(0)\rangle$, using the formula from Eq. (13.45) for the function $g(\theta) = \cos\theta$. We get

$$\frac{d}{dt}\langle \cos\theta\rangle = -2D_R \iint \Psi \sin\theta \cos\theta \, d\theta \, d\phi. \qquad (13.49)$$

Recognising that the right-hand side of Eq. (13.49) is proportional to the mean cosine, we thus obtain a differential equation,

$$\left(\frac{d}{dt} + 2D_R\right)\langle \cos\theta\rangle = 0. \qquad (13.50)$$

The particle is oriented initially along $\theta = 0$ so $\langle \cos\theta\rangle(0) = 1$, leading to

$$\langle \cos\theta\rangle(t) = e^{-2D_R t}, \qquad (13.51)$$

and the particle losses its orientation exponentially with rate $2D_R$. Note that a Taylor expansion of Eq. (13.51) at short times using $\cos\theta \approx 1 - \theta^2/2$ and $e^{-2D_R t} \approx 1 - 2D_R t$ allows us to recover the identity in Eq. (13.47).

13.2 Cells vs. Noise

Having characterised the thermal diffusion of a spherical particle, we now focus on the impact of noise on cells. We first consider the Brownian motion of the smallest organisms, bacteria, in the absence of motility. We next address the consequences of noise for locomotion and show how Brownian reorientation leads to long-time diffusive behaviour for self-propelling cells.

13.2.1 Brownian Motion of Bacteria

Translation

To address the thermal motion of a bacterium, we model it as a rod of length L and width $2a$ (with $a \ll L$). The cell has thus two different directions, denoted by $\parallel$ along the length and $\perp$ perpendicular to it, which are associated with drag coefficients $c_\parallel$ and $c_\perp$ (see Chapter 6) and mobilities in translation $M_{\parallel,\perp} = (c_{\parallel,\perp}L)^{-1}$.

The molecular timescale characterising fluctuations in the fluid, t_1, is unchanged from the case of a spherical particle. Since the drag coefficients $c_\parallel$ and $c_\perp$ are within a factor of 2 of each other, the momentum-relaxation timescales for the bacterium, $t_{2;\parallel}$ and $t_{2;\perp}$, where $t_{2;\parallel,\perp} = M_{\parallel,\perp}m$, are of the same order of magnitude in both directions. Geometrically we have

$$t_{2;\parallel,\perp} = \frac{m}{c_{\parallel,\perp}L} \approx \frac{\rho(2a)^2 L}{c_{\parallel,\perp}L} = \frac{4\rho a^2}{c_{\parallel,\perp}}. \qquad (13.52)$$

We consider the case of E. coli, a typical bacterium of length $L \approx 8\ \mu m$ and width $2a \approx 1\ \mu m$. For such a cell modelled as a rod of aspect ratio $L/(2a) = 8$, we get from Eq. (6.32) that $c_\perp \approx 4 \times 10^{-3}$ Pa s, which leads to $t_2 = O(10^{-7}$ s), much larger than t_1.

The thermal diffusivities of the cell, $D_\parallel$ and $D_\perp$, where $D_{\parallel,\perp} = M_{\parallel,\perp} k_B T$, also have (to within a factor of 2) the same magnitude in both directions of the organism. Putting numbers in, we have $D_{\parallel,\perp} = M_{\parallel,\perp} k_B T = k_B T / (c_{\parallel,\perp} L) = O(0.1\ \mu\mathrm{m}^2/\mathrm{s})$. This agrees with the theoretical estimate of Berg (1993) and with experimental measurements for the passive diffusion of *E. coli* near surfaces giving $D \approx 0.1\ \mu\mathrm{m}^2/\mathrm{s}$ (Tavaddod et al., 2011). Note that these numbers are robust to the exact shape and size of the cell. For example, for the bacterium *Leptospira*, whose whole body has a helical shape and is about 9 μm long, measurements and computations have reported $D \approx 0.2\ \mu\mathrm{m}^2/\mathrm{s}$ (Butenko et al., 2012; Koens and Lauga, 2014).

Associated with the cell's Brownian motion, we may evaluate the typical timescales, t_3, for the cell to diffuse a distance comparable to its own size. Since the cell is elongated and t_3 scales as the square of the relevant length scale, the timescales $t_{3;\perp}$ and $t_{3;\parallel}$ along the two directions of the cell have very different magnitudes. In the direction perpendicular to the cell we obtain $t_{3;\perp} = (2a)^2 / D_\perp \approx 8$ s, while along its elongated direction we have $t_{3;\parallel} = L^2 / D_\parallel \approx 4$ min.

Rotation

In order to characterise the rotational diffusivity of the cell, we need to calculate the value of its mobility in rotation, M_R. This is easily done using resistive-force theory. Assuming that the elongated cell rotates with angular velocity Ω about an axis perpendicular to its long axis and passing through its centre, the velocity of each point on the cell is $u_\perp = \Omega r$ where r is the distance to the centre. This leads to a hydrodynamic force density $f_\perp = -c_\perp \Omega r$ and therefore a hydrodynamic torque

$$L = \int_{-L/2}^{L/2} r u_\perp \, dr = -\int_{-L/2}^{L/2} c_\perp \Omega r^2 \, dr = -\frac{\Omega}{M_R}, \quad M_R = \frac{12}{c_\perp L^3}. \tag{13.53}$$

Using these numbers, we obtain the theoretical estimate $D_R = M_R k_B T \approx 0.07\ \mathrm{rad}^2/\mathrm{s}$, consistent with the prediction of $D_R \approx 0.062\ \mathrm{rad}^2/\mathrm{s}$ from Berg (1993), and leading to the relevant timescale for thermal reorientation of the long axis of the cell as $t_4 \approx 15$ s.

These theoretical estimates agree with experimental values. Measurements with non-tumbling *E. coli* reported thermal diffusivities of $D_R \approx 0.057\ \mathrm{rad}^2/\mathrm{s}$ (Drescher et al., 2011), while the rotational diffusion coefficient of paralysed *E. coli* was measured to be $D_R \approx 0.03\ \mathrm{rad}^2/\mathrm{s}$ near surfaces (Tavaddod et al., 2011). Here again, the numbers are also robust to the detailed shape and size of the organisms. For example, the bacterium *Leptospira* has $D_R \approx 0.02\ \mathrm{rad}^2/\mathrm{s}$ (Butenko et al., 2012; Koens and Lauga, 2014).

13.2.2 Diffusion of Straight-Swimming Cells

In Section 13.1 we saw how to compute the first two moments for the trajectory of a passive particle subject to fluctuating thermal forces. We turn here to the impact of Brownian motion on swimming cells, and in particular bacteria. We demonstrate that the combination of rotational diffusion with self-propulsion allows cells to display long-time diffusion enhanced by orders of magnitude above that given by thermal forces alone. We focus first on the case where the bacterium has no intrinsic rotation and introduce rotation in the next section.

Model

As a simple model of a self-propelling bacterium we consider the same setup as in Chapter 10, where a spherical cell is assumed to swim at constant speed in a body-fixed frame, $\mathbf{U}_s = U\mathbf{p}$, where $\mathbf{p}$ is a unit vector. Using the Brownian dynamics framework from Section 13.1.2, we denote by $\mathbf{X}(t)$ the position of the cell. Its instantaneous velocity is given by the sum of swimming along the direction $\mathbf{p}$ and thermal forces, i.e.

$$\frac{d\mathbf{X}}{dt} = U\mathbf{p}(t) + M\mathbf{F}_B, \qquad (13.54)$$

where the noise term in Eq. (13.54) is the same as in Eq. (13.13). Due to rotational diffusion, the unit vector $\mathbf{p}$ has zero mean direction at long times, $\langle\mathbf{p}\rangle(t \to \infty) = \mathbf{0}$, and therefore an ensemble average of Eq. (13.54) leads to $\langle\mathbf{X}\rangle(t \to \infty) = \mathbf{0}$. Despite the fact that the cell is swimming, its thermal loss of orientation means that it is eventually going nowhere on average.

Mean Square Displacements

Similar to diffusion in orientation, the cell undergoes rms motion, however, and has a nonzero value of $\langle\mathbf{X}\cdot\mathbf{X}\rangle$. For a spherical cell, $\mathbf{F}_B$ and $\mathbf{p}$ are uncorrelated, and therefore the rms displacement of the organism is the sum of the passive diffusion in Eq. (13.38) and the motion obtained by ignoring the random force $\mathbf{F}_B$ in Eq. (13.54), which has a solution denoted by $\tilde{\mathbf{X}}$. That solution is found by integrating Eq. (13.54) once, leading to

$$\tilde{\mathbf{X}}(t) = U \int_0^t \mathbf{p}(t')\,dt', \qquad (13.55)$$

and a dot product with Eq. (13.54) leads to

$$\frac{1}{2}\frac{d(\tilde{\mathbf{X}}\cdot\tilde{\mathbf{X}})}{dt} = \frac{d\tilde{\mathbf{X}}}{dt}\cdot\tilde{\mathbf{X}} = U^2 \int_0^t \mathbf{p}(t')\cdot\mathbf{p}(t)\,dt'. \qquad (13.56)$$

In order to compute the ensemble average $\langle\tilde{\mathbf{X}}\cdot\tilde{\mathbf{X}}\rangle(t)$ we therefore need to evaluate the orientation correlation $\langle\mathbf{p}(t')\cdot\mathbf{p}(t)\rangle$ for times $t \geq t'$. This is equal to the expected value of the cosine of the angle between the vectors $\mathbf{p}(t')$ and $\mathbf{p}(t)$. Since the origin of times is irrelevant, this is therefore given by the exponential decay found in Eq. (13.51), so that

$$\langle \mathbf{p}(t') \cdot \mathbf{p}(t) \rangle = \langle \mathbf{p}(0) \cdot \mathbf{p}(t - t') \rangle = e^{-2D_R(t-t')}, \quad t \geq t'. \tag{13.57}$$

Substituting this result in Eq. (13.56) we then obtain

$$\frac{d\langle \tilde{\mathbf{X}} \cdot \tilde{\mathbf{X}} \rangle}{dt} = 2U^2 \int_0^t e^{-2D_R(t-t')} dt' = \frac{U^2}{D_R} \left(1 - e^{-2D_R t} \right), \tag{13.58}$$

which may be integrated once to get the exact result

$$\langle \tilde{\mathbf{X}} \cdot \tilde{\mathbf{X}} \rangle(t) = \frac{U^2}{D_R} \left(t + \frac{1}{2D_R} e^{-2D_R t} - \frac{1}{2D_R} \right). \tag{13.59}$$

Effective Diffusion

At short times, $t \ll D_R^{-1}$, Eq. (13.59) leads to the expected ballistic scaling $\langle \tilde{\mathbf{X}} \cdot \tilde{\mathbf{X}} \rangle(t) \approx (Ut)^2$. In contrast, at long times we obtain a diffusive behaviour. Specifically, adding back the Brownian diffusion of Eq. (13.38), we see that for $t \gg D_R^{-1}$ the swimming cell displays a diffusive scaling in three dimensions, $\langle \mathbf{X} \cdot \mathbf{X} \rangle \approx 6D_{\text{eff}} t$, with an effective diffusion constant given by

$$D_{\text{eff}} = D + \frac{U^2}{6D_R}. \tag{13.60}$$

The result is similar in two dimensions, with the factor of 6 replaced by 4.

For a bacterium swimming with speed $U \approx 30\ \mu\text{m/s}$ and recalling that $D_R = O(0.06\ \text{rad}^2/\text{s})$ we obtain $D_{\text{eff}} = O(10^3\ \mu\text{m}^2/\text{s})$ and thus an enhancement over Brownian motion by four orders of magnitude.

In order to illustrate this diffusion enhancement visually, we show in Fig. 13.1(b) five different trajectories over a time interval of 30 s for a 1 μm spherical organism swimming with constant speed $U = 1\ \mu\text{m/s}$ at room temperature. While the organism still explores its environment in a random fashion as a result of rotational diffusion, its locomotion allows it to diffuse much faster than what is obtained without swimming (Fig. 13.1(a)).

13.2.3 Diffusion of Cells with Circular Trajectories

Another situation of interest in the case of bacteria is their diffusing dynamics when they are swimming near surfaces. As seen in Section 11.2, hydrodynamic interactions with surfaces lead to circular cell trajectories, which can thus be expected to decrease the ability of cells to explore their environment.

Model

To compute the effective diffusion of bacteria with circular trajectories we assume that the cell always remains near the surface, and consider a probabilistic description for the cell orientation in two dimensions quantified by the orientation probability distribution function $\Psi(\theta, t)$, where $0 \leqslant \theta < 2\pi$ is the polar angle between the cell swimming direction, $\mathbf{p}$, and the direction on the surface along which the cell is aligned at $t = 0$. Conservation of probability states

$$\frac{\partial \Psi}{\partial t} + \frac{\partial}{\partial \theta}(\Omega \Psi) = 0. \tag{13.61}$$

If hydrodynamic interactions between the cell and the surface lead to rotation with angular velocity ω perpendicular to surface, the instantaneous rotation rate of the cell, Ω, is given by

$$\Omega = \omega + M_R L, \tag{13.62}$$

where the thermodynamic torque is the gradient of the free energy,

$$L = -\frac{\partial \mathcal{U}}{\partial \theta}, \quad \mathcal{U} = k_B T \ln \Psi, \tag{13.63}$$

so that Eq. (13.61) becomes an advection-diffusion equation for Ψ:

$$\frac{\partial \Psi}{\partial t} + \omega \frac{\partial \Psi}{\partial \theta} = D_R \frac{\partial^2 \Psi}{\partial \theta^2}. \tag{13.64}$$

Mean Square Displacements

From Eq. (13.64), we may compute orientation correlations for the direction of the cell, $\langle \cos \theta \rangle$ and $\langle \sin \theta \rangle$, using integration by parts. Evaluating first the rate of change of $\langle \cos \theta \rangle$, we have

$$\frac{d}{dt}\langle \cos \theta \rangle = \int_0^{2\pi} \cos \theta \frac{\partial \Psi}{\partial t} \, d\theta = \int_0^{2\pi} \cos \theta \left(D_R \frac{\partial^2 \Psi}{\partial \theta^2} - \omega \frac{\partial \Psi}{\partial \theta} \right) d\theta. \tag{13.65}$$

Writing the integrand as

$$\cos \theta \left(D_R \frac{\partial^2 \Psi}{\partial \theta^2} - \omega \frac{\partial \Psi}{\partial \theta} \right) = \frac{\partial}{\partial \theta} \left(-\Psi \omega \cos \theta + D_R \cos \theta \frac{\partial \Psi}{\partial \theta} + \Psi D_R \sin \theta \right)$$
$$- \omega \Psi \sin \theta - \Psi D_R \cos \theta, \tag{13.66}$$

the exact derivative integrates to zero by periodicity, leading to

$$\frac{d}{dt}\langle \cos \theta \rangle = \int_0^{2\pi} (-\omega \Psi \sin \theta - \Psi D_R \cos \theta) \, d\theta = -\omega \langle \sin \theta \rangle - D_R \langle \cos \theta \rangle. \tag{13.67}$$

Using similar algebra we also obtain

$$\frac{d}{dt}\langle \sin\theta \rangle = \omega\langle\cos\theta\rangle - D_R\langle\sin\theta\rangle. \tag{13.68}$$

The system in Eqs. (13.67)–(13.68) can be solved exactly with initial condition of $\theta = 0$ leading to the orientation correlations

$$\langle\cos\theta\rangle(t) = \cos\omega t\, e^{-D_R t}, \tag{13.69a}$$

$$\langle\sin\theta\rangle(t) = \sin\omega t\, e^{-D_R t}. \tag{13.69b}$$

To compute the effective diffusion constant of the cell we may then use the same setup as in the previous section and evaluate Eq. (13.56) using the orientation correlation, $\langle \mathbf{p}(t')\cdot\mathbf{p}(t)\rangle$, $t \geq t'$, applicable in the circular case. This correlation is equal to the mean cosine of the angle between two different times, which is only a function of the difference in time, $t - t'$, and is therefore given by Eq. (13.69a) as

$$\langle\mathbf{p}(t')\cdot\mathbf{p}(t)\rangle = \langle\mathbf{p}(0)\cdot\mathbf{p}(t - t')\rangle = \cos\left(\omega(t - t')\right)e^{-D_R(t-t')}. \tag{13.70}$$

As a result, the mean square displacement in the absence of Brownian motion in translation, $\langle\tilde{\mathbf{X}}\cdot\tilde{\mathbf{X}}\rangle$, is obtained using Eq. (13.56) as

$$\frac{d\langle\tilde{\mathbf{X}}\cdot\tilde{\mathbf{X}}\rangle}{dt} = 2U^2\int_0^t\left[\cos\left(\omega(t - t')\right)e^{-D_R(t-t')}\right]dt'$$

$$= \frac{2U^2}{\omega^2 + D_R^2}\left[D_R + e^{-D_R t}\left(\omega\sin(\omega t) - D_R\cos(\omega t)\right)\right]. \tag{13.71}$$

Effective Diffusion

Since the cell is assumed to remain near the surface, it is only allowed to diffuse in two dimensions. In that case the two-dimensional version of the diffusive scaling for Brownian motion is similar to Eq. (13.38) but with a factor of 4 on the right-hand side instead of 6 (a contribution of $2Dt$ from each direction). Adding this Brownian contribution to Eq. (13.71) we see that, in the long-time limit, the cell displays long-time effective diffusive behaviour, $\langle\mathbf{X}\cdot\mathbf{X}\rangle = 4D_{\text{eff}}t$ for $t \gg D_R^{-1}$, with the effective diffusion constant

$$D_{\text{eff}} = D + \frac{U^2 D_R}{2(D_R^2 + \omega^2)}. \tag{13.72}$$

In the limit where $\omega = 0$, the result of Eq. (13.72) becomes the two-dimensional version of Eq. (13.60). Otherwise, we see that increasing the value of the cell angular velocity, ω, always decreases the effective diffusivity, with a significant change in magnitude as soon as $\omega \gtrsim D_R$.

In the case of swimming bacteria such as *E. coli*, the resulting value of D_{eff} near a surface falls in between the values for non-motile cells and for free-swimming cells in the bulk fluid. The angular velocity of a cell swimming in circles as in Fig. 11.6 is given by $\omega \approx U/R$, where U is the swimming speed and R the radius of curvature of the circular trajectory, leading to $\omega = O(1 \text{ rad/s})$. Given that $D_R = O(0.06 \text{ rad}^2/\text{s})$, we see that the denominator in Eq. (13.72) is dominated by the ω^2 term, and leads to an effective diffusivity about two orders of magnitude above the Brownian one, D.

13.3 Run-and-Tumble

As first introduced in Chapter 8, the strategy used by many flagellated bacteria to sample their environment consists in alternating periods of approximately straight swimming at constant speed ('runs', about 1 sec long) with quick reorientation events ('tumbles', typically one tenth of a second long) (Berg, 2004). Here we first introduce the argument, due to Purcell, rationalising this run-and-tumble search strategy as a way to outrun nutrient diffusion, and then show how stochastic run-and-tumble motion with known statistical properties can be described as a diffusive process.

13.3.1 Outrunning Diffusion

In his lecture 'Life at Low Reynolds Number', Purcell put forward a physical rationale behind the run-and-tumble strategy. Bacteria are very small and thus chemical transport around them, in particular that of nutrients, is dominated by diffusion. Even at high swimming speeds, bacteria cannot create flows that modify this transport appreciably – something that, in the language of transport science, would be referred to as a small Péclet number transport problem (Leal, 2007). Unlike Pac-Man, who can gobble extra Pac-Dots when he is moving faster, the swimming of bacteria cannot impact the rate at which they feed.

In that case, why swim at all? Purcell proposed that bacteria swim in order to outrun the diffusion of nutrients, allowing the organisms to compare chemical concentrations at different locations, which in turn can allow chemotaxis towards higher concentrations. Using the video-game analogy, this would be equivalent to Pac-Man moving around the game in order to detect where the density of Pac-Dots is highest.

To understand why this leads naturally to a locomotion strategy à la run-and-tumble, we follow Purcell's simple scaling argument. Consider first a cell swimming with constant speed U during a time interval t. The distance travelled by the swimmer, L_{swim}, scales as $L_{\text{swim}} \sim Ut$. During that same interval, a particular nutrient molecule travels by a rms distance, L_{diff}, following a diffusing scaling $L_{\text{diff}} \sim \sqrt{Dt}$, where D is the molecular diffusion constant of the substance of

interest. Considering the race between the swimmer and the nutrient, it is apparent that at very short times the diffusion process always outruns the swimmer since
$t^{1/2} \gg t$ when t is small. In order for the cell to win the race, swimming has to be be
sustained for a minimum amount of time, t_c, satisfying the criterion $L_{swim} > L_{diff}$,
which means

$$Ut > \sqrt{Dt} \rightarrow t > t_c = \frac{D}{U^2}. \tag{13.73}$$

Swimming for significantly less than t_c would not allow the cell to compare concentrations at different locations (Purcell, 1977, 'If you don't swim that far, you
haven't gone anywhere'), whereas locomotion for much longer than t_c would waste
the cell's opportunity to try another (potentially even better) direction. It therefore
appears that it is best for the bacterium to swim for a time just above t_c, and then
tumble, which is precisely the run-and-tumble mechanism. For nutrients with standard molecular diffusivities $D = O(10^{-9} \text{ m}^2/\text{s})$ and cells with swimming speed on
the order of $U \approx 30 \ \mu\text{m/s}$ we obtain a critical time given by $t_c = O(1 \text{ s})$, fully
consistent with the average duration of runs for *E. coli* (Berg, 2004).

13.3.2 Orientation Correlation

The stochastic nature of the run-and-tumble motion of *E. coli* has been characterised by a number of experimental studies. In particular, the angle change
between two successive runs has a mean value of about $68°$ and a probability distribution function that peaks at approximately at $63°$, indicating reorientations with
positive correlations (Berg and Brown, 1972; Turner et al., 2000; Saragosti et al.,
2012). We now build a stochastic model for the motion of run-and-tumble bacteria that allows us to compute this orientation correlation and to describe run-and-
tumble as an effective diffusive process.

   In order to construct the model, we follow Lovely and Dahlquist (1975) and
adapt the freely rotating chain model of polymer physics (Flory, 1969). The geometrical description of the run-and-tumble motion is illustrated in Fig. 13.2. The
cell undergoes straight runs followed by reorientation events (tumbles) assumed to
be instantaneous, and is subject to no other source of noise. When reorientation
i takes place, the unit vector in the swimming direction changes from $\mathbf{p}_i$ to a new
value $\mathbf{p}_{i+1}$. We use the two spherical coordinate angles $0 \leqslant \theta_i \leqslant \pi$ and $0 \leqslant \phi_i < 2\pi$
to parametrise the direction of the vector $\mathbf{p}_{i+1}$ relative to that of $\mathbf{p}_i$, with ϕ_i assumed
to be distributed uniformly.

   We aim to compute the orientation correlation, $\langle \mathbf{p}(0) \cdot \mathbf{p}(t) \rangle$, for $t \geqslant 0$. We first
calculate the correlation $\langle \mathbf{p}(0) \cdot \mathbf{p}(t) \rangle_n$ conditional upon the number of reorientations, n, having taken place in the time interval between 0 and t. Using the theorem

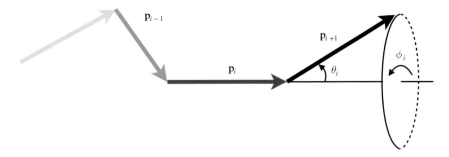

Figure 13.2 Geometrical model of run-and-tumble motion. A cell undergoes straight runs followed by instantaneous reorientation events (tumbles). The unit vector along run i is denoted by $\mathbf{p}_i$ and we use the spherical coordinates angles θ_i and ϕ_i to parametrise the direction of $\mathbf{p}_{i+1}$ relative to that of $\mathbf{p}_i$. We neglect any additional sources of noise.

of total probability, we next sum over all possible values of n to determine the expected value of the total correlation.

If $n = 0$ then there is no reorientation, and therefore $\langle \mathbf{p}(0) \cdot \mathbf{p}(t) \rangle_0 = 1$. If $n = 1$, computing an angular average over ϕ_1 for fixed value of θ_1 leads to

$$\langle \mathbf{p}_0 \cdot \mathbf{p}_1 \rangle_{\phi_1} = \cos \theta_1. \tag{13.74}$$

Averaging over all possible values of θ_1 we then obtain

$$\langle \mathbf{p}(0) \cdot \mathbf{p}(t) \rangle_1 = \langle \cos \theta_1 \rangle \equiv \alpha, \tag{13.75}$$

where α denotes the mean reorientation cosine, a characteristic of run-and-tumble motion assumed to be known and identical for all tumbles.

For the case with $n = 2$, we can carry out successive averages to obtain a similar result. First, we compute the average over the angle ϕ_2 for the rotation of the vector $\mathbf{p}_2$ around $\mathbf{p}_1$, with everything else fixed, as

$$\langle \mathbf{p}_0 \cdot \mathbf{p}_2 \rangle_{\phi_2} = \mathbf{p}_0 \cdot \langle \mathbf{p}_2 \rangle_{\phi_2} = \cos \theta_2 \mathbf{p}_0 \cdot \mathbf{p}_1, \tag{13.76}$$

so that

$$\langle \mathbf{p}_0 \cdot \mathbf{p}_2 \rangle_{\phi_2, \theta_2} = \langle \cos \theta_2 \rangle \mathbf{p}_0 \cdot \mathbf{p}_1 = \alpha \mathbf{p}_0 \cdot \mathbf{p}_1. \tag{13.77}$$

Using Eq. (13.75), the average of Eq. (13.77) over ϕ_1 and θ_1 leads to

$$\langle \mathbf{p}(0) \cdot \mathbf{p}(t) \rangle_2 = \alpha^2. \tag{13.78}$$

Clearly, the procedure leading to Eqs. (13.75) and (13.78) could be extended for $n > 2$ reorientations and we would obtain similarly

$$\langle \mathbf{p}(0) \cdot \mathbf{p}(t) \rangle_n = \alpha^n. \tag{13.79}$$

The expected value of the orientation correlation is then obtained using the theorem of total probability as the sum of all the possible reorientations,

$$\langle \mathbf{p}(0) \cdot \mathbf{p}(t) \rangle = \sum_{n=0}^{\infty} \tilde{p}(n;t)\langle \mathbf{p}(0) \cdot \mathbf{p}(t) \rangle_n, \tag{13.80}$$

where $\tilde{p}(n;t)$ is the probability of having exactly n reorientations during a time interval of size t. Since reorientation events are known to be Poisson-distributed (Berg, 2004), this probability is given by

$$\tilde{p}(n;t) = \frac{(t/T)^n}{n!} e^{-(t/T)}, \tag{13.81}$$

where T is mean duration of a run, $T \approx 1$ s. Using Eq. (13.81) we can evaluate exactly the sum in Eq. (13.79) as

$$\langle \mathbf{p}(0) \cdot \mathbf{p}(t) \rangle = \sum_{n=0}^{\infty} \frac{(t/T)^n}{n!} e^{-(t/T)} \alpha^n = e^{-(t/T)} \sum_{n=0}^{\infty} \frac{(\alpha t/T)^n}{n!}, \tag{13.82}$$

and therefore obtain finally

$$\langle \mathbf{p}(0) \cdot \mathbf{p}(t) \rangle = e^{-t/\tau_c}, \quad \tau_c = \frac{T}{1 - \alpha}. \tag{13.83}$$

13.3.3 Effective Diffusion

Even though run-and-tumble is not a thermal diffusion process, the orientation correlation is found to be exponential, similar in form to Eq. (13.51), which was obtained for Brownian motion. The characteristic timescale, τ_c, is related to the mean duration of a run (T), increased to account for the orientation correlation between each run. Note that for random reorientations we have $\alpha = 0$ and the correlation time is exactly T. For *E. coli*, a mean run time of $T \approx 1$ s and $\langle \cos \theta \rangle \approx 0.33$ (Berg, 1993) lead to the correlation time $\tau_c \approx 1.5$ s.

The mean square displacement of a cell undergoing run-and-tumble motion is calculated from the integral in Eq. (13.56). Since the loss of orientation is exponential, we get the same result as in Eq. (13.59) but with $2D_R$ replaced by τ_c^{-1}. Ignoring Brownian motion in translation, the effective diffusion constant for the running-and-tumbling bacterium in three dimensions is then

$$D_{\text{eff}} = \frac{U^2 T}{3(1 - \alpha)}. \tag{13.84}$$

For cells swimming with velocity $U \approx 30$ μm/s and $\langle \cos \theta \rangle \approx 0.33$, the prediction from Eq. (13.84) leads to effective diffusion of swimming *E. coli* of $D \approx 450$ μm^2/s (Berg, 1993).

Further Reading

This chapter was written to present a concise pedagogical introduction to Brownian motion and the impact of a fluctuating environment on the swimming of bacteria. Stochastic processes have long been a focus of research in physics (Chandrasekhar, 1943), and we refer to classical texts for an appreciation of some of the subtle aspects of thermal noise we did not address (Batchelor, 1976; Ermak and McCammon, 1978; Mazo, 2002; Graham, 2018), including its impact on active particles (Romanczuk et al., 2012). Beyond thermal fluctuations, modelling of the stochastic behaviour of molecular motors has been reviewed by Jülicher et al. (1997).

Returning to results derived in the chapter, we note that the enhancement in diffusion above Brownian motion obtained for swimmers is still valid even if their locomotion is reciprocal, i.e. if their swimming speed is periodic in time with zero average motion (Lauga, 2011a; see Exercises). The fluctuating dynamics of swimmers was also characterised in the presence of external flows (Ten Hagen et al., 2011; Sandoval et al., 2014). In the case of run-and-tumble bacteria, quick tumbles can be modelled as an effective rotational diffusion with diffusivity $D_R \approx 2.1$–3.5 rad^2/s (Saragosti et al., 2012). Since the mean timescale for the duration of a tumble is $\tau \approx 0.14$ s (Saragosti et al., 2012), this is in agreement with the simple estimate $D_R \approx 1/(2\tau) \approx 3.6$ rad^2/s.

Exercises

1. In order to compute the diffusive dynamics of a spherical particle, we argued in Section 13.1.1 that the ensemble average of the product between the Brownian force and the position of the particle in Eq. (13.7) is zero. Use an exponential factor in Eq. (13.3) to relate $\mathbf{X}(t)$ to $\mathbf{F}_B(t')$, with $t' < t$, and deduce that $\langle \mathbf{F}_B \cdot \mathbf{X} \rangle = 0$ in the Langevin framework.

2. Starting from the diffusion equation satisfied by the probability distribution function of a diffusing particle, $\Psi(\mathbf{r}, t)$, i.e.

$$\frac{\partial \Psi}{\partial t} = D \nabla^2 \Psi,$$

show by calculating ensemble averages and their derivatives explicitly that $\langle \mathbf{X} \cdot \mathbf{X} \rangle = 6Dt$.

3. Although a body undergoing reciprocal deformation cannot swim (scallop theorem from Chapter 2), a reciprocal motion can increase the effective diffusion coefficient of the body in a thermal environment. Consider a spherical particle swimming with periodic speed $U(t) = U_0 \cos(\omega t)$ along a direction $\mathbf{p}(t)$. The

particle is subject to Brownian motion both in translation and in rotation. Compute the long-time effective diffusivity of the active particle in three dimensions, D_{eff}, and show that it is always larger than its Brownian diffusivity.

4. Show that the Fokker–Planck equation describing the probability distribution function of a particle moving along a straight line and subject to a position-dependent velocity, $U(x)$, is an advection–diffusion equation.

5. An active particle walks along a straight line using discrete steps. It moves either to the right with velocity v^+ for a time τ^+ or to the left using velocity v^- for time τ^-. At every step, the probability that the particle moves to the right is p^+ and that to the left is p^-. Calculate the long-time velocity of the particle. Show that a sufficient condition to obtain a nonzero mean speed is that one parameter characterising motion to the right (v^+, τ^+, p^+) is different from the corresponding parameter for moving to the left.

6. A motile cell swims in the gap between two parallel rigid surfaces separated by a distance H. The cell always remains parallel to the surfaces and is modelled as an axisymmetric force dipole of strength $\mathcal{P}$. Applying results from Chapter 11 on the impact of rigid walls on force dipoles and adding the effects of both surfaces in the far field, use a balance between cell diffusion and wall attraction to compute the steady-state probability distribution function for the cell between the surfaces (the final result is the one plotted as a solid line in Fig. 11.1, right).

7. Bottom-heavy spherical cells swim with speed U in a vertical pipe of radius R. A downward Poiseuille flow of velocity $V(r) = V_0(1 - r^2/R^2)$ is established in the pipe, where r is the radial coordinate measured from the centre of the pipe. As explained in Section 10.3.3, the combination of downward flow with bottom-heaviness leads to the reorientation of the cells and their focusing at the centre of the pipe. By balancing the diffusion of swimming cells with their advection by the flow and their own locomotion, show that the steady-state probability distribution function of the cells is a Gaussian centred at $r = 0$.

14

Hydrodynamics of Collective Locomotion

In this chapter we move beyond the individual motion of swimming microorganisms and turn to the collective dynamics of cell populations such as bacterial colonies. Specifically, we use the results from Chapters 9 and 10 on hydrodynamic interactions to adapt the discrete and continuum frameworks introduced in Chapter 13 for the case of collective cell locomotion. Swimming cells create flows, which advect and rotate neighbouring organisms, and since the flow induced by each cell depends on its location and orientation, this coupling leads to complex nonlinear swimming dynamics. In the discrete case, we derive a first-principle model of cells interacting in the dilute limit, demonstrate the different ways in which two swimming microorganisms affect each other hydrodynamically, and show how the model can be used to explain clustering instabilities of swimming algae. We then develop a continuum approach coupling the dynamics of the fluid with the distribution in position and orientation of the cell population. After relating the model to alternative phenomenological descriptions based on symmetry arguments, we use this continuum framework to capture collective cell instabilities.

14.1 Discrete Model of Active Suspensions

14.1.1 Mathematical Model

The discrete framework for collective cell locomotion is an intuitive modelling approach based on extending models for the swimming of individual cells. Consider a collection of cells located in an unbounded fluid, with each cell characterised by its own body-fixed unit vector $\mathbf{p}^{(i)}(t)$ along which it swims (we use the same notation as in Chapters 10 and 13 for consistency). The setup is sketched in Fig. 14.1 and we use superscripts to label the cell of interest.

We assume that the suspension is in the dilute limit, i.e. that the typical size of each cell, a, is much smaller than the distance to its nearest neighbour, r. In that

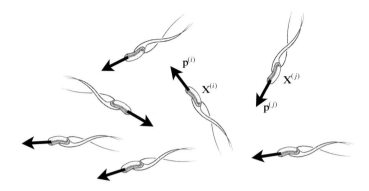

Figure 14.1 Discrete model for the collective dynamics of swimmers. Cell i located at $\mathbf{X}^{(i)}(t)$ and swimming along the body-fixed unit vector $\mathbf{p}^{(i)}(t)$ experiences flows from all other swimmers located at $\mathbf{X}^{(j)}(t)$ and swimming along $\mathbf{p}^{(j)}(t)$ ($j \neq i$). The flows from all neighbouring organisms advect and rotate the swimmer, leading to coupled nonlinear dynamics. Drawings courtesy of Jacques Lauga.

case, the dynamics of organism i is given by the sum of: (a) its intrinsic swimming (linear velocity $\mathbf{U}^{(i)}_{\text{swim}}$ and angular velocity $\mathbf{\Omega}^{(i)}_{\text{swim}}$); (b) thermal noise as in Chapter 13 (or other source of noise characterised by an effective temperature, with linear velocity $\mathbf{U}^{(i)}_{\text{noise}}$ and angular velocity $\mathbf{\Omega}^{(i)}_{\text{noise}}$); and (c) the hydrodynamic interactions from all cells $j \neq i$, which we can compute at leading order in a/r since the suspension is dilute. Denoting the instantaneous position of cell i by $\mathbf{X}^{(i)}(t)$, its velocity is thus written as

$$\frac{d\mathbf{X}^{(i)}}{dt} = \mathbf{U}^{(i)}_{\text{swim}} + \mathbf{U}^{(i)}_{\text{noise}} + \sum_{j \neq i} \mathbf{u}^{(j)}(\mathbf{X}^{(i)} - \mathbf{X}^{(j)}; \mathbf{p}^{(j)}), \qquad (14.1)$$

where $\mathbf{u}^{(j)}$ is the flow velocity created by cell j as it swims along the direction $\mathbf{p}^{(j)}$ and is a function of the relative position between the cells, $\mathbf{X}^{(i)} - \mathbf{X}^{(j)}$. We saw in Chapter 9 that swimmers induce dipolar flows, and thus we follow Eq. (9.30) to write the flow created by each cell at a location $\mathbf{r}$ relative to it as

$$\mathbf{u}^{(j)}(\mathbf{r}; \mathbf{p}) = \frac{\mathcal{P}}{8\pi\mu} \left[-\frac{1}{r^3} + \frac{3(\mathbf{p} \cdot \mathbf{r})^2}{r^5} \right] \mathbf{r}, \quad r = |\mathbf{r}|, \qquad (14.2)$$

where $\mathcal{P}$ is the dipole strength, assumed here to be constant and identical for all cells (recall that when $\mathcal{P} > 0$ the cell is a pusher). While more realistic unsteady flows containing higher-order singularities could be used, the minimal model in Eq. (14.2) allows us to capture generic hydrodynamic interactions.

In addition to its position, the orientation of the cell, $\mathbf{p}^{(i)}(t)$, evolves in time as

$$\frac{d\mathbf{p}^{(i)}}{dt} = \mathbf{\Omega}^{(i)} \times \mathbf{p}^{(i)}, \tag{14.3}$$

where the cell rotation rate, $\mathbf{\Omega}^{(i)}$, is given, similar to Eq. (14.1), as

$$\mathbf{\Omega}^{(i)} = \mathbf{\Omega}_{\text{swim}}^{(i)} + \mathbf{\Omega}_{\text{noise}}^{(i)} + \sum_{j \neq i} \mathbf{\Omega}^{(j)}(\mathbf{X}^{(i)} - \mathbf{X}^{(j)}; \mathbf{p}^{(j)}; \mathbf{p}^{(i)}). \tag{14.4}$$

The last term in Eq. (14.4) quantifies the reorientation of swimmer i due to interactions with swimmer j. In the dilute limit, the flows from other cells can be approximated as locally linear so that, when modelling each swimmer as a prolate spheroid, we may use Jeffery's equation from Chapter 10 giving the angular velocity in a linear flow, and write

$$\mathbf{\Omega}^{(j)}(\mathbf{r}; \mathbf{p}; \tilde{\mathbf{p}}) = \frac{1}{2}\boldsymbol{\omega}^{(j)}(\mathbf{r}; \mathbf{p}) + B\tilde{\mathbf{p}} \times \left(\mathbf{E}^{(j)}(\mathbf{r}; \mathbf{p}) \cdot \tilde{\mathbf{p}} \right), \tag{14.5}$$

where $B \geqslant 0$ is the shape factor from Eq. (10.64) for the spheroidal shape. The vorticity, $\boldsymbol{\omega}^{(j)}$, and rate of strain, $\mathbf{E}^{(j)}$, are given as

$$\boldsymbol{\omega}^{(j)}(\mathbf{r}; \mathbf{p}) = \nabla \times \mathbf{u}^{(j)}\Big|_{(\mathbf{r};\mathbf{p})}, \quad \mathbf{E}^{(j)}(\mathbf{r}; \mathbf{p}) = \frac{1}{2}\left(\nabla\mathbf{u}^{(j)} + \left(\nabla\mathbf{u}^{(j)}\right)^{T} \right)\Big|_{(\mathbf{r};\mathbf{p})}, \tag{14.6}$$

where the velocity field $\mathbf{u}^{(j)}$ is taken as the dipolar flow from Eq. (14.2).

14.1.2 Interactions between Two Swimming Cells

Equipped with a method to evaluate the sums in Eqs. (14.1) and (14.4), the discrete model above provides an intuitive approach to studying the coupled dynamics of cell populations in the dilute (i.e. far-field) limit. Physical insight can be obtained by considering the case either of two cells or of regular arrangements of swimmers.

Attraction vs. Repulsion

We consider a first cell located at $\mathbf{X}$ and swimming along $\mathbf{p}$ and characterise its hydrodynamic influence on a second organism located at $\tilde{\mathbf{X}}$ and swimming along $\tilde{\mathbf{p}}$. Denoting by $\mathbf{r} = \tilde{\mathbf{X}} - \mathbf{X}$ the relative position between the two cells, the velocity induced at the second organism by the first is given in Eq. (14.2). Some characteristics of this flow were already noted in Chapter 9. Its velocity field, axisymmetric around the direction of the first dipole and decaying as $1/r^2$, is directed in the radial direction everywhere and is independent of the orientation and geometry of the swimmer it is affecting (i.e. independent of $\tilde{\mathbf{p}}$).

We sketch in Fig. 14.2 the direction of the flow induced by a pusher cell and a puller. When the relative position between the swimmers is perpendicular to the

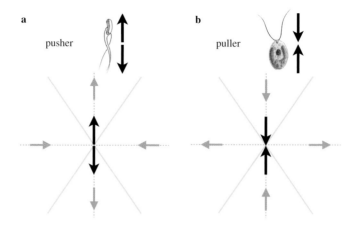

Figure 14.2 Direction of the axisymmetric flow induced by a steady force dipole. (a) A pusher cell such as a flagellated bacterium ($\mathcal{P} > 0$) creates an attractive flow in the direction perpendicular to its symmetry axis and repulsive in the front and back directions. (b) The direction of the flow is reversed for a puller such as the algal genus *Chlamydomonas* ($\mathcal{P} < 0$). Drawings courtesy of Jacques Lauga.

symmetry axis of the first dipole, the flow is directed towards the cell for a pusher, resulting in cell–cell attraction (see Fig. 14.2(a)) and away from the cell for a puller, leading to cell-cell repulsion (Fig. 14.2(b)). The flow direction is reversed in front of and behind the cell. Denoting by θ the angular position of the second swimmer away from the axis of the dipole (i.e. $\cos\theta = \mathbf{p} \cdot \mathbf{r}/r$), the transition between the attractive and repulsive regions occurs at the critical angles $\cos^2\theta_c = 1/3$, i.e. $\theta_c \approx 54.7°$ and $125.3°$, indicated by grey solid lines in Fig. 14.2.

Reorientation

The reorientation of the cell swimming along $\tilde{\mathbf{p}}$ and located at $\mathbf{r}$ relative to the first swimmer occurs with angular velocity $\mathbf{\Omega}$ given by Jeffery's equation, Eqs. (14.5) and (14.6). From the dipolar flow in Eq. (14.2), we can use vector calculus to compute these two terms, each decaying in space as $1/r^3$, as

$$\omega(\mathbf{r}; \mathbf{p}) = \frac{3\mathcal{P}}{4\pi\mu} \frac{(\mathbf{p}\cdot\mathbf{r})(\mathbf{p}\times\mathbf{r})}{r^5}, \tag{14.7a}$$

$$\tilde{\mathbf{p}} \times (\mathbf{E}(\mathbf{r}; \mathbf{p}) \cdot \tilde{\mathbf{p}}) = \frac{3\mathcal{P}}{8\pi\mu r^5}\left(1 - 5\frac{(\mathbf{p}\cdot\mathbf{r})^2}{r^2}\right)(\tilde{\mathbf{p}}\cdot\mathbf{r})(\tilde{\mathbf{p}}\times\mathbf{r}) \tag{14.7b}$$

$$+ \frac{3\mathcal{P}}{8\pi\mu r^5}(\mathbf{p}\cdot\mathbf{r})\left[(\tilde{\mathbf{p}}\cdot\mathbf{p})(\tilde{\mathbf{p}}\times\mathbf{r}) + (\tilde{\mathbf{p}}\cdot\mathbf{r})(\tilde{\mathbf{p}}\times\mathbf{p})\right],$$

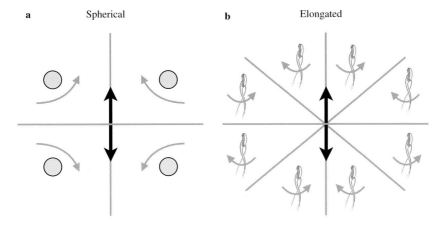

Figure 14.3 Reorientation of swimmers by a pusher dipole ($\mathcal{P} > 0$). (a) In the case of spherical cells ($B = 0$), there are four regions with alternating signs of angular velocity, as illustrated using curved arrows, and the reorientation rate is independent of the direction of the cell, $\tilde{\mathbf{p}}$. (b) For parallel elongated swimmers (illustrated here for $B = 1$), there are now eight different regions with alternating reorientation directions. All results are reversed for a puller. Drawings courtesy of Jacques Lauga.

and thus from Eq. (14.5) we obtain the rotation rate of the second swimmer as

$$
\boldsymbol{\Omega} = \frac{3\mathcal{P}}{8\pi\mu r^5}\left[(\mathbf{p}\cdot\mathbf{r})(\mathbf{p}\times\mathbf{r}) + B\left(1 - 5\frac{(\mathbf{p}\cdot\mathbf{r})^2}{r^2}\right)(\tilde{\mathbf{p}}\cdot\mathbf{r})(\tilde{\mathbf{p}}\times\mathbf{r}) \right.
$$
$$
\left. + B(\mathbf{p}\cdot\mathbf{r})(\tilde{\mathbf{p}}\cdot\mathbf{p})(\tilde{\mathbf{p}}\times\mathbf{r}) + B(\mathbf{p}\cdot\mathbf{r})(\tilde{\mathbf{p}}\cdot\mathbf{r})(\tilde{\mathbf{p}}\times\mathbf{p}) \right]. \qquad (14.8)
$$

The formula in Eq. (14.8) is hard to interpret in general so we focus now on two special cases. First, for a spherical swimmer the shape factor is $B = 0$ so the angular velocity is governed only by vorticity, and we obtain

$$
\boldsymbol{\Omega} = \frac{3\mathcal{P}}{8\pi\mu r^5}(\mathbf{p}\cdot\mathbf{r})(\mathbf{p}\times\mathbf{r}). \qquad (14.9)
$$

The reorientation of the second swimmer is independent of its orientation ($\tilde{\mathbf{p}}$) and is dictated by its position relative to the orientation of the first swimmer. The result is illustrated in Fig. 14.3(a) for a pusher dipole ($\mathcal{P} > 0$), using curved arrows to indicate the direction of rotation of the second swimmer. The space is split into four quadrants according to the signs of $\mathbf{p}\cdot\mathbf{r}$ and $\mathbf{p}\times\mathbf{r}$. For a puller ($\mathcal{P} < 0$), all directions in Fig. 14.3(a) are reversed.

The more complex case of an elongated swimmer, $B > 0$, can be simplified by making the additional assumption that $\mathbf{p} = \tilde{\mathbf{p}}$ to focus on the configuration where both swimmers are initially parallel to each other. In that case, the angular velocity in Eq. (14.8) becomes

$$\mathbf{\Omega} = \frac{3\mathcal{P}}{8\pi\mu r^5} \left[(1 + 2B) - 5B\frac{(\mathbf{p} \cdot \mathbf{r})^2}{r^2} \right] (\mathbf{p} \cdot \mathbf{r})(\mathbf{p} \times \mathbf{r}). \qquad (14.10)$$

When the cells are in a side-by-side configuration (i.e. small values of $\mathbf{p} \cdot \mathbf{r}$), the sign of the relative rotation is that of $(\mathbf{p} \cdot \mathbf{r})(\mathbf{p} \times \mathbf{r})$, similar to the spherical case. A transition in reorientation happens at a critical angle $\bar{\theta}_c$ satisfying

$$\cos^2 \bar{\theta}_c = \frac{1 + 2B}{5B}, \qquad (14.11)$$

which has a solution as soon as its right-hand side is less than 1. This happens for $B > 1/3$ and corresponds to a swimmer of aspect ratio greater than $\sqrt{2} \approx 1.41$. For elongated cells with $B \approx 1$, the critical angles are $\bar{\theta}_c \approx 39.2°$ and $140.8°$. The space surrounding the dipole is split in this case into eight different regions, as illustrated in Fig. 14.3(b). As above, all rotations are reversed for pullers.

Unstable Angular Modes

The expressions for the relative reorientation of swimmers can be used to examine the stability of symmetric cell arrangements. We focus on the case of parallel spherical swimmers and treat the elongated case in the Exercises. When two identical spherical cells interact, fixed points of their angular dynamics are found by setting $\mathbf{\Omega} = \mathbf{0}$ in Eq. (14.9), leading to two lines of fixed points, namely swimmers in a side-by-side ($\mathbf{p} \perp \mathbf{r}$) or front-back ($\mathbf{p} \parallel \mathbf{r}$) configuration.

Using the information on cell reorientation shown in Fig. 14.3(a), we can deduce that all fixed points are unstable. This is illustrated graphically in Fig. 14.4. On the left of the figure we consider the side-by-side configuration. If the two swimmers are tilted away from their parallel configuration in an antisymmetric (i.e. mirror-image) way, the hydrodynamically induced reorientations (sketched using curved arrows) bring them back to parallel swimming, indicating stability. In contrast, when the two swimmers are perturbed in a symmetric way and remain parallel to each other, the induced angular velocities act to amplify the perturbation and lead to instability.

The front-back setup, shown on the right of Fig. 14.4, leads to different unstable modes. In this case, using the information from the sketches in Fig. 14.3(a), we see that when the cells are perturbed symmetrically and remain parallel to each other, hydrodynamic interactions bring them back to their original configuration, while in the antisymmetric mode the reorientations created by the dipolar flows now amplify the disturbance and lead to instability.

Although we deduced these results graphically, they can be made rigorous by computing the growth rates of the different modes (see Exercises). Importantly,

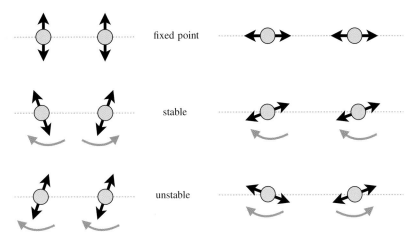

Figure 14.4 Orientation instabilities of two interacting spherical pushers ($B = 0$, $\mathcal{P} > 0$). Left: instability of the side-by-side configuration. If the orientations of the two swimmers are perturbed in a mirror-image way, hydrodynamic interactions shown using curved arrows act to bring them back to their original orientations, whereas if they are perturbed while remaining parallel, the induced flows act to amplify the initial perturbation. Right: in the front-back configuration, parallel perturbations are stable but mirror-image ones are unstable.

since each fixed point (either side-by-side and front-back) has both stable and unstable modes in the case of pushers, we deduce that the same is true for pullers, and therefore aligned configurations of any types of swimmers are always unstable. This remains true for elongated cells, which have additional fixed points (see Exercises), a result that we re-derive using the continuum approach below (see Section 14.3.2).

14.1.3 Clustering of Focused Cells

The previous section highlighted instabilities in the orientations of cells driven by hydrodynamic interactions. We now discuss instabilities in the positions of organisms, such as those occurring in the light-focused swimmers of Section 10.3.3. In that case, we assume that the orientations of the swimmers remain fixed due to their phototactic bias and consider a collection of identical cells swimming along a line (Fig. 14.5(a)). The resulting instabilities were reported numerically by Jibuti et al. (2014) and explained theoretically by Lauga and Nadal (2017).

To describe the dynamics of the swimmers mathematically, we assume they are all sufficiently far from each other that they can be modelled as identical dipoles, which we take as having constant magnitude $\mathcal{P}$ (Fig. 14.5(b)). The cells are distributed uniformly in the x direction, along which they swim with constant speed

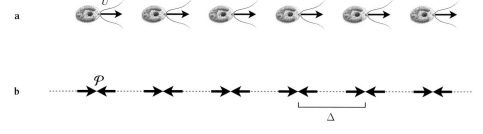

Figure 14.5 Clustering instability of focused cells. (a) Collection of light-focused algae swimming with speed U along a straight line with fixed orientation. (b) Mathematical model in the swimming frame where the algae are replaced by identical puller dipoles ($\mathcal{P} < 0$) separated by distance Δ. Drawings in (a) courtesy of Jacques Lauga.

U. In that case, since the organisms are aligned with their swimming direction, the dipolar flow from Eq. (14.2) instantaneously induced at location x by a swimmer at x_0 is given by

$$\mathbf{u}(x - x_0) = u(x - x_0)\mathbf{e}_x, \quad u(x - x_0) = \frac{\mathcal{P}}{4\pi\mu|x - x_0|^3}(x - x_0). \tag{14.12}$$

Using $x^{(n)}(t)$ to denote the position of swimmer n and considering only nearest-neighbour interactions, the velocity of cell n is given as the sum of its intrinsic swimming plus the velocity induced by its two neighbours as

$$\frac{dx^{(n)}}{dt} = U + u\left(x^{(n)} - x^{(n+1)}\right) + u\left(x^{(n)} - x^{(n-1)}\right). \tag{14.13}$$

The configuration where the distance between each cell and its nearest neighbours is constant, $x^{(n)} - x^{(n-1)} = \Delta$, is a fixed point of the dynamics, with all swimmers moving together with speed U. To quantify the stability of such an arrangement, we perturb the position of each cell as $x^{(n)}(t) = n\Delta + Ut + \epsilon^{(n)}(t)$ and calculate the evolution equation for the perturbation, $\epsilon^{(n)}(t)$. For $\delta = \pm 1$ we first use a Taylor expansion of Eq. (14.12) as

$$u(\delta\Delta + \epsilon) = \frac{\mathcal{P}\delta}{4\pi\mu\Delta^2}\left(1 - 2\frac{\delta\epsilon}{\Delta}\right) + O(\epsilon^2), \tag{14.14}$$

which allows us to evaluate Eq. (14.13) at leading order in the $\epsilon^{(n)}$'s as

$$\frac{d\epsilon^{(n)}}{dt} = u\left(-\Delta + \epsilon^{(n)} - \epsilon^{(n+1)}\right) + u\left(\Delta + \epsilon^{(n)} - \epsilon^{(n-1)}\right) \tag{14.15a}$$

$$\approx \frac{\mathcal{P}}{2\pi\mu\Delta^3}\left(\epsilon^{(n+1)} + \epsilon^{(n-1)} - 2\epsilon^{(n)}\right). \tag{14.15b}$$

This discrete system is characterised by the tridiagonal symmetric Toeplitz matrix. To compute the eigenvalues of Eq. (14.15b), we may look for solutions written as Fourier modes along the discrete spatial coordinate Δ of the form

$$\epsilon^{(n)}(t) = \epsilon e^{\sigma t + ikn\Delta}, \tag{14.16}$$

where k is a discrete wavenumber. Note that for periodic perturbations we have $k\Delta = 2r\pi$ where r is a rational number. Substituting the form of the perturbation into Eq. (14.15b), the growth rate of the mode, σ, is obtained as

$$\sigma = \frac{\mathcal{P}}{\pi\mu\Delta^3}[\cos(k\Delta) - 1]. \tag{14.17}$$

Importantly, since $\cos(k\Delta) \leqslant 1$, then σ and $\mathcal{P}$ always have opposite signs. A line of pushers with $\mathcal{P} > 0$ is therefore always stable ($\sigma \leqslant 0$) whereas a line of pullers is unstable ($\sigma \geqslant 0$), consistent with the results obtained numerically for model algae (Jibuti et al., 2014). This analysis is robust to an extension beyond the nearest-neighbour approximation (Lauga and Nadal, 2017) and can also be recovered using the continuum model developed in the next section (see Exercises).

The physical interpretation for the instability arising in a line of pullers originates from the spatial dependence of the flow in Eq. (14.12), which, when $\mathcal{P} < 0$, is attractive and increases in magnitude when two cells approach each other. If the position of a swimmer is perturbed, it becomes closer to one of its neighbours and further away from the other one. Since the flow magnitude in Eq. (14.12) decreases with separation distance, the net effect of this perturbation is to increase the attraction between the approaching swimmers and to decrease the attraction between the separating ones. This results in a net flow bringing the two nearby swimmers even closer together.

In the case of pushers, the effect is reversed and the flows bring the swimmers back to their original relative position. Using the same analysis, we see that if the swimmers were aligned perpendicularly to their swimming direction, it would instead be lines of pushers that would be unstable, since the flow perpendicular to the swimming direction of a pusher is attractive (see Fig. 14.2(a)).

14.2 Continuum Model of Active Suspensions

In this second part of the chapter, we turn to an alternative modelling approach for the hydrodynamics of active suspensions. Similar to the mathematical framework developed in Section 13.1.3 for diffusing particles, we now propose a continuum probabilistic model of collective motion. This approach, which in its many forms has gained much popularity in the physics community, adapts the classical probabilistic formalism developed for polymer dynamics (Doi and Edwards, 1988) and passive suspensions (Koch and Shaqfeh, 1989) to the context of fluids driven by self-propelled cells (Saintillan and Shelley, 2008a,b).

14.2.1 Probability Distribution Function

Similar to Section 13.1.3, the continuum model consists in deriving a conservation equation for the distribution of cells. We denote by $\Psi(\mathbf{r}, \mathbf{p}, t)$ the probability distribution function for a swimmer located at position $\mathbf{r}$ with orientation $\mathbf{p}$ at time t. We denote the total volume of fluid by V and the orientation space for the swimmer by Ω. The concentration of swimmers, $c(\mathbf{r}, t)$, is given by an average of Ψ in orientation space and is normalised for a total number of swimmers N in the fluid as

$$c(\mathbf{r}, t) \equiv \iint_{\Omega} \Psi \, d\mathbf{p}, \qquad \iiint_{V} c(\mathbf{r}, t) \, d\mathbf{r} = N, \qquad (14.18)$$

where we have used the shorthand notation $\iint_{\Omega} \ldots d\mathbf{p}$ and $\iiint_{V} \ldots d\mathbf{r}$ to indicate, respectively, integration over the two-dimensional orientation space and over the three-dimensional position space.

Conservation of probability is written as

$$\frac{\partial \Psi}{\partial t} + \nabla_{\mathbf{r}} \cdot (\mathbf{u_r} \Psi) + \nabla_{\mathbf{p}} \cdot (\mathbf{u_p} \Psi) = 0, \qquad (14.19)$$

where we have used $\nabla_{\mathbf{r}}$ to denote the gradient in position ($\mathbf{r}$) space while $\nabla_{\mathbf{p}} \equiv (\mathbf{1} - \mathbf{pp}) \cdot \partial / \partial \mathbf{p}$ is the gradient on the unit sphere (recall that the swimming direction, $\mathbf{p}$, is a unit vector).

Assuming first that thermal (Brownian) fluctuations are the only source of noise in the collection of swimming cells, the flux in position can be modelled as the sum of the deterministic rate of change in the position of the cells ($\dot{\mathbf{r}}$) and the thermodynamic Brownian term from Section 13.1.3, so that

$$\mathbf{u_r} = \dot{\mathbf{r}} + M\mathbf{F_r}, \quad \mathbf{F_r} = -\nabla_{\mathbf{r}} \mathcal{U}, \quad \mathcal{U} = k_B T \ln \Psi. \qquad (14.20)$$

Similarly, the probability flux in orientation ($\mathbf{p}$) space is the sum of the deterministic rate of change in the swimmer orientation ($\dot{\mathbf{p}}$) and the Brownian contribution

$$\mathbf{u_p} = \dot{\mathbf{p}} + M_R \mathbf{F_p}, \quad \mathbf{F_p} = -\nabla_{\mathbf{p}} \mathcal{U}. \qquad (14.21)$$

With Eqs. (14.20)–(14.21), the conservation equation, Eq. (14.19), becomes an advection-diffusion equation,

$$\frac{\partial \Psi}{\partial t} + \nabla_{\mathbf{r}} \cdot (\dot{\mathbf{r}} \Psi) + \nabla_{\mathbf{p}} \cdot (\dot{\mathbf{p}} \Psi) = D \nabla_{\mathbf{r}}^2 \Psi + D_R \nabla_{\mathbf{p}}^2 \Psi, \qquad (14.22)$$

where the diffusion coefficients in position (D) and rotation (D_R) are given by the Stokes–Einstein relationships, Eqs. (13.35) and (13.39), respectively. Other sources of noise in the system could be included in the model by modifying the effective values of both D and D_R.

14.2.2 Probability Fluxes

In order to close Eq. (14.22), we need to know the values of $\dot{\mathbf{r}}$ and $\dot{\mathbf{p}}$. In the dilute limit, the former is the sum of the intrinsic swimming velocity of the cell ($U\mathbf{p}$, assuming that all cells have identical speeds) and the fluid velocity, denoted by $\mathbf{u}$, induced at the cell location by all the other swimmers in the suspension:

$$\dot{\mathbf{r}} = U\mathbf{p} + \mathbf{u}(\mathbf{r}, t). \tag{14.23}$$

In the dilute limit, the cells are sufficiently small that we may approximate flows around them as locally linear. Modelling the cells as identical prolate spheroids (shape factor $0 \leqslant B < 1$) that do not undergo any intrinsic rotation, the rate of change of their orientation is therefore given by Jeffery's equation, Eq. (10.68), as

$$\dot{\mathbf{p}} = (\mathbf{1} - \mathbf{p}\mathbf{p}) \cdot (\mathbf{W} + B\mathbf{E}) \cdot \mathbf{p}, \tag{14.24}$$

where we recall that $2\mathbf{W} = (\nabla\mathbf{u})^T - \nabla\mathbf{u}$ and $2\mathbf{E} = (\nabla\mathbf{u})^T + \nabla\mathbf{u}$.

The flow field appearing in Eqs. (14.23)–(14.24) arises from the activity of the swimming cells. The total hydrodynamic stress tensor, σ, satisfies the Cauchy equation for mechanical equilibrium of an incompressible fluid, Eq. (2.13):

$$\nabla \cdot \sigma = \mathbf{0}, \quad \nabla \cdot \mathbf{u} = 0. \tag{14.25}$$

The hydrodynamic stress tensor in the active fluid is the sum of the hydrodynamic stress from the background flow and the deviatoric active stress induced by the swimmers, $\Sigma^{s,\text{dev}}$. Denoting the pressure in the fluid by q and the shear viscosity by μ, we have a stress written as

$$\sigma = -q\mathbf{1} + 2\mu\mathbf{E} + \Sigma^{s,\text{dev}}. \tag{14.26}$$

The value of the swimmer stress was derived in Chapter 9. For an isolated organism inducing a stresslet $\mathbf{S}$ in a fluid of volume V, it is given by

$$\Sigma^{s,\text{dev}} = \frac{\mathbf{S}}{V}. \tag{14.27}$$

For a dilute suspension of swimmers described in the continuum framework by the probability distribution function Ψ, the effect of all swimmers can be superposed linearly and the expected value of the swimmer stress is given by an orientational average,

$$\Sigma^{s,\text{dev}} = \iint_{\Omega} \mathbf{S}\Psi(\mathbf{r}, \mathbf{p}, t)\, d\mathbf{p}, \tag{14.28}$$

where we recall that $\iint_\Omega \Psi(\mathbf{r}, \mathbf{p}, t)\, d\mathbf{p}$ has dimensions of number of cells per unit volume. Here we take the value of $\mathbf{S}$ for a steady, axisymmetric swimmer from Eq. (9.18),

$$\mathbf{S} = \mathcal{P}\left(\frac{1}{3}\mathbf{1} - \mathbf{pp}\right), \tag{14.29}$$

where $\mathcal{P}$ is constant and identical for all cells. Denoting averages in orientation space by $\langle\ldots\rangle(\mathbf{r}, t) \equiv \iint_\Omega \ldots \Psi(\mathbf{r}, \mathbf{p}, t)\, d\mathbf{p}$, Eq. (14.28) then becomes

$$\boldsymbol{\Sigma}^{s,\text{dev}} = \mathcal{P}\iint_\Omega \left(\frac{1}{3}\mathbf{1} - \mathbf{pp}\right)\Psi(\mathbf{r}, \mathbf{p}, t)\, d\mathbf{p} = \mathcal{P}\left(\frac{1}{3}c(\mathbf{r}, t)\mathbf{1} - \langle\mathbf{pp}\rangle\right). \tag{14.30}$$

Using the definitions of Eq. (14.26) and the swimmer stress from Eq. (14.30), the equations for instantaneous mechanical equilibrium of the active fluid, Eq. (14.25), become the incompressible Stokes equations forced by the swimmer stresses,

$$-\nabla q + \mu\nabla^2\mathbf{u} = \mathcal{P}\nabla \cdot \langle\mathbf{pp}\rangle, \quad \nabla \cdot \mathbf{u} = 0, \tag{14.31}$$

where we have absorbed the first term in $\boldsymbol{\Sigma}^{s,\text{dev}}$ into the fluid pressure. Note that, notation-wise, we have used ∇ instead of $\nabla_\mathbf{r}$ here because the continuum fluid velocity and pressure, $\mathbf{u}$ and q, are functions of $\mathbf{r}$ and t only, and therefore it is no longer necessary to consider gradients in orientation space.

14.2.3 Probabilistic Equations for an Active Fluid

In summary, the system of equations satisfied by the probability distribution function for the elongated cells (Ψ) and the flow velocity ($\mathbf{u}$) is given for the probabilistic continuum model by

$$\frac{\partial\Psi}{\partial t} + \nabla_\mathbf{r} \cdot (\dot{\mathbf{r}}\Psi) + \nabla_\mathbf{p} \cdot (\dot{\mathbf{p}}\Psi) = D\nabla_\mathbf{r}^2\Psi + D_R\nabla_\mathbf{p}^2\Psi, \tag{14.32a}$$

$$\dot{\mathbf{r}} = U\mathbf{p} + \mathbf{u}, \tag{14.32b}$$

$$\dot{\mathbf{p}} = (\mathbf{1} - \mathbf{pp}) \cdot (\mathbf{W} + B\mathbf{E}) \cdot \mathbf{p}, \tag{14.32c}$$

$$-\nabla q + \mu\nabla^2\mathbf{u} = \mathcal{P}\nabla \cdot \langle\mathbf{pp}\rangle, \tag{14.32d}$$

$$\nabla \cdot \mathbf{u} = 0. \tag{14.32e}$$

The dynamics of the cells and the flow field are strongly coupled. Organisms create stresses due to their swimming activity, which generate fluid flow; in turn, the flows affect the positions and orientations of the cells. While these equations were derived in the dilute limit, they are valid generically for all dipolar swimmers and can be supplemented by rules for near-field dynamics and collisions specific to an organism of interest.

14.2.4 Other Modelling Approaches to Active Matter

The model given by Eq. (14.32) is the result of mutiple approximations and assumptions about the nature of the fluid, the distribution of active particles in it and their interactions. Historically, it belongs to the realm of 'active matter', a subfield of physics that considers the collective dynamics of individual units able to convert an internal source of energy into local movement. Classical examples include schools of fish, the dynamics of the cellular cytoskeleton and the collective motion of bacteria. The field of active matter was kick-started at the end of the twentieth century with the discovery of non-equilibrium phase transitions in a model of flocks where moving particles attempt to align their velocity with that of their neighbours (Vicsek et al., 1995). Over the following 20 years, active matter has generated a wide range of new physical models and has matured into an important research area in its own right (Marchetti et al., 2013; Jülicher et al., 2018). Though it is beyond the scope of this book to present a thorough review, it is relevant to appreciate the variety of modelling approaches for active matter, and how they might differ from the one presented in the previous section.

Models for the physics of active matter can be first categorised according to the nature of the broken symmetries in the system (Marchetti et al., 2013). In polar active matter, the individual unit has a distinguishable front and back; this includes, for example, swimming bacteria characterised by a cell body (front) propelled by helical flagella (back). Order in polar active matter can then be measured by a vector order parameter, which is a local average of orientations, $\mathbf{n} = \frac{1}{c} \iint_{\Omega} \mathbf{p} \Psi \, d\mathbf{p}$ (using notation from the previous section). In contrast, in apolar active matter the front and back of the individual particles cannot be distinguished, and alignment is measured by the tensor order parameter, $\mathbf{Q} = \frac{1}{c} \iint_{\Omega} \left(\mathbf{pp} - \frac{1}{3}\mathbf{1} \right) \Psi \, d\mathbf{p}$, similar to what is done for nematic liquid crystals (Chaikin et al., 1995).

The second important distinction among models of active matter is the extent to which they include the dynamics of the surrounding fluid and conserve momentum. So-called wet models, such as that in Eq. (14.32), consider the dynamics of both the active units and that of the surrounding fluid, which mediates their interaction, and as a result the total momentum of the system is conserved. In contrast, dry active matter models include the fluid only as a source of friction that acts as a sink of momentum for the dynamics of the active elements.

Broadly speaking, active matter models that couple the dynamics of the concentration, c, the order parameters, $\mathbf{n}$ and $\mathbf{Q}$, and potentially also the fluid velocity, $\mathbf{u}$, have resulted from two distinct mathematical approaches. The first starts from a specific microscopic description, which is then coarse-grained to obtain continuum equations. This is the modelling framework adopted in this chapter where we considered the full probabilistic equations, and historical approaches along these lines are reviewed by Marchetti et al. (2013). The final system in Eq. (14.32) can then be turned into equations for c, $\mathbf{n}$ and $\mathbf{Q}$ by taking successive moments in orientational space for Ψ and postulating closure relationships for higher moments (Saintillan and Shelley, 2013).

The second modelling approach consists in postulating phenomenological hydrodynamic equations for the various fields of interest by combining the principles of statistical physics with new terms allowed by the symmetry of the system. This approach is often attributed to Toner and Tu (1998) for flocks and Simha and Ramaswamy (2002) for swimming cells. While this is not the approach taken in this book, it has proven popular in the physics community (Marchetti et al., 2013; Jülicher et al., 2018). Furthermore, phenomenological models are consistent with the microscopic approach and their predictions often agree. For example, ordered suspensions of swimming cells have been predicted phenomenologically to be unstable to long-wavelength fluctuations in orientation (Simha and Ramaswamy, 2002), a result that is recovered using the microscopic model in Section 14.3.2. It should be noted that phenomenological models of this second type come with a variety of names resulting from different modelling assumptions on potentially similar biological systems. For example, both active gels (Kruse et al., 2004, 2005) and active nematics (Doostmohammadi et al., 2018) have been proposed to capture the dynamics of cellular tissues, the cytoskeleton of living cells and *in-vitro* mixtures of microtubules and motor proteins.

14.3 Collective Instabilities

14.3.1 Experimental Observations

One of the hallmarks of collective cell locomotion is its display of chaotic and intermittent dynamics. Most experimental work in the area has been carried out for suspensions of swimming bacteria, with early quantitative measurements showing patterns of unsteady vortices and jets (Mendelson et al., 1999). The dynamic state of dense bacterial suspensions, often called bacterial turbulence, is characterised by velocity correlation lengths larger than the size of individual cells (Soni et al., 2003; Sokolov et al., 2007), as well as strong inhomogeneities in the density and the orientation of swimmers, with large clumps and zones of alignment (Wu and Libchaber, 2000; Dombrowski et al., 2004; Cisneros et al., 2007). An example of such dynamics is illustrated in Fig. 14.6(a), where we show an experimental snapshot of the collective motion of *Bacillus subtilis* bacteria near a water/air/glass contact line. A detailed analysis of the instantaneous vector field of swimming velocities of the cells is plotted in Fig. 14.6(b), with associated vorticity shown in Fig. 14.6(c).

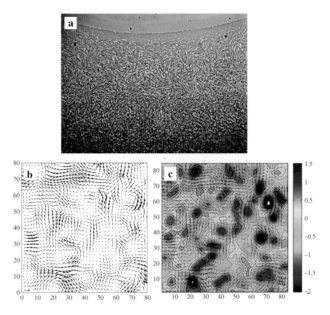

Figure 14.6 Collective locomotion in suspensions of swimming *Bacillus subtilis* bacteria (reprinted by permission from Springer Nature: Cisneros et al. (2007)). (a) Concentration of swimming cells near a water/air/glass contact line. (b) Vector field of instantaneous bacterial swimming velocity; the axis units are μm and the velocity scale is indicated by the lower-left horizontal arrow of magnitude $50\,\mu$m/s. (c) Vorticity of the swimming velocity field in (b), with scale in unit of 1/s.

Similar to the approach for tackling the transition to turbulence in traditional hydrodynamics, one route to understanding the physics of bacterial turbulence from first principles is to discover steady states of the active-fluid equations from Section 14.2 and study their stability. In what follows we consider two of these steady states, first the situation where all cells are aligned with each other and then the homogenous, isotropic state where all cells are equally likely to be located anywhere, and to be pointing in any direction.

14.3.2 Instability of Aligned State

We first consider the situation where all cells point initially in the same direction (polar order). This was addressed using a phenomenological approach by Simha and Ramaswamy (2002), and shown numerically to be unstable for all types of swimmers by Saintillan and Shelley (2007) (Fig. 14.7).

Setup

To characterise the stability of the aligned state, we use the continuum probabilistic model from Section 14.2. We assume that the cells are all aligned locally with a

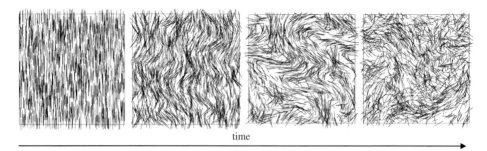

time

Figure 14.7 Numerical simulations showing the time evolution of a suspension of 2500 elongated pusher swimmers, initially aligned (left panel) and progressively becoming unstable (left to right) (reprinted figure with permission from Saintillan and Shelley (2007), Copyright 2007 by the American Physical Society).

preferred vector **n** and thus write the probability distribution function as

$$\Psi(\mathbf{r}, \mathbf{p}, t) = c(\mathbf{r}, t)\delta[\mathbf{p} - \mathbf{n}(\mathbf{r}, t)]. \tag{14.33}$$

Dynamics of Concentration and Orientation

If we further assume that diffusion can be neglected (i.e. $D = 0$ and $D_R = 0$), then we may derive closed evolution equations for the fields c and **n** (Saintillan and Shelley, 2008a). Computing the orientational average of Eq. (14.32a) assuming Eq. (14.33) and using the fact that an integral in **p** of a divergence in **p** is zero, we obtain

$$\frac{\partial c}{\partial t} + \nabla \cdot [c(U\mathbf{n} + \mathbf{u})] = 0, \tag{14.34}$$

where we have replaced $\nabla_{\mathbf{r}}$ by ∇ since the fields c and **n** are only functions of **r** and t. Similarly, when multiplying Eq. (14.32a) by **p** and computing its orientational average, we obtain, after one integration by parts in **p**,

$$\frac{\partial (c\mathbf{n})}{\partial t} + \nabla \cdot [c(U\mathbf{n} + \mathbf{u})\mathbf{n}] = c(\mathbf{1} - \mathbf{n}\mathbf{n}) \cdot (\mathbf{W} + B\mathbf{E}) \cdot \mathbf{n}. \tag{14.35}$$

Finally, combining Eq. (14.35) with Eq. (14.34) leads to an equation for **n** as

$$\frac{\partial \mathbf{n}}{\partial t} + (U\mathbf{n} + \mathbf{u}) \cdot \nabla \mathbf{n} = (\mathbf{1} - \mathbf{n}\mathbf{n}) \cdot (\mathbf{W} + B\mathbf{E}) \cdot \mathbf{n}. \tag{14.36}$$

Next, using the locally aligned assumption, Eq. (14.33), the equations for the incompressible fluid, Eqs. (14.32d)–(14.32e), become

$$-\nabla q + \mu \nabla^2 \mathbf{u} = \mathcal{P}\nabla \cdot (c\mathbf{n}\mathbf{n}), \quad \nabla \cdot \mathbf{u} = 0. \tag{14.37}$$

Linearisation

The aligned state where all cells point in the same direction, e.g. $\mathbf{n} = \mathbf{e}_z$, with uniform concentration, $c = c_0$, is a steady solution of the equations of motion, Eqs. (14.34), (14.36) and (14.37), with zero flow and pressure. To characterise its stability, we assume a small perturbation around it as

$$
\begin{pmatrix} c(\mathbf{r},t) \\ \mathbf{n}(\mathbf{r},t) \\ \mathbf{u}(\mathbf{r},t) \\ q(\mathbf{r},t) \end{pmatrix} = \begin{pmatrix} c_0 \\ \mathbf{e}_z \\ \mathbf{0} \\ 0 \end{pmatrix} + \epsilon \begin{pmatrix} c'(\mathbf{r},t) \\ \mathbf{n}'(\mathbf{r},t) \\ \mathbf{u}'(\mathbf{r},t) \\ q'(\mathbf{r},t) \end{pmatrix}, \tag{14.38}
$$

where ϵ is a small dimensionless parameter. Note that since $\mathbf{n}$ is a unit vector, we have $\mathbf{n}' \cdot \mathbf{e}_z = 0$. Substituting Eq. (14.38) into Eqs. (14.34), (14.36) and (14.37) and linearising lead to the equations for the perturbation at order ϵ as

$$
\left(\frac{\partial}{\partial t} + U \frac{\partial}{\partial z} \right) c' + U c_0 \nabla \cdot \mathbf{n}' = 0, \tag{14.39a}
$$

$$
\left(\frac{\partial}{\partial t} + U \frac{\partial}{\partial z} \right) \mathbf{n}' = (1 - \mathbf{e}_z \mathbf{e}_z) \cdot (\mathbf{W}' + B\mathbf{E}') \cdot \mathbf{e}_z, \tag{14.39b}
$$

$$
-\nabla q' + \mu \nabla^2 \mathbf{u}' = \mathcal{P} \nabla \cdot (c' \mathbf{e}_z \mathbf{e}_z + c_0 \mathbf{e}_z \mathbf{n}' + c_0 \mathbf{n}' \mathbf{e}_z), \tag{14.39c}
$$

$$
\nabla \cdot \mathbf{u}' = 0, \tag{14.39d}
$$

where $\mathbf{W}'$ and $\mathbf{E}'$ are the gradient tensors for the perturbed flow $\mathbf{u}'$.

Unstable Modes

We proceed to look for Fourier modes of Eq. (14.38) as

$$
\begin{pmatrix} c'(\mathbf{r},t) \\ \mathbf{n}'(\mathbf{r},t) \\ \mathbf{u}'(\mathbf{r},t) \\ q'(\mathbf{r},t) \end{pmatrix} = \begin{pmatrix} \tilde{c} \\ \tilde{\mathbf{n}} \\ \tilde{\mathbf{u}} \\ \tilde{q} \end{pmatrix} \exp(\sigma t + i\mathbf{k} \cdot \mathbf{r}), \tag{14.40}
$$

where we now have $\tilde{\mathbf{n}} \cdot \mathbf{e}_z = 0$. The solution to the flow problem in Fourier space, Eqs. (14.39c)–(14.39d), is obtained as

$$
\tilde{\mathbf{u}} = \frac{-i\mathcal{P}}{\mu k^2} (1 - \hat{\mathbf{k}}\hat{\mathbf{k}}) \cdot (\tilde{c}\mathbf{e}_z \mathbf{e}_z + c_0 \mathbf{e}_z \tilde{\mathbf{n}} + c_0 \tilde{\mathbf{n}} \mathbf{e}_z) \cdot \mathbf{k}, \tag{14.41}
$$

where $k = |\mathbf{k}|$ and $\hat{\mathbf{k}} = \mathbf{k}/k$. Noticing that if $\mathbf{k} \cdot \mathbf{e}_z = \mathbf{k} \cdot \tilde{\mathbf{n}} = 0$ then $\tilde{\mathbf{u}} = \mathbf{0}$, we may assume that the flow perturbation is in the $(\mathbf{e}_z, \tilde{\mathbf{n}})$ plane and write

$$
\mathbf{k} = k(\cos \theta \mathbf{e}_z + \sin \theta \tilde{\mathbf{n}}/\tilde{n}), \tag{14.42}
$$

where $\tilde{n} = |\tilde{\mathbf{n}}|$. Evaluating Eq. (14.39a) in Fourier space leads next to the first relationship between $\tilde{c}$ and $\tilde{n}$ as

$$(\sigma + iUk\cos\theta)\tilde{c} = -c_0 Uik\sin\theta\tilde{n}. \tag{14.43}$$

To compute the second one we need to examine Eq. (14.39b). With Eq. (14.41) we can compute the right-hand side of Eq. (14.39b), leading to

$$\tilde{\mathbf{E}} = \frac{i}{2}(\mathbf{k}\tilde{\mathbf{u}} + \tilde{\mathbf{u}}\mathbf{k}), \quad \tilde{\mathbf{W}} = \frac{i}{2}(\tilde{\mathbf{u}}\mathbf{k} - \mathbf{k}\tilde{\mathbf{u}}). \tag{14.44}$$

After some elementary algebra and trigonometry we obtain

$$(1 - \mathbf{e}_z\mathbf{e}_z)\cdot(\mathbf{k}\tilde{\mathbf{u}})\cdot\mathbf{e}_z = \frac{i\mathcal{P}}{\mu}\sin^2\theta\,(c_0\tilde{n}\cos 2\theta - \tilde{c}\sin\theta\cos\theta)\,\frac{\tilde{\mathbf{n}}}{\tilde{n}}, \tag{14.45a}$$

$$(1 - \mathbf{e}_z\mathbf{e}_z)\cdot(\tilde{\mathbf{u}}\mathbf{k})\cdot\mathbf{e}_z = -\frac{i\mathcal{P}}{\mu}\cos^2\theta\,(c_0\tilde{n}\cos 2\theta - \tilde{c}\sin\theta\cos\theta)\,\frac{\tilde{\mathbf{n}}}{\tilde{n}}, \tag{14.45b}$$

so that Eq. (14.39b) leads to the second relationship,

$$(\sigma + iUk\cos\theta)\tilde{n} = \frac{\mathcal{P}}{2\mu}(B\cos 2\theta + 1)(c_0\tilde{n}\cos 2\theta - \tilde{c}\sin\theta\cos\theta). \tag{14.46}$$

While the full eigenvalue analysis for the growth rate σ was carried out by Saintillan and Shelley (2008a), we focus here on the long-wavelength limit $k \to 0$. Setting $k = 0$ in Eqs. (14.43) and (14.46) allows us to obtain the eigenvalue system

$$\sigma\begin{pmatrix}\tilde{c}\\\tilde{n}\end{pmatrix} = \frac{\mathcal{P}}{2\mu}(B\cos 2\theta + 1)\begin{pmatrix}0 & 0\\-\sin\theta\cos\theta & c_0\cos 2\theta\end{pmatrix}\begin{pmatrix}\tilde{c}\\\tilde{n}\end{pmatrix}. \tag{14.47}$$

Since the matrix on the right-hand side of Eq. (14.47) is triangular, its eigenvalues are obtained directly from its diagonal elements. One eigenvalue is $\sigma = 0$ indicating a neutral mode and the other is obtained as

$$\sigma = \frac{\mathcal{P}c_0}{2\mu}(B\cos 2\theta + 1)\cos 2\theta. \tag{14.48}$$

Discussion

The result in Eq. (14.48) is the prediction from the continuum model for the growth rate of the long-wavelength mode for aligned swimmers. Since $0 \leqslant B < 1$ for an elongated swimmer, the term in parentheses is always positive as θ varies, while the second $\cos 2\theta$ term changes sign with θ. For any sign of $\mathcal{P}$, we can therefore always find values of θ leading to $\sigma > 0$. All elongated swimmer populations, be they pushers or pullers, are thus unstable in the aligned state. The unstable modes have nonzero amplitude of both concentration and orientation, so any orientational instability away from the aligned configuration is also associated with concentration instabilities. Notably, the final result for σ does not depend on the value of the swimming speed, U. The instability is thus purely a result of the active stresslets and is also present in non-motile, active systems.

14.3.3 Instability of Uniform Isotropic State

The next situation of interest is one where the cells in the suspension are distributed equally in both position ('uniform') and orientation ('isotropic').

Setup

If c_0 denotes the cell concentration in this base state, its probability distribution function is $\Psi_0 = c_0/(4\pi)$, leading to an isotropic tensor for the average swimmer stress, $\langle \mathbf{pp} \rangle \propto \mathbf{1}$, and thus a fixed point of Eq. (14.32) with zero associated flow and pressure. Using Ψ' to denote the dimensionless perturbation to the probability distribution function, we then write

$$\Psi = \frac{c_0}{4\pi} \left[1 + \epsilon \Psi'(\mathbf{r}, \mathbf{p}, t) \right], \tag{14.49}$$

where ϵ is a small dimensionless number. In a similar fashion to the previous section, we write the perturbations in velocity and pressure as $\mathbf{u} = \epsilon \mathbf{u}'$ and $q = \epsilon q'$.

Linearised Dynamics

We first consider the perturbed version of Eq. (14.32a) in the absence of diffusion. Using Eqs. (14.32b) and (14.32e), the second term on the left-hand side of Eq. (14.32a) is, at order ϵ, proportional to

$$\nabla_\mathbf{r} \cdot \left[(1 + \epsilon \Psi') (U\mathbf{p} + \mathbf{u}) \right] \approx \epsilon \nabla_\mathbf{r} \cdot (\Psi' U\mathbf{p}) = \epsilon U\mathbf{p} \cdot \nabla_\mathbf{r} \Psi'. \tag{14.50}$$

In order to compute the third term in Eq. (14.32a) we need to evaluate Eq. (14.32c) using vector calculus. Since $\nabla_\mathbf{p} \equiv (\mathbf{1} - \mathbf{pp}) \cdot \partial/\partial \mathbf{p}$, and since the tensors $\mathbf{W}'$ and $\mathbf{E}'$ are trace-free and, respectively, antisymmetric and symmetric, we obtain

$$\nabla_\mathbf{p} \cdot \left[(\mathbf{1} - \mathbf{pp}) \cdot \mathbf{W}' \cdot \mathbf{p} \right] = 0, \quad \nabla_\mathbf{p} \cdot \left[(\mathbf{1} - \mathbf{pp}) \cdot \mathbf{E}' \cdot \mathbf{p} \right] = -3\mathbf{p} \cdot \mathbf{E}' \cdot \mathbf{p}. \tag{14.51}$$

We may then write Eq. (14.32a) at order ϵ as

$$\frac{\partial \Psi'}{\partial t} + U\mathbf{p} \cdot \nabla_\mathbf{r} \Psi' = 3B\mathbf{p} \cdot \mathbf{E}' \cdot \mathbf{p}. \tag{14.52}$$

Finally, the perturbation to Eq. (14.32d) leads, at order ϵ, to the flow equation

$$-\nabla q' + \mu \nabla^2 \mathbf{u}' = \frac{\mathcal{P}c_0}{4\pi} \nabla \cdot \left(\iint_\Omega \Psi' \mathbf{pp} \, d\mathbf{p} \right). \tag{14.53}$$

Unstable modes

We look for Fourier modes of the perturbed equations, Eqs. (14.52)–(14.53),

$$
\begin{pmatrix} \Psi'(\mathbf{r}, \mathbf{p}, t) \\ \mathbf{u}'(\mathbf{r}, t) \\ q'(\mathbf{r}, t) \end{pmatrix} = \begin{pmatrix} \tilde{\Psi}(\mathbf{p}) \\ \tilde{\mathbf{u}} \\ \tilde{q} \end{pmatrix} \exp\left(\sigma t + i\mathbf{k} \cdot \mathbf{r}\right).
\tag{14.54}
$$

The solution to Eq. (14.53) in Fourier space is first obtained as

$$
\tilde{\mathbf{u}} = -\frac{i\mathcal{P}c_0}{4\pi k\mu} \left(\mathbf{1} - \hat{\mathbf{k}}\hat{\mathbf{k}}\right) \cdot \left(\iint_\Omega \tilde{\Psi}\mathbf{p}\mathbf{p} \, d\mathbf{p}\right) \cdot \hat{\mathbf{k}},
\tag{14.55}
$$

where $k = |\mathbf{k}|$ and $\hat{\mathbf{k}} = \mathbf{k}/k$.

In order to compute Eq. (14.52) in Fourier space we need to evaluate $\tilde{\mathbf{E}}$, i.e. the Fourier transform of $\mathbf{E}'$. Using Eqs. (14.44) and (14.55) we obtain

$$
\mathbf{p} \cdot \tilde{\mathbf{E}} \cdot \mathbf{p} = \frac{\mathcal{P}c_0}{4\pi\mu}(\mathbf{p} \cdot \hat{\mathbf{k}})\mathbf{p} \cdot \left(\mathbf{1} - \hat{\mathbf{k}}\hat{\mathbf{k}}\right) \cdot \left(\iint_\Omega \tilde{\Psi}\mathbf{p}\mathbf{p} \, d\mathbf{p}\right) \cdot \hat{\mathbf{k}},
\tag{14.56}
$$

which allows us to evaluate Eq. (14.52) in Fourier space as

$$
(\sigma + iU\mathbf{p} \cdot \mathbf{k})\,\tilde{\Psi}(\mathbf{p}) = \frac{3\mathcal{P}Bc_0}{4\pi\mu}(\mathbf{p} \cdot \hat{\mathbf{k}})\mathbf{p} \cdot \left(\mathbf{1} - \hat{\mathbf{k}}\hat{\mathbf{k}}\right) \cdot \left(\iint_\Omega \tilde{\Psi}\mathbf{p}\mathbf{p} \, d\mathbf{p}\right) \cdot \hat{\mathbf{k}}.
\tag{14.57}
$$

The full eigenvalue analysis was carried out by Saintillan and Shelley (2008a), and here again we focus on the long-wavelength limit, $k \to 0$. Since the vector on the right-hand side of Eq. (14.55) is perpendicular to $\hat{\mathbf{k}}$, along a direction denoted $\hat{\mathbf{k}}_\perp$, the long-wavelength solution to Eq. (14.57) can be written in the form

$$
\tilde{\Psi}(\mathbf{p}) = \Gamma(\mathbf{p} \cdot \hat{\mathbf{k}})(\mathbf{p} \cdot \hat{\mathbf{k}}_\perp).
\tag{14.58}
$$

This ansatz allows us to evaluate the tensor

$$
\iint_\Omega \tilde{\Psi}(\mathbf{p})\mathbf{p}\mathbf{p} \, d\mathbf{p} = \Gamma\hat{\mathbf{k}} \cdot \left(\iint_\Omega \mathbf{p}\mathbf{p}\mathbf{p}\mathbf{p} \, d\mathbf{p}\right) \cdot \hat{\mathbf{k}}_\perp = \frac{4\pi}{15}\Gamma\left(\hat{\mathbf{k}}_\perp\hat{\mathbf{k}} + \hat{\mathbf{k}}\hat{\mathbf{k}}_\perp\right),
\tag{14.59}
$$

where we have used the result of Eq. (9.52b) for an isotropic fourth-rank tensor. Substituting this result in Eq. (14.57) leads to

$$
\sigma\Gamma(\mathbf{p} \cdot \hat{\mathbf{k}})(\mathbf{p} \cdot \hat{\mathbf{k}}_\perp) = \frac{\mathcal{P}Bc_0}{5\mu}\Gamma(\mathbf{p} \cdot \hat{\mathbf{k}})(\mathbf{p} \cdot \hat{\mathbf{k}}_\perp),
\tag{14.60}
$$

and thus the growth rate of the Fourier mode, σ, is given by

$$
\sigma = \frac{\mathcal{P}Bc_0}{5\mu}.
\tag{14.61}
$$

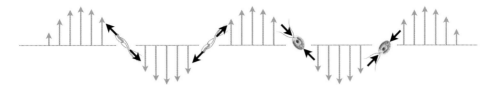

Figure 14.8 Physical interpretation of the instability of the uniform isotropic state (Subramanian and Koch (2009), reproduced with permission). Cells near the nodes of a Fourier mode of the flow field align preferably with the local extensional axis. The local flows created by active pushers reinforce the perturbation (left) while pullers act to decrease the perturbation amplitude (right). Drawings courtesy of Jacques Lauga.

Discussion

The simple result in Eq. (14.61) makes an important prediction. Spherical cells with $B = 0$ are neutrally stable, but elongated ($B > 0$) cells with a negative dipole strength, $\mathcal{P} < 0$, have a negative growth rate, $\sigma < 0$, and therefore suspensions of pullers are stable. In contrast, elongated pusher cells have $\mathcal{P} > 0$, so that $\sigma > 0$ and thus uniform isotropic suspensions of flagellated bacteria, for example, are unstable to long-wavelength perturbations. This theoretical prediction remains valid for finite wavelengths (which have smaller growth rates) and was confirmed by numerical simulations (Saintillan and Shelley, 2008a; Hohenegger and Shelley, 2010; Saintillan and Shelley, 2013). Notably, the long-wavelength growth rate in Eq. (14.61) does not depend on the value of U, and therefore active non-motile elongated cells would display the same instability.

A physical interpretation of the mechanism for this instability was put forward by Subramanian and Koch (2009) and is illustrated in Fig. 14.8. A sinusoidal perturbation in the velocity of the fluid (light arrows) leads to anisotropy in the orientation distribution of cells. Near the nodes of the flow perturbation, the elongated swimmers have a preferred nematic alignment along the extensional axis of the flow. Examining the direction of the local forcing from the cells on the surrounding fluid (dark arrows) we see that pushers act to amplify the flow perturbation whereas pullers act to decrease it. This instability can also be interpreted in terms of a negative apparent viscosity for the active suspension (Subramanian and Koch, 2009).

A final important feature of the unstable mode in Eq. (14.58) is the absence of associated concentration instabilities. This can be seen by evaluating the Fourier amplitude of the cell concentration, $\tilde{c}$, given by

$$\tilde{c} = \iint_{\Omega} \tilde{\Psi} \, d\mathbf{p} = \Gamma \hat{\mathbf{k}} \cdot \left(\iint_{\Omega} \mathbf{p}\mathbf{p} \, d\mathbf{p} \right) \cdot \hat{\mathbf{k}}_{\perp}. \qquad (14.62)$$

Since the tensor $\iint_\Omega \mathbf{pp}\, d\mathbf{p}$ is isotropic, $\iint_\Omega \mathbf{pp}\, d\mathbf{p} \propto \mathbf{1}$, and due to the orthogonality of $\hat{\mathbf{k}}_\perp$ and $\hat{\mathbf{k}}$, we obtain $\tilde{c} = 0$. At leading order in ϵ, the model predicts that the orientational instability is not accompanied by any instabilities in concentration. Numerical computations later showed that concentration fluctuations originate from nonlinearities in the model (Saintillan and Shelley, 2013).

Further Reading

Hydrodynamic interactions between swimming cells were addressed in this chapter in the far-field limit and we focused on the instantaneous physical picture of how flows from one cell affect others. Time-varying cell-cell interactions can also be important, arising either from time variations in the swimmer flows or from the fact that they move along trajectories and therefore their relative positions and orientations change in time. The fully coupled dynamics of two swimmers was addressed computationally for flagellated bacteria by Ishikawa et al. (2007). The spherical squirmer model from Chapter 4 can be used to map out the full relationship between the initial position and orientations of two cells and their trajectories (Ishikawa et al., 2006) and to also make analytical progress beyond the dilute limit (Papavassiliou and Alexander, 2017). For swimmers with confined trajectories, the far-field limit can be used to derive the long-time dynamics of two coupled cells (Michelin and Lauga, 2010b). Unsteady hydrodynamic interactions can also be exploited to induce collective locomotion in non-swimmers (Lauga and Bartolo, 2008; see Exercises).

The instability of uniform isotropic suspensions of elongated pusher cells was derived in this chapter in the three-dimensional, unbounded case. For confined swimmers, hydrodynamic interactions change in nature and become independent of the microscopic details of the cellular propulsion mechanism, leading to instabilities that depend on the polarity in the swimmer shape and induce spontaneous large-scale directed motion (Brotto et al., 2013). For suspensions of run-and-tumble bacteria, the tumbling motion provides an additional orientation decorrelation mechanism, which, although it cannot be described as an effective thermal process, can be included in the model by modifying the conservation of probability in Eq. (14.32a). As a result, the instability of the uniform isotropic state occurs in this case only for cell concentrations above a critical value (Subramanian and Koch, 2009; Koch and Subramanian, 2011).

Beyond stability characteristics, the nonlinear dynamics of swimmer suspensions leads to a wide range of pattern formation. Spermatozoa swimming along circular trajectories can self-organise on surfaces along vortices with local hexagonal order (Riedel et al., 2005). The collective motion of swimming bacteria shows strong spatiotemporal coherence in a manner that depends on the concentration of cells (Sokolov et al., 2007) and is governed ultimately by the balance between long-range hydrodynamic interactions and short-range cell-cell collisions (Sokolov and Aranson, 2012). Bacteria that have differentiated to an elongated and hyper-flagellated 'swarming' state also show collective modes of locomotion at very high density (Copeland and Weibel, 2009; Kearns, 2010; Darnton et al., 2010), while chemically active particles can be used to create active crystals (Palacci et al., 2013). From a physical point of view, locomotion can lead to a new type of non-equilibrium phase transition, termed motility-induced phase separation, in which active swimmers slowing down in high-density regions induce a separation between dense and dilute phases (Cates and Tailleur, 2015).

Another class of collective processes does not require hydrodynamic interactions but involves the cooperative action of large numbers of microorganisms. The classical example is bioconvection, in which upward-swimming bottom-heavy algae gather at the surface of a liquid. When the cells are slightly heavier than the surrounding fluid, accumulation in large numbers leads to an overturning instability in the fluid accompanied by downward plumes and flow recirculation (Childress et al., 1975; Pedley and Kessler, 1992). Other examples involve the cooperative motion of bacteria powering microscopic gears (Sokolov et al., 2010), creating recirculation flows around dense swarms (Wu et al., 2011; Dauparas and Lauga, 2016) and enabling the transport and chemotaxis of synthetic particles to which they are attached (Zhuang and Sitti, 2016; Zhuang et al., 2017).

Exercises

1. Consider the orientation stability of a pair of parallel swimmers in the (x, y) plane. The cells are identical, separated by $\mathbf{r} = r\mathbf{e}_x$, and they interact hydrodynamically as axisymmetric force dipoles in the far field, so their reorientation is described by Jeffery's equation with shape factor B. Both swimmers, of orientations $\mathbf{p}$ and $\tilde{\mathbf{p}}$, are tilted by small angles away from their swimming directions. Compute the growth rates of the parallel and mirror-image modes for the side-by-side ($\mathbf{p} = \tilde{\mathbf{p}}$, $\mathbf{p} \cdot \mathbf{r} = 0$) and front-back ($\mathbf{p} = \tilde{\mathbf{p}}$, $\mathbf{p} \times \mathbf{r} = \mathbf{0}$) configurations. Recover the results obtained graphically in Section 14.1.2 in the case of spherical swimmers ($B = 0$) and characterise the stability map of elongated swimmers as the value of B increases.

2. Two identical elongated cells swim in the (x, y) plane and interact hydrodynamically in the far field as axisymmetric force dipoles. Their orientation obeys Jeffery's equation with shape factor $B > 0$. Assuming that the swimmers always remain parallel, show that there exists, for a range of values of B, a new fixed point in the orientation dynamics different from side-by-side and front-back, and characterise its stability.

3. For the continuum model of an active fluid Section 14.2 with probability distribution function $\Psi(\mathbf{r}, \mathbf{p}, t)$, we define the total entropy of the system, S, as an integral over position and orientation spaces:

$$S = \iiint_V \iint_\Omega \frac{\Psi}{\Psi_0} \ln\left(\frac{\Psi}{\Psi_0}\right) d\mathbf{p}\, d\mathbf{r},$$

with $\Psi_0 = 1/(4\pi)$. The cells in the active fluid are modelled as axisymmetric force dipoles of strength $\mathcal{P}$ and are elongated with shape parameter B. Show that the rate of change of entropy, $\dot{S}$, is given by

$$\frac{\dot{S}}{4\pi} = \frac{6B}{\mathcal{P}} \iiint_V \mathbf{E} : \mathbf{E}\, d\mathbf{r} - \iiint_V \iint_\Omega \left(D|\nabla_\mathbf{r} \ln \Psi|^2 + D_R|\nabla_\mathbf{p} \ln \Psi|^2\right) d\mathbf{p}\, d\mathbf{r},$$

where $\mathbf{E}$ is the rate of strain tensor for the fluid and D and D_R are the diffusivities in translation and rotation.

4. The clustering instability of focused cells was analysed in Section 14.1.3 using a discrete model for the cells and their interactions. Show that the continuum model in Section 14.2 can be used to recover the same result.

5. Using the framework of active matter in Section 14.2, show that the swimmer concentration, $c(\mathbf{r}, t) \equiv \int_\Omega \Psi\, d\mathbf{p}$, satisfies an advection-diffusion equation with a forcing term involving the vector order parameter, $\mathbf{n} \equiv \frac{1}{c} \int_\Omega \mathbf{p}\Psi\, d\mathbf{p}$.

6. Although an isolated reciprocal swimmer cannot swim at low Reynolds number, multiple reciprocal swimmers can exploit hydrodynamic interactions to self-propel. Consider two reciprocal swimmers modelled as parallel axisymmetric force dipoles $\mathbf{p}$ and $\tilde{\mathbf{p}}$, with $\mathbf{p} = \tilde{\mathbf{p}} = \mathbf{e}_x$. The dipoles are aligned along x so that the flow remains axisymmetric. The strengths of the dipoles are $\mathcal{P}_1(t) = \mathcal{P}_0 \cos(\omega t)$ and $\mathcal{P}_2(t) = \mathcal{P}_0 \cos(\omega t + \phi)$. Assuming that dipole i moves at the velocity set up by dipole $j \neq i$ in its absence, show that the two-dipole system undergoes a net collective translation, and calculate the average speed at leading order in $\mathcal{P}_0$.

7. Consider two dumbbells aligned with the x direction. The first one is composed of two spheres of radii a_1 and a_2 separated by the time-varying length $\ell_a(t)$, while the spheres in the second dumbbell have radii b_1 and b_2 and are separated by $\ell_b(t)$. Both dumbbells undergo periodic oscillations of amplitudes δ_a and δ_b with angular frequency ω, and they are separated by a distance $d(t)$, with $d \gg a_i, b_i, \ell_j$. Calculate the time-averaged swimming speed of the two-dumbbell system at leading order in the oscillation amplitudes.

15

Locomotion and Transport in Complex Fluids

In many situations of biological significance, self-propelled cells and slender appendages interact with complex fluids that display non-Newtonian behaviour. Important examples include the locomotion of mammalian spermatozoa in cervical mucus, the motion of bacteria in biofilms and the clearance of mucus by airway cilia. The hydrodynamic principles reviewed in Chapter 2 and used in the rest of this book focused solely on Newtonian fluids. In this final chapter, we revisit these results for fluids characterised by a nonlinear, or non-instantaneous, relationship between stress and deformation. We first consider linear viscoelastic fluids. We show how the presence of memory in the fluid affects the hydrodynamic forces on, and the energy expended by, slender filaments, and compare a modified active-filament model of spermatozoa to experiments in viscoelastic fluids. We next address fluids with nonlinear rheological properties and their impact on the locomotion of, and transport by, Taylor's waving sheet. When the wave kinematics are prescribed, the nonlinear fluid leads to decreases in both speed and energy expenditure, which compares favourably to experiments using small nematodes. If instead the wave results from a balance between activity, elasticity and fluid forces, a transition to enhanced motion is possible for sufficiently flexible swimmers. We finish by addressing heterogeneous fluids and show how the multiscale nature of the fluid systematically enhances flow and transport.

15.1 Locomotion and Transport in Linear Viscoelastic Fluids

Motivated by the importance of polymers in biology, we first consider in this section the case of linear viscoelastic fluids. Classical references on the physics and structure of polymers are the textbooks by Doi and Edwards (1988) and Larson (1999), while the mathematical approach to rheology is reviewed in the books by Bird et al. (1987a) and Morrison (2001).

15.1.1 Viscoelasticity

To facilitate physical intuition for viscoelasticity, the simplest approach is to consider the one-dimensional shear deformation of a fluid bounded by two infinite, rigid plates separated by a distance h. The top plate is displaced by a distance $\gamma(t)h$ relative to the bottom plate in the direction parallel to the plates, where $\gamma > 0$ is the dimensionless shear strain. If the fluid is Newtonian, the shear stress, σ, required to move the plate is

$$\sigma = \eta\dot{\gamma}, \tag{15.1}$$

where $\dot{\gamma} \equiv d\gamma/dt$ is the shear rate, and η the shear (or dynamic) viscosity with dimension of stress times time (we use the notation η as done conventionally for complex fluids, instead of μ).

A Newtonian fluid is characterised by a constant shear viscosity and therefore a shear stress proportional to the rate of deformation. In contrast, a viscoelastic fluid has a history-dependent relationship between stress and shear rate. For polymer solutions, the physical origin of this time dependence is the entropic elasticity of polymer molecules, which confers a mix of viscous and elastic properties to the fluid.

Springs and Dashpots

It has been a long-standing tradition in the rheology community to illustrate elementary rheological behaviour of fluids visually. For example, the Newtonian fluid in the example above may be represented by a dashpot, indicating a viscous response for all times (Fig. 15.1(a)). If instead the material between the two plates is purely elastic, the shear stress is proportional to the displacement and is given by

$$\sigma = G\gamma, \tag{15.2}$$

where G is the shear (strain) modulus, with dimension of stress. In contrast to a Newtonian fluid that responds instantaneously to velocities, a purely elastic material responds instantaneously to displacements, and it is illustrated as a spring (Fig. 15.1(b)).

Maxwell Fluid and Stress Relaxation

Viscoelastic materials exhibit a combination of viscous and elastic behaviours. Since we are interested in materials that behave as fluids in the long-time limit, the simplest linear superposition of viscous and elastic responses consists of a spring and dashpot connected in series. This is the classical Maxwell fluid (Fig. 15.1(c)).

To derive the constitutive relationship for such a fluid, we denote by σ_1 the stress in the spring and σ_2 that in the dashpot. Since the spring and dashpot are connected in series, applying Newton's third law to each element in the sequence

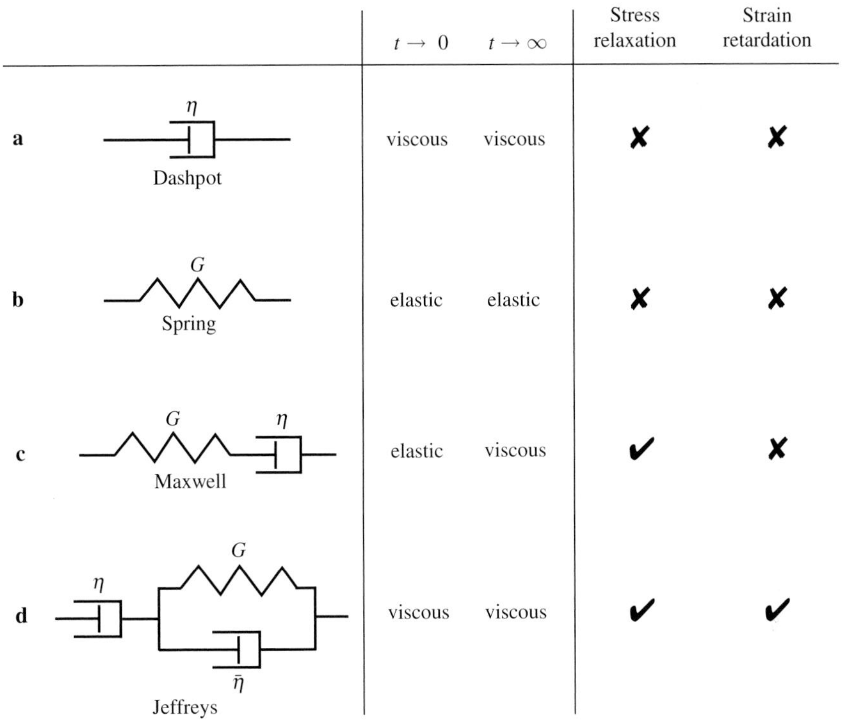

		$t \to 0$	$t \to \infty$	Stress relaxation	Strain retardation
a	Dashpot (η)	viscous	viscous	✘	✘
b	Spring (G)	elastic	elastic	✘	✘
c	Maxwell (G, η)	elastic	viscous	✔	✘
d	Jeffreys (η, G, $\bar{\eta}$)	viscous	viscous	✔	✔

Figure 15.1 Viscoelastic fluids as springs and dashpots. (a) A viscous fluid is represented by a dashpot. (b) An elastic material is represented by a spring. (c) A Maxwell fluid, composed of a spring and dashpot in series, is a viscoelastic fluid with short-time elastic and long-time viscous responses with stress relaxation. (d) A Jeffreys viscoelastic fluid, composed of a connection in series between a dashpot and an element made up of a parallel spring and dashpot connection, features both stress relaxation and strain retardation.

leads to equality of the stresses, $\sigma_1 = \sigma_2 \equiv \sigma$. In contrast, the displacements are added, so the total strain for the fluid is given by the sum $\gamma \equiv \gamma_1 + \gamma_2$. Since $\sigma = \sigma_1 = G\gamma_1$ and $\sigma = \sigma_2 = \eta\dot{\gamma}_2$, we have

$$\dot{\gamma} = \dot{\gamma}_1 + \dot{\gamma}_2 = \frac{\dot{\sigma}}{G} + \frac{\sigma}{\eta}, \qquad (15.3)$$

which may be rewritten classically as

$$\sigma + \lambda\frac{\partial\sigma}{\partial t} = \eta\dot{\gamma}, \quad \lambda = \frac{\eta}{G}. \qquad (15.4)$$

Note that we have used partial derivatives in Eq. (15.4) because later in this chapter stresses will also be allowed to vary spatially.

The timescale, λ, introduced in Eq. (15.4) is called the relaxation time for the fluid. Over short times, $t \ll \lambda$, the time derivative in Eq. (15.4) dominates and the complex fluid responds like an elastic solid. In contrast, at long times, $t \gg \lambda$, the time derivative may be neglected and the response is that of a viscous fluid with viscosity η. The limit of an elastic solid is therefore $\lambda \to \infty$ while that of a viscous fluid is $\lambda \to 0$.

The term 'relaxation' arises from the stress variation in the fluid when motion is stopped. Indeed, consider a shearing motion under constant stress, σ_0. If zero displacements are imposed starting at $t = 0^+$, then the solution to Eq. (15.4) is given for $t > 0$ by $\sigma = \sigma_0 e^{-t/\lambda}$ and the stress decays exponentially to zero over the characteristic timescale λ. The relaxation timescale is therefore the relevant timescale for stresses in the fluid to adapt to changes in deformation.

Jeffreys Fluid and strain Retardation

For the Maxwell fluid above, stresses adapt with some delay to changes in kinematics. In contrast, changes to stresses on the left-hand side of Eq. (15.4) are experienced immediately by the deformation in the fluid with no delay. In particular, a stress-free fluid is brought to rest immediately.

In order to model the ability of the fluid deformation to have memory of the stress state, we introduce the second classical model of linear viscoelasticity, the Jeffreys fluid. This fluid, illustrated in Fig. 15.1(d), has a dashpot (viscosity η) connected in series with an element made up of a spring (shear modulus G) and a dashpot (viscosity $\bar{\eta}$) connected in parallel.

The Jeffreys constitutive relationship may be derived by noting that in the isolated dashpot, the stress is given by $\sigma_1 = \eta \dot{\gamma}_1$, while in the parallel spring-dashpot element, the stress is written as the sum $\sigma_2 = G\gamma_2 + \bar{\eta}\dot{\gamma}_2$. Since both elements are connected in series, the total stress is $\sigma \equiv \sigma_1 = \sigma_2$ while displacements add up as $\gamma = \gamma_1 + \gamma_2$, and in particular $\dot{\gamma} = \dot{\gamma}_1 + \dot{\gamma}_2$. Applying the differential operator $G + \bar{\eta}\partial/\partial t$ to this last equality leads to

$$\left(1 + \lambda_1 \frac{\partial}{\partial t}\right)\sigma = \eta\left(1 + \lambda_2 \frac{\partial}{\partial t}\right)\dot{\gamma}, \quad \lambda_1 = \frac{\eta + \bar{\eta}}{G}, \quad \lambda_2 = \frac{\bar{\eta}}{G}. \tag{15.5}$$

The relationship between stresses and strains in Eq. (15.5) now involves two different timescales: λ_1 is the relaxation timescale with the same physical interpretation as for the Maxwell fluid; in contrast, λ_2 is termed the strain retardation time, and characterises the relevant timescale for deformation in the fluid to adapt to changes in stresses. In particular, the stress-free state of a Jeffreys fluid has residual rates of strain that decay exponentially as e^{-t/λ_2}. Importantly, we note that $\lambda_2 < \lambda_1$.

On short timescales, i.e. for deformation rates faster than both λ_1^{-1} and λ_2^{-1}, the behaviour in Eq. (15.5) is viscous with a viscosity of $(1/\eta + 1/\bar{\eta})^{-1}$, resulting from the two dashpots being in series, while at long times (small deformation rates), the dynamics is also Newtonian but governed by the isolated dashpot with viscosity η.

Deborah Numbers

In the context of cell propulsion, the relevant timescales for the viscoelastic fluid have to be compared to the characteristic timescale for the oscillatory motion of the microorganism. Denoting by ω the frequency characterising the motion of cellular appendages (flagella or cilia), the importance of viscoelasticity is quantified by the two dimensionless Deborah numbers

$$\text{De}_1 = \lambda_1 \omega, \quad \text{De}_2 = \lambda_2 \omega, \tag{15.6}$$

and we note that, for nonzero frequencies, we always have $\text{De}_2 < \text{De}_1$ since $\lambda_2 < \lambda_1$. Swimming cells with small Deborah numbers do not feel the viscoelastic nature of the fluid, whereas as soon as De_1 is of order one, or more, we expect viscoelasticity to play an important role. Biologically, the typical relaxation timescale of cervical mucus is at least on the order of seconds (Katz and Berger, 1980). Spermatozoa flagella, which oscillate with frequencies of order 10 Hz, are thus in the high De_1 regime. Similarly, the relaxation time for airway mucus is on the order of tens of seconds (Gilboa and Silberberg, 1976), and therefore the non-Newtonian nature of the fluid should impact ciliary transport strongly.

Linear Viscoelastic Fluid

The two model fluids considered above for unidirectional shear flows are special one-dimensional cases of the broad category of incompressible (i.e. satisfying $\nabla \cdot \mathbf{u} = 0$) linear viscoelastic fluids. They obey a linear constitutive relationship between the deviatoric stress tensor (i.e. the non-isotropic part of the stress), $\boldsymbol{\tau}(\mathbf{r}, t)$, and the shear rate tensor, $\dot{\boldsymbol{\gamma}}(\mathbf{r}, t) = 2\mathbf{E} = \nabla \mathbf{u} + (\nabla \mathbf{u})^T$. For a Maxwell fluid, this relationship takes the form of a tensorial version of Eq. (15.4),

$$\left(1 + \lambda \frac{\partial}{\partial t}\right) \boldsymbol{\tau} = \eta \dot{\boldsymbol{\gamma}}, \tag{15.7}$$

while for a Jeffreys fluid, Eq. (15.5) becomes

$$\left(1 + \lambda_1 \frac{\partial}{\partial t}\right) \boldsymbol{\tau} = \eta \left(1 + \lambda_2 \frac{\partial}{\partial t}\right) \dot{\boldsymbol{\gamma}}. \tag{15.8}$$

The most general relationship for a linear viscoelastic fluid is of the form

$$\boldsymbol{\tau}(\mathbf{r}, t) = \int_{-\infty}^{t} G(t - t') \dot{\boldsymbol{\gamma}}(\mathbf{r}, t') \, dt', \tag{15.9}$$

with a memory kernel, $G(u)$, termed the relaxation modulus and nonzero only for $\tau \geqslant 0$ (principle of causality). For a Newtonian fluid, G is a half-delta function, $G(u) = 2\eta \delta(u)|_{u \geqslant 0}$, and the stress only depends on the instantaneous value of the rate of strain. For a Maxwell fluid, it is straightforward to turn Eq. (15.7) into

Eq. (15.9) using an exponential integration factor, and to obtain $G(u) = (\eta/\lambda)e^{-u/\lambda}$. Therefore, a Maxwell fluid has an exponential loss of memory (see Exercises for a Jeffreys fluid).

15.1.2 Viscoelastic Resistive-Force Theory

How do relaxation and retardation affect hydrodynamic forces on slender filaments? For any fluid with a nonlinear constitutive relationship, it is expected that the rheological behaviour is linear for small-amplitude motion, and we can exploit this linearity to compute hydrodynamic forces at leading order (Fulford et al., 1998).

Consider a viscoelastic fluid whose rheological behaviour is described by the linear viscoelastic model of Eq. (15.9). Since cell locomotion always involves oscillatory motion, let us assume a periodic time dependence and use a Fourier series with lowest frequency ω. We may thus write for the shear rate and deviatoric stress

$$\dot{\gamma}(\mathbf{r}, t) = \sum_{n=-\infty}^{\infty} \dot{\gamma}^{(n)}(\mathbf{r})e^{-in\omega t}, \quad \tau(\mathbf{r}, t) = \sum_{n=-\infty}^{\infty} \tau^{(n)}(\mathbf{r})e^{-in\omega t}. \tag{15.10}$$

Substituting into Eq. (15.9), we obtain the relationship

$$\tau^{(n)}(\mathbf{r}) = \eta^{(n)}(\omega)\dot{\gamma}^{(n)}(\mathbf{r}), \tag{15.11}$$

where the complex viscosity, $\eta^{(n)}(\omega)$, is related to the Fourier transform of the relaxation modulus as

$$\eta^{(n)}(\omega) = \int_0^\infty G(u)e^{in\omega u}\, du. \tag{15.12}$$

The total stress in the fluid, σ, is the sum of the isotropic pressure and the deviatoric stress, $\sigma = -p\mathbf{1} + \tau$. The equation for mechanical equilibrium is $\nabla \cdot \sigma = \mathbf{0}$ (see Eq. (2.13)) and therefore we always have

$$\nabla p = \nabla \cdot \tau. \tag{15.13}$$

Using Eq. (15.11), we obtain in Fourier space the Stokes equations,

$$\nabla p^{(n)} = \eta^{(n)}(\omega)\nabla^2 \mathbf{u}^{(n)}, \tag{15.14}$$

with a complex, frequency-dependent viscosity at each Fourier mode n.

In order to be able to use this result to compute hydrodynamic forces on slender filaments, we have to be allowed to Fourier transform not just the governing equation but also its boundary conditions. The boundary conditions for the fluid velocity, $\mathbf{u}$, near a moving slender filament with a centreline located at $\mathbf{x}(s,t)$ are written in real space as

$$\mathbf{u}\big(\mathbf{x}(s,t),t\big) = \frac{\partial \mathbf{x}}{\partial t}. \tag{15.15}$$

We cannot use Fourier series directly yet, because the boundary condition in Eq. (15.15) is evaluated on a filament whose position is a function of time, so there are contributions from higher-order nonlinear terms (similar to the squirming boundary-perturbation problem from Chapter 4). In order to carry out this calculation rigorously, we have to linearise the problem. Consider, thus, small-amplitude motion where we write $\mathbf{x}(s,t) = \mathbf{x}_0(s) + \epsilon \mathbf{x}_1(s,t)$ with $\epsilon \ll 1$. At leading order, Eq. (15.15) becomes

$$\mathbf{u}_1\big(\mathbf{x}_0(s),t\big) = \frac{\partial \mathbf{x}_1}{\partial t}. \tag{15.16}$$

At linear order in the dimensionless amplitude ϵ, the governing equation and boundary conditions for each Fourier mode, Eqs. (15.14) and (15.16), are identical to the Newtonian ones. To solve the viscoelastic problem, we can thus use the Newtonian solution and, for each Fourier mode, just replace the Newtonian viscosity by the mode-dependent complex value for the viscosity, $\eta^{(n)}(\omega)$.

All results from the previous chapters can therefore be extended formally at leading order in the deformation using a complex viscosity. In particular, we can extend the resistive-force theory derived in Chapter 6 for slender filaments. Recall that in the Newtonian limit we have the linear relationship between the hydrodynamic force per unit length and velocity of the filament given by Eq. (6.40). For a complex fluid, we decompose the leading-order forces in Fourier modes as

$$\mathbf{f}(s,t) = \sum_{n=-\infty}^{\infty} \mathbf{f}^{(n)}(s) e^{-in\omega t}, \tag{15.17}$$

and using the mathematical analogy with Eq. (6.40) we can state directly a viscoelastic resistive-force theory for each Fourier mode as

$$\mathbf{f}^{(n)}(s) = -\left[c_{\parallel}^{(n)} \mathbf{tt} + c_{\perp}^{(n)} (\mathbf{1} - \mathbf{tt}) \right] \cdot \mathbf{u}^{(n)}(s), \tag{15.18}$$

where $\mathbf{t}$ is the leading-order tangent vector, $\mathbf{t} = \mathbf{t}_0$, and the complex drag coefficients are given in Fourier space by

$$c_{\perp}^{(n)} = 2c_{\parallel}^{(n)}, \quad c_{\parallel}^{(n)} = \frac{2\pi\eta^{(n)}(\omega)}{\ln(L/a)}. \tag{15.19}$$

It is important to emphasise that the viscoelastic resistive-force theory in Eq. (15.18) is valid only at leading order (order ϵ) in the amplitude of the motion and cannot be used to compute the swimming speeds of waving filaments.

Indeed, not only does swimming always scale quadratically with the waving amplitude but, as will be explained in Section 15.2, a fluid can never be truly linearly viscoelastic, and therefore any quadratic term in the constitutive relationship is bound to also contribute to the swimming speed.

15.1.3 Rate of Work for Locomotion in Viscoelastic Fluid

While it cannot be used to calculate swimming speeds, the viscoelastic resistive-force theory derived in Eq. (15.18) can be exploited to compute the energy expenditure of oscillating filaments because its leading-order value is given explicitly by the product of two order-ϵ quantities. Indeed, as first seen in Chapter 5, the rate of work of a slender filament on the fluid, $\dot{W}$, is a quadratic quantity equal to the product of the filament velocity and hydrodynamic force

$$\dot{W}(t) = -\int \mathbf{f} \cdot \mathbf{u}\,ds. \qquad (15.20)$$

For small-amplitude motion, the leading-order value of $\dot{W}$ (order ϵ^2) can thus be obtained by using the linearised values of both $\mathbf{u}$ and $\mathbf{f}$ in Eq. (15.20).

Consider a slender filament undergoing waving motion with fundamental frequency ω. As in Chapter 5, the location of the filament centreline is denoted by $y(x, t)$ in Cartesian coordinates with x along the filament and y perpendicular to it. The waveform is a superposition of Fourier modes as

$$y(x, t) = \sum_{n=-\infty}^{\infty} y^{(n)} e^{in(kx-\omega t)}, \qquad (15.21)$$

so that the velocity at leading order is

$$u_y(x, t) = \frac{\partial y}{\partial t} = \sum_{n=-\infty}^{\infty} u^{(n)} e^{in(kx-\omega t)}, \quad u^{(n)} = -in\omega y^{(n)}. \qquad (15.22)$$

Similarly, the y component of the leading-order hydrodynamic force on the filament is written using Fourier modes as

$$f_y(x, t) = \sum_{n=-\infty}^{\infty} f^{(n)} e^{in(kx-\omega t)}. \qquad (15.23)$$

The instantaneous rate of work by the filament over one wavelength $2\pi/k$, Eq. (15.20), is given at leading order by

$$\dot{W}(t) = -\int_0^{2\pi/k} f_y(x, t) u_y(x, t)\,dx. \qquad (15.24)$$

Using the Fourier decompositions in Eqs. (15.21) and (15.23), we obtain that the time average of the rate of work per wavelength, $\langle \dot{W} \rangle$, is given by the infinite sum of Fourier modes

$$\langle \dot{W} \rangle = -\frac{2\pi}{k} \sum_{n=-\infty}^{\infty} f^{(n)} u^{(-n)} = -\frac{2\pi}{k} i\omega \sum_{n=-\infty}^{\infty} n f^{(n)} y^{(-n)}. \tag{15.25}$$

To calculate the rate of work in the case of a non-Newtonian fluid, $\dot{W}_{NN}$, we use the leading-order viscoelastic resistive-force theory in Eq. (15.18),

$$f^{(n)} = -c_\perp^{(n)} u^{(n)} = i n \omega c_\perp^{(n)} y^{(n)}. \tag{15.26}$$

Considering the case of a Jeffreys fluid, the complex viscosity from Eq. (15.12) is obtained by evaluating Eq. (15.5) for each Fourier mode as

$$\eta^{(n)}(\omega) = \eta \frac{1 - i n \text{De}_2}{1 - i n \text{De}_1}, \tag{15.27}$$

where η is the long-time viscosity of the fluid. Writing Eq. (15.19) as $c_\perp^{(n)}/c_\perp = \eta^{(n)}/\eta$, where $c_\perp$ is the Newtonian drag coefficient, we obtain

$$f^{(n)} = i n \omega c_\perp \frac{1 - i n \text{De}_2}{1 - i n \text{De}_1} y^{(n)}, \tag{15.28}$$

and therefore, using the fact that $y^{(n)} y^{(-n)} = \left| y^{(n)} \right|^2$, Eq. (15.25) becomes

$$\dot{W}_{NN} = \frac{2\pi}{k} c_\perp \omega^2 \sum_{n=-\infty}^{\infty} \frac{1 - i n \text{De}_2}{1 - i n \text{De}_1} n^2 \left| y^{(n)} \right|^2. \tag{15.29}$$

The imaginary part of this equation is zero by symmetry, and we get finally

$$\dot{W}_{NN} = \frac{2\pi}{k} c_\perp \omega^2 \sum_{n=-\infty}^{\infty} \frac{1 + n^2 \text{De}_1 \text{De}_2}{1 + n^2 \text{De}_1^2} n^2 \left| y^{(n)} \right|^2. \tag{15.30}$$

Comparing this formula with the Newtonian case ($\text{De}_1 = \text{De}_2 = 0$), we see that the rate of work of the waving filament is given by a sum of positive modes, each with a decreased contribution compared to the Newtonian result, $\dot{W}_N$, since $\text{De}_2 < \text{De}_1$. In particular, if we assume the filament to wave with a single mode $n = 1$, we have explicitly the ratio

$$\frac{\dot{W}_{NN}}{\dot{W}_N} = \frac{1 + \text{De}_1 \text{De}_2}{1 + \text{De}_1^2}, \tag{15.31}$$

a result identical to the one obtained later using a nonlinear constitutive model in Eq. (15.61). The energy expenditure of the filament is therefore always less in a viscoelastic fluid than in a Newtonian fluid (at leading order). The physical origin of this decrease lies in the elastic nature of the fluid, which stores, rather than dissipates, some of the work done by the filament.

15.1.4 Beating Patterns in Viscoelastic Fluids

The hydrodynamic forces predicted by the linear viscoelastic resistive-force theory can also be used to characterise the beating shapes for eukaryotic flagella, as done by Fu et al. (2008).

Consider an active slender filament driven internally, as in Section 7.4. Using Cartesian coordinates, the leading-order elasto-hydrodynamic force balance in the y direction for a filament in a Newtonian fluid is given by Eq. (7.53),

$$c_\perp \frac{\partial y}{\partial t} + B\frac{\partial^4 y}{\partial x^4} = f_a(x,t),\tag{15.32}$$

where f_a is the density of internal active forces.

For small-amplitude motion, the complex fluid can be modelled appropriately as linearly viscoelastic, and linear viscoelastic resistive-force theory allows us to compute the change in hydrodynamic forces. For each Fourier mode, we have Eq. (15.26) and therefore the force balance becomes

$$-in\omega c_\perp^{(n)} y^{(n)} + B\frac{d^4 y^{(n)}}{dx^4} = f_a^{(n)},\tag{15.33}$$

where $f_a^{(n)}$ is the nth Fourier component of the active force. In the specific case of a Jeffreys fluid, we can use Eq. (15.27) to evaluate the drag coefficient so that, when focusing for simplicity on the fundamental mode $n = 1$,

$$-i\omega\frac{1 - iDe_2}{1 - iDe_1}c_\perp y^{(1)} + B\frac{d^4 y^{(1)}}{dx^4} = f_a^{(1)}.\tag{15.34}$$

With a non-dimensionalisation as in Section 7.4, using the frequency ω and wavenumber k, the waveform is characterised by four dimensionless numbers, namely the sperm number, Sp, from Eq. (7.55), quantifying the balance between hydrodynamics and flexibility (and based on the long-time viscosity of the fluid), the dimensionless length of the filament, kL, and the two Deborah numbers, De_1 and De_2.

15.1.5 Comparison with Experiments: Human Spermatozoa

The comparison between the predictions of this model and experiments is shown in Fig. 15.2. The experiments, carried out by Ishijima et al. (1986), show the beating waveforms of human spermatozoa whose cell bodies are held in place by pipettes (Fig. 15.2(a) and (b)). The comparison between experiments and theory is displayed in Fig. 15.2(c), assuming clamped boundary conditions on the end of the flagellum held by the pipette (Fig. 15.2, left) and free boundary conditions on the

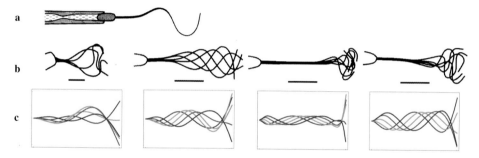

Figure 15.2 (a) Human spermatozoon with cell body held by a pipette. (b) Flagellar waveforms of human spermatozoa in synthetic Newtonian fluids of increasing viscosity (first two panels), in a synthetic non-Newtonian fluid (third panel; relaxation time of about 10^{-2} s) and in cervical mucus (fourth panel; relaxation time of about 1 s) (scale bars: 10 μm) (Ishijima et al., 1986, reproduced with permission). (c) Time sequence of the theoretical model for a clamped active filament with $kL = 3\pi$. The four panels have sperm numbers Sp $= 5.6, 13.6, 37.3$ and 45. The last two panels are for a non-Newtonian fluid and have Deborah numbers $\mathrm{De}_1 = 0.5$, $\mathrm{De}_2/\mathrm{De}_1 = 1/4000$ and $\mathrm{De}_1 = 50$, $\mathrm{De}_2/\mathrm{De}_1 = 1/4360$, respectively (reprinted figure with permission from Fu et al. (2008), Copyright 2008 by the American Physical Society).

other end. The panels in Fig. 15.2(b) and (c) show waveforms for a human spermatozoon in synthetic Newtonian fluids of increasing viscosity (first two panels), in a synthetic non-Newtonian fluid with a relaxation time of about 10^{-2} s (third panel) and in cervical mucus, which has a relaxation time of about 1 s (fourth panel), with dimensionless parameters given in the caption. The beating patterns predicted by the simple one-dimensional linearised theory appear to be in qualitative agreement with the experimental observations. In particular, beating is concentrated at the distal end of the flagellum, a mathematical consequence of the boundary-layer structure of the solution of Eq. (15.34) in the case of a high-viscosity fluid (Fu et al., 2008).

15.2 Locomotion and Transport in Nonlinear Viscoelastic Fluids

15.2.1 Nonlinear Viscoelasticity

While linear viscoelastic fluids have a simple mathematical structure, which makes them appealing for calculations, they are subject to a well-known fundamental problem. Specifically, the constitutive relationship in Eq. (15.9) is not independent of the frame in which it is evaluated (it is said to be 'not objective'). To see this, consider the time derivatives appearing in the Jeffreys fluid, Eq. (15.8). Both the stress and rate of strain tensors are expressed in a specific basis, and time derivatives of each tensor also include time derivatives of the basis vectors themselves. In the case where the basis vectors are modified by the flow (through rotation due to velocity gradients), the value of the time derivative then depends on the frame in which it is evaluated. The classical demonstration of this result consists in calculating the viscosity of a linear viscoelastic fluid as in Eq. (15.9) for the shear flow between two plates on a rotating table, and showing that the value of the viscosity computed in the lab frame depends on the rotation rate of the table (see Exercises).

This issue of frame dependence can be resolved by defining objective tensorial derivatives that render the constitutive relationships frame invariant. This can be done in a number of ways, and references to textbooks on the subject are listed in the Further Reading section. We use in this book the most common objective derivative, termed the upper-convected derivative and defined for any second-rank tensor $\mathbf{A}$ as

$$\overset{\triangledown}{\mathbf{A}} \equiv \frac{\partial \mathbf{A}}{\partial t} + \mathbf{u} \cdot \nabla \mathbf{A} - (\nabla \mathbf{u})^T \cdot \mathbf{A} - \mathbf{A} \cdot (\nabla \mathbf{u}). \tag{15.35}$$

This derivative quantifies the rate of $\mathbf{A}$ calculated while translating and deforming with the flow $\mathbf{u}$. Importantly, we see that Eq. (15.35) is bilinear in $\mathbf{u}$ and $\mathbf{A}$, which gives rise naturally to a nonlinear relationship between stresses and flow and allows us to capture the nonlinear phenomenology of complex fluids.

An important and widely used complex fluid model is the Oldroyd-B fluid, which is the upper-convected version of the Jeffreys fluid in Eq. (15.8). Using the definition in Eq. (15.35), the Oldroyd-B constitutive relationship is

$$\tau + \lambda_1 \overset{\triangledown}{\tau} = \eta(\dot{\gamma} + \lambda_2 \overset{\triangledown}{\dot{\gamma}}). \tag{15.36}$$

Importantly, this fluid can not only be understood phenomenologically following the arguments presented in this chapter, but it can also be derived mathematically from a kinetic theory for a dilute polymer solution whose individual molecules are modelled as elastic dumbbells (Hohenegger and Shelley, 2011).

15.2.2 Waving with Fixed Kinematics

In a fluid with a nonlinear constitutive relationship such as Eq. (15.36), nonlinearities are expected to lead to irreversibility in fluid transport, and thus to impact locomotion. In the next two sections we quantify this effect using Taylor's waving sheet model introduced in Chapter 3. Similar to what was done in Chapters 11 and 12 to address locomotion near surfaces and synchronisation, the waving sheet model allows us to rigorously quantify swimming and ciliary transport in a complex fluid. We assume that the fluid is described by the Oldroyd-B relationship from Eq. (15.36) and consider in this first section the case where the kinematics of the waving motion are prescribed (Lauga, 2007).

Waving Sheet in Oldroyd-B Fluid

The setup and notation are identical to those in Section 3.3 and Fig. 3.1, the only change being the nature of the fluid. The two-dimensional sheet has wavelength

$2\pi/k$, frequency ω and wave speed $V = \omega/k$. We solve the problem in the swimming frame of the sheet, (x, y). The flow velocity at infinity is $U\mathbf{e}_x$ and, as in Chapter 3, can be viewed either as the flow induced by the envelope of a dense ciliary array or minus the velocity of a swimming sheet.

We consider only the case where the sheet undergoes transverse deformation with amplitude B, so that the material points (x_s, y_s) on the surface of the sheet in the (x, y) frame are given by Eq. (3.1) with $A = 0$, i.e.

$$x_s = x, \tag{15.37a}$$

$$y_s = B \sin(kx - \omega t). \tag{15.37b}$$

We non-dimensionalise the problem using k^{-1} and ω^{-1} as characteristic length and timescales, so that material points follow Eq. (3.6) with $a = 0$,

$$x_s = x, \tag{15.38a}$$

$$y_s = \epsilon b \sin(x - t), \tag{15.38b}$$

where we use the same symbols for notation convenience.

The flow problem is solved using a streamfunction, ψ, defined as in Eq. (3.7),

$$u_x = \frac{\partial \psi}{\partial y}, \quad u_y = -\frac{\partial \psi}{\partial x}. \tag{15.39}$$

The boundary conditions are identical to the Newtonian case, i.e. on the sheet

$$u_x(x_s, y_s) = \frac{\partial \psi}{\partial y}\Big|_{(x_s, y_s)} = 0, \tag{15.40a}$$

$$u_y(x_s, y_s) = -\frac{\partial \psi}{\partial x}\Big|_{(x_s, y_s)} = -\epsilon b \cos(x - t), \tag{15.40b}$$

while at infinity

$$\lim_{y \to \infty} \nabla \psi = U\mathbf{e}_y. \tag{15.41}$$

The statement of mechanical equilibrium in Eq. (15.13) together with the constitutive relationship from Eq. (15.36) replace the Stokes equations. Using $\eta\omega$ and ω to non-dimensionalise stresses and shear rates, respectively, and using the same symbols for notation convenience, Eq. (15.36) becomes

$$\tau + De_1 \overset{\triangledown}{\tau} = \dot{\gamma} + De_2 \overset{\triangledown}{\dot{\gamma}}, \tag{15.42}$$

with the Deborah numbers defined in Eq. (15.6). We aim to determine U for given values of ϵb, De_1 and De_2. Note that the rate of strain tensor, $\dot{\gamma}$, is only a function of the second-order derivatives of ψ.

To make analytical progress, we use regular perturbation expansions in ϵ as in Chapter 3. We thus write for the streamfunction, the swimming velocity, the deviatoric stress and rate of strain tensors

$$\psi = \epsilon \psi_1 + \epsilon^2 \psi_2 + \cdots, \tag{15.43a}$$

$$U = \epsilon U_1 + \epsilon^2 U_2 + \cdots, \tag{15.43b}$$

$$\boldsymbol{\tau} = \epsilon \boldsymbol{\tau}_1 + \epsilon^2 \boldsymbol{\tau}_2 + \cdots, \tag{15.43c}$$

$$\dot{\boldsymbol{\gamma}} = \epsilon \dot{\boldsymbol{\gamma}}_1 + \epsilon^2 \dot{\boldsymbol{\gamma}}_2 + \cdots, \tag{15.43d}$$

and as usual we expect swimming to occur at order ϵ^2. Since we no longer have the biharmonic equation for ψ, the procedure to obtain the non-Newtonian equation consists in taking the curl of Eq. (15.13) in order to eliminate the pressure, which leads to a fourth-order partial differential equation for the streamfunction. The boundary conditions for ψ remain unchanged from the Newtonian case and are given at order ϵ by

$$\left. \frac{\partial \psi_1}{\partial y} \right|_{(x,0)} = 0, \tag{15.44a}$$

$$\left. \frac{\partial \psi_1}{\partial x} \right|_{(x,0)} = b \cos(x - t) = \mathcal{R} \left\{ b e^{i(x-t)} \right\}, \tag{15.44b}$$

$$\lim_{y \to \infty} \nabla \psi_1 = U_1 \mathbf{e}_y, \tag{15.44c}$$

and at order ϵ^2 by

$$\left. \frac{\partial \psi_2}{\partial y} \right|_{(x,0)} = -b \sin(x - t) \left. \frac{\partial^2 \psi_1}{\partial y^2} \right|_{(x,0)}, \tag{15.45a}$$

$$\left. \frac{\partial \psi_2}{\partial x} \right|_{(x,0)} = -b \sin(x - t) \left. \frac{\partial^2 \psi_1}{\partial x \partial y} \right|_{(x,0)}, \tag{15.45b}$$

$$\lim_{y \to \infty} \nabla \psi_2 = U_2 \mathbf{e}_y. \tag{15.45c}$$

Order-ϵ Solution

At order ϵ, Eq. (15.42) is linearised as the Jeffreys fluid

$$\left(1 + \mathrm{De}_1 \frac{\partial}{\partial t} \right) \boldsymbol{\tau}_1 = \left(1 + \mathrm{De}_2 \frac{\partial}{\partial t} \right) \dot{\boldsymbol{\gamma}}_1. \tag{15.46}$$

To make further progress we use the identity, valid for any symmetric second-rank tensor field $\mathbf{A}(x, y)$ that does not depend on z,

$$\mathbf{e}_z \cdot [\nabla \times (\nabla \cdot \mathbf{A})] = \left(\frac{\partial^2}{\partial x^2} - \frac{\partial^2}{\partial y^2} \right) A_{xy} + \frac{\partial^2}{\partial x \partial y} (A_{yy} - A_{xx}), \tag{15.47}$$

where we use subscripts to denote the components of the tensor. In particular, in the case where $\mathbf{A} = \dot{\boldsymbol{\gamma}}$ this becomes

$$\mathbf{e}_z \cdot [\nabla \times (\nabla \cdot \dot{\boldsymbol{\gamma}})] = -\nabla^4 \psi. \tag{15.48}$$

This result can be used to derive the equation for ψ_1 as follows. First, take the divergence of Eq. (15.46). The resulting left-hand side of the equation is linear in the pressure gradient, through Eq. (15.13), and thus if we proceed to compute the curl of that equation, the left-hand side becomes identically zero. The value of the right-hand side is obtained by using the result in Eq. (15.48) and leads to the equation for the streamfunction at order ϵ

$$\left(1 + \mathrm{De}_2 \frac{\partial}{\partial t}\right) \nabla^4 \psi_1 = 0. \tag{15.49}$$

The boundary conditions for ψ_1 associated with this equation are the same as for the Newtonian case, namely

$$\left.\frac{\partial \psi_1}{\partial y}\right|_{(x,0)} = 0, \tag{15.50a}$$

$$\left.\frac{\partial \psi_1}{\partial x}\right|_{(x,0)} = b\cos(x - t) = \mathcal{R}\left\{be^{i(x-t)}\right\}, \tag{15.50b}$$

$$\lim_{y \to \infty} \nabla\psi_1 = U_1 \mathbf{e}_y, \tag{15.50c}$$

where $\mathcal{R}$ denotes the real part of a complex number. Since Eq. (15.49) is linear and since the boundary conditions in Eq. (15.50b) only involve one Fourier mode, at long times (i.e. after any decaying transients), the solution to Eq. (15.49) is periodic in time. As a result, ψ_1 is also a solution to $\nabla^4\psi_1 = 0$, and therefore it is equal to the Newtonian solution from Eq. (3.21). At leading order, the non-Newtonian flow problem is thus identical to the Newtonian one, and in particular we have $U_1 = 0$. Since we use complex notation in this chapter, we write this leading-order solution as

$$\psi_1(x, y, t) = \mathcal{R}\left\{\tilde{\psi}_1(x, y)e^{-it}\right\}, \quad \tilde{\psi}_1 = -ib(1 + y)e^{-y}e^{ix}, \tag{15.51}$$

with similar notation for all other fields in this section. It will be useful for later calculations to note that the Fourier amplitude of the velocity field, using this notation, is

$$\tilde{\mathbf{u}}_1 = b\begin{pmatrix} iy \\ -(1+y) \end{pmatrix} e^{-y}e^{ix}. \tag{15.52}$$

Rate of Work

We can use the first-order flow solution to evaluate the rate of work done by the waving sheet. This is given by the surface integral in Eq. (2.18), written for a two-dimensional sheet as

$$\dot{W} = -\int_S \mathbf{u} \cdot \boldsymbol{\sigma} \cdot \mathbf{n} \, dS. \tag{15.53}$$

This integral may be transformed into a volume integral using the divergence theorem to become the first equality in Eq. (2.20), which is for a two-dimensional sheet

$$\dot{W}(t) = \iint_V \frac{\partial u_i}{\partial x_j} \sigma_{ij} \, dV. \tag{15.54}$$

Note that we have used the notation $\int_S \dots dS$ and $\iint_V \dots dV$ to emphasise that, for a two-dimensional sheet, the surface and volume integrals are, respectively, one- and two-dimensional integrals.

Using the identity $\sigma_{ij} = -p\delta_{ij} + \tau_{ij}$ and the fact that the fluid is incompressible, we obtain

$$\dot{W} = \iint_V \frac{\partial u_i}{\partial x_j} \tau_{ij} \, dV = \frac{1}{2} \iint_V \tau_{ij}\dot{\gamma}_{ij} \, dV = \frac{1}{2} \iint_V \boldsymbol{\tau} : \dot{\boldsymbol{\gamma}} \, dV. \tag{15.55}$$

This is a quadratic quantity, so the perturbation expansion is $\dot{W} = \epsilon^2 \dot{W}_2 + \cdots$, and the leading-order term $\dot{W}_2$ is half the volume integral of $\boldsymbol{\tau}_1 : \dot{\boldsymbol{\gamma}}_1$.

Using Fourier notation for the stress and rate of strain tensors, similar to the streamfunction in Eq. (15.51), we may transform the linearised constitutive relationship from Eq. (15.46) in Fourier space as

$$\tilde{\boldsymbol{\tau}}_1 = \frac{1 - i\mathrm{De}_2}{1 - i\mathrm{De}_1} \tilde{\dot{\boldsymbol{\gamma}}}_1. \tag{15.56}$$

Since terms of the form $e^{2i(x-t)}$ average to zero over one wavelength of the sheet, the mean value of $\boldsymbol{\tau}_1 : \dot{\boldsymbol{\gamma}}_1$ per wavelength is

$$\langle \boldsymbol{\tau}_1 : \dot{\boldsymbol{\gamma}}_1 \rangle = \frac{1}{2} \mathcal{R}\{\tilde{\boldsymbol{\tau}}_1 : \tilde{\dot{\boldsymbol{\gamma}}}_1^*\} = \frac{1}{2} \left(\frac{1 + \mathrm{De}_1 \mathrm{De}_2}{1 + \mathrm{De}_1^2} \right) \left(\tilde{\dot{\boldsymbol{\gamma}}}_1 : \tilde{\dot{\boldsymbol{\gamma}}}_1^* \right), \tag{15.57}$$

where * is used to indicate complex conjugate. The total rate of work in the Newtonian case, $\dot{W}_{2,N}$, was calculated in Chapter 3 per wavelength as

$$\dot{W}_{2,N} = \frac{1}{2} \iint_V \boldsymbol{\tau}_1 : \dot{\boldsymbol{\gamma}}_1 \, dV, \tag{15.58}$$

which using Eq. (15.57) becomes

$$\dot{W}_{2,N} = \frac{\pi}{2} \int_0^\infty \tilde{\dot{\boldsymbol{\gamma}}}_1 : \tilde{\dot{\boldsymbol{\gamma}}}_1^* \, dy = 2\pi b^2. \tag{15.59}$$

Since Eq. (15.57) indicates a simple proportionality between the non-Newtonian and Newtonian energy densities, we can deduce the rate of work done by the sheet per wavelength in the Oldroyd-B fluid, $\dot{W}_{2,NN}$, at leading order as

$$\dot{W}_{2,NN} = 2\pi b^2 \frac{1 + \mathrm{De}_1 \mathrm{De}_2}{1 + \mathrm{De}_1^2}. \tag{15.60}$$

Using the Newtonian result, Eq. (15.59), we can rewrite Eq. (15.60) as the ratio

$$\frac{\dot{W}_{2,NN}}{\dot{W}_{2,N}} = \frac{1 + \mathrm{De}_1 \mathrm{De}_2}{1 + \mathrm{De}_1^2}, \tag{15.61}$$

which is the same result as the one obtained in Eq. (15.31) for a waving filament. Compared to a Newtonian fluid with the same viscosity, the waving sheet experiences, at leading order in its amplitude, a systematic decrease in its rate of work.

Order-ϵ^2 Solution

In order to calculate U_2, we first evaluate the order-ϵ^2 component of Eq. (15.42), using the upper-convected derivative definition in Eq. (15.35), as

$$\left(1 + \mathrm{De}_1 \frac{\partial}{\partial t}\right) \tau_2 - \left(1 + \mathrm{De}_2 \frac{\partial}{\partial t}\right) \dot{\gamma}_2 \tag{15.62}$$

$$= \mathrm{De}_1 \left((\nabla \mathbf{u}_1)^T \cdot \tau_1 + \tau_1 \cdot \nabla \mathbf{u}_1 - \mathbf{u}_1 \cdot \nabla \tau_1\right)$$

$$- \mathrm{De}_2 \left((\nabla \mathbf{u}_1)^T \cdot \dot{\gamma}_1 + \dot{\gamma}_1 \cdot \nabla \mathbf{u}_1 - \mathbf{u}_1 \cdot \nabla \dot{\gamma}_1\right).$$

Since the right-hand side of Eq. (15.62) is quadratic in ψ_1, the solution at order ϵ^2 involves products of Fourier terms $\propto \exp i(x - t)$ with themselves and with their complex conjugates, and thus ψ_2 must take the form

$$\psi_2(x, y, t) = \mathcal{R}\left\{\hat{\psi}_2(y) e^{2i(x-t)}\right\} + f(y). \tag{15.63}$$

Since the first term in Eq. (15.63) is periodic, the function $f(y)$ is the mean value of the streamfunction over one wavelength, so we can write

$$\psi_2(x, y, t) = \mathcal{R}\left\{\tilde{\psi}_2(x, y) e^{-2it}\right\} + \langle \psi_2 \rangle(y), \quad \tilde{\psi}_2(x, y) = \hat{\psi}_2(y) e^{2ix}. \tag{15.64}$$

Computing next the mean value of Eq. (15.62) over one wavelength and noting that the resulting terms no longer depend on time lead to

$$\langle \tau_2 \rangle - \langle \dot{\gamma}_2 \rangle = \mathcal{R}\left\{\frac{\mathrm{De}_1 - \mathrm{De}_2}{2(1 - i\mathrm{De}_1)} \left(\nabla \tilde{\mathbf{u}}_1^{T*} \cdot \tilde{\dot{\gamma}}_1 + \tilde{\dot{\gamma}}_1 \cdot \nabla \tilde{\mathbf{u}}_1^* - \tilde{\mathbf{u}}_1^* \cdot \nabla \tilde{\dot{\gamma}}_1\right)\right\}. \tag{15.65}$$

From the velocity in Eq. (15.52) we evaluate the tensors as

$$\nabla \tilde{\mathbf{u}}_1 = b \begin{pmatrix} -y & -i(1+y) \\ i(1-y) & y \end{pmatrix} e^{-y} e^{ix}, \quad \tilde{\dot{\gamma}}_1 = b \begin{pmatrix} -2y & -2iy \\ -2iy & 2y \end{pmatrix} e^{-y} e^{ix},$$

$$(15.66)$$

and can thus calculate the right-hand side of Eq. (15.65) as

$$\langle \tau_2 \rangle - \langle \dot{\gamma}_2 \rangle = b^2 \left(\frac{\mathrm{De}_1 - \mathrm{De}_2}{1 + \mathrm{De}_1^2} \right) e^{-2y} \left(\begin{array}{cc} 6y^2 - 2y - 1 & \mathrm{De}_1(1 + 2y - 2y^2) \\ \mathrm{De}_1(1 + 2y - 2y^2) & 2y^2 + 2y + 1 \end{array} \right).$$

$$(15.67)$$

To obtain an equation for $\langle \psi_2 \rangle$ we proceed as we did at order ϵ. We first take the divergence of Eq. (15.67). Invoking mechanical equilibrium from Eq. (15.13), we then take the curl to eliminate the pressure and therefore the first term on the left-hand side of Eq. (15.67). The second term is related to the streamfunction at order ϵ^2 through Eq. (15.48). Using Eq. (15.47) to evaluate the right-hand side of Eq. (15.67) and noting that both sides now only depend on y, we obtain the equation for the mean streamfunction

$$\frac{\mathrm{d}^4}{\mathrm{d}y^4} \langle \psi_2 \rangle = b^2 \frac{\mathrm{De}_1(\mathrm{De}_2 - \mathrm{De}_1)}{1 + \mathrm{De}_1^2} \frac{\mathrm{d}^2}{\mathrm{d}y^2} \left[e^{-2y}(1 + 2y - 2y^2) \right]. \qquad (15.68)$$

The boundary conditions for ψ_2 in real space are given by Eq. (3.22), i.e.

$$\left. \frac{\partial \psi_2}{\partial y} \right|_{(x,0)} = -b \sin(x - t) \left. \frac{\partial^2 \psi_1}{\partial y^2} \right|_{(x,0)}, \qquad (15.69a)$$

$$\left. \frac{\partial \psi_2}{\partial x} \right|_{(x,0)} = -b \sin(x - t) \left. \frac{\partial^2 \psi_1}{\partial x \partial y} \right|_{(x,0)}, \qquad (15.69b)$$

$$\lim_{y \to \infty} \nabla \psi_2 = U_2 \mathbf{e}_y, \qquad (15.69c)$$

and they become for the mean streamfunction

$$\frac{\mathrm{d}\langle \psi_2 \rangle}{\mathrm{d}y}(0) = \frac{b^2}{2}, \qquad (15.70a)$$

$$\frac{\mathrm{d}\langle \psi_2 \rangle}{\mathrm{d}y}(\infty) = U_2. \qquad (15.70b)$$

Using the boundary condition in Eq. (15.70) we can therefore integrate Eq. (15.68) three times to obtain

$$\frac{\mathrm{d}}{\mathrm{d}y} \langle \psi_2 \rangle = \frac{b^2}{2} \left(\frac{1 + \mathrm{De}_1 \mathrm{De}_2}{1 + \mathrm{De}_1^2} \right) + b^2 \left(\frac{\mathrm{De}_1(\mathrm{De}_2 - \mathrm{De}_1)}{1 + \mathrm{De}_1^2} \right) \left(y^2 - \frac{1}{2} \right) e^{-2y}.$$

$$(15.71)$$

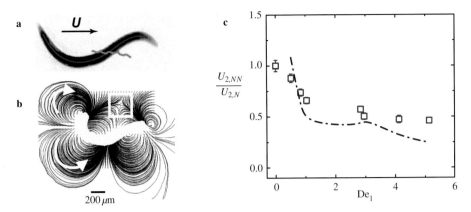

Figure 15.3 Locomotion of nematode *C. elegans* in Newtonian and synthetic polymer solutions (reprinted figure with permission from Shen and Arratia (2011), Copyright 2011 by the American Physical Society). (a) Snapshot of the worm, approximately 1 mm long and 80 μm wide, with a fit to its instantaneous centreline and tracking of its centre of mass. (b) Streamlines for instantaneous flow field created by the waving of the worm in a polymer solution. (c) Measured ratio between the non-Newtonian and Newtonian swimming speeds, $U_{2,NN}/U_{2,N}$, as a function of the Deborah number, De_1, in experiments (symbols) and prediction from the theory (line) using measured kinematic and rheological data.

Swimming Velocity

Comparing Eq. (15.71) with the condition at infinity, Eq. (15.70b), we obtain finally the swimming speed at order two in the non-Newtonian fluid as

$$U_{2,NN} = \frac{b^2(1 + De_1 De_2)}{2(1 + De_1^2)}. \tag{15.72}$$

Denoting by $U_{2,N}$ the leading-order swimming velocity in a Newtonian fluid, $U_{2,N} = b^2/2$, we can then rewrite Eq. (15.72) as the ratio

$$\frac{U_{2,NN}}{U_{2,N}} = \frac{1 + De_1 De_2}{1 + De_1^2}. \tag{15.73}$$

At any finite frequency, $De_2 < De_1$, so $U_{2,NN} < U_{2,N}$ and a waving sheet with prescribed kinematics always swims (or transports fluid) at a rate smaller than in a Newtonian fluid. The same leading-order result applies to locomotion powered by a slender flagellum (Fu et al., 2009) and remains valid for a large range of constitutive models (Lauga, 2007). Note that the speed ratio in Eq. (15.73) is the same as for the rate of work, Eq. (15.61).

Comparison with Experiments: C. elegans Locomotion

The result in Eq. (15.72) has been compared successfully with experiments for the small nematode *C. elegans* swimming in synthetic polymer solutions (Shen and Arratia, 2011). The results are illustrated in Fig. 15.3. In Fig. 15.3(a) we show a snapshot of the worm, approximately 1 mm long and 80 μm wide, together with the path of its body centre and the fit to the worm centreline used to extract its waving kinematics. Instantaneous streamlines for the flow induced by the organism in a viscoelastic fluid are shown in Fig. 15.3(b), highlighting large vortical structures and extensional regions. The comparison between the measurements and the predicted swimming speed in Eq. (15.72) is shown in Fig. 15.3(c) as a function of De_1 using, for each data point, measured kinematic data for the waving motion and rheological data for the fluid. The good agreement between the two shows that the decrease in the worm speed observed experimentally can be explained successfully by the small-amplitude theory.

15.2.3 Flexible Waving Swimmer

The calculations in the previous section showed that when the characteristics of the waves are prescribed, the induced flow speed (interpreted as either swimming or pumping) always decreases in the Oldroyd-B fluid compared to the Newtonian case. In contrast, an enhancement in locomotion has been reported experimentally for *C. elegans* worms self-propelling in concentrated polymer solutions (Gagnon et al., 2013). Numerical simulations of finite waving sheets driven elastically around a preferred shape also showed an increase in swimming speeds for order-one Deborah numbers (Teran et al., 2010). These results, at odds with the theory from the previous section, challenge us therefore to propose a different theoretical explanation.

One physical mechanism that predicts a swimming enhancement is the dynamic feedback between the fluid stresses and the waving shape via the flexibility of the swimmer. The change in waveform due to modifications of the forces in the fluid was summarised in Section 15.1.4 for slender filaments at order ϵ. Here we carry out the calculation quantifying the interplay between shape change and locomotion at order ϵ^2 (Riley and Lauga, 2014).

Active Sheet in Oldroyd-B Fluid

The setup is similar to that in Section 15.2.2, but instead of prescribing the amplitude of the waving sheet, we now impose the internal forces. The waveform is then obtained by balancing the internal forcing, the hydrodynamic stresses and the flexibility of the swimmer. As before, we solve the problem asymptotically as a perturbation expansion, but now the small parameter ϵ is the dimensionless amplitude of the internal force distribution.

The modelling for active swimmers was introduced in Section 7.4 for slender filaments and was adapted to the case of two-dimensional sheets in Section 12.2 to address synchronisation. Following this, the leading-order dimensional waveform $y_1(x,t)$ satisfies the active force balance in the y direction from Eq. (12.122), written

$$B\frac{\partial^4 y_1}{\partial x^4} - \mathbf{e}_y \cdot \boldsymbol{\sigma} \cdot \mathbf{n}|_{(x,0)} = f_a(x,t), \tag{15.74}$$

where $\mathbf{e}_y \cdot \boldsymbol{\sigma} \cdot \mathbf{n}$ is the y component of hydrodynamic stresses acting on the sheet at leading order and f_a is the prescribed active stress. If we assume that the fluid is above the sheet, then in the limit where the wave amplitude is small compared to the wavelength we have $\mathbf{n} \approx \mathbf{e}_y$ and thus $-\mathbf{e}_y \cdot \boldsymbol{\sigma} \cdot \mathbf{n} \approx p_1$. The linear response of the waving the sheet, Eq. (15.74), then becomes

$$B\frac{\partial^4 y_1}{\partial x^4} + p_1|_{(x,0)} = f_a(x,t). \tag{15.75}$$

If the active force f_a is a pure travelling wave with wavenumber k and frequency ω, we non-dimensionalise lengths with k^{-1}, hydrodynamic pressure with $\eta\omega$ and active stresses with the elastic term, Bk^3 (since in the limit of large sperm number, we know from Section 7.4 that these two balance). Keeping the same symbols for simplicity, the dimensionless version of Eq. (15.75) becomes

$$\frac{\partial^4 y_1}{\partial x^4} + \mathrm{Sp}^3 p_1|_{(x,0)} = f_a(x,t), \tag{15.76}$$

where we have used the two-dimensional sperm number from Eq. (12.134),

$$\mathrm{Sp}^3 = \frac{\eta\omega}{Bk^3}. \tag{15.77}$$

Shape and Flow at Order ϵ

At order ϵ, the streamfunction for the flow, ψ_1, is still a solution to the non-Newtonian biharmonic equation in Eq. (15.49) with the same no-penetration boundary condition, Eq. (15.50a). However, the normal boundary condition of prescribed displacement, Eq. (15.50b), is replaced by the force balance for the active sheet, Eq. (15.76). In order to determine the shape, we thus need to relate the fluid pressure p to the waveform y.

We use Fourier notation for the active force and write

$$f_a(x,t) = \mathcal{R}\{\tilde{f}_a e^{-it}\}, \quad \tilde{f}_a(x) = \hat{f}_a e^{ix}, \tag{15.78}$$

where $\hat{f}_a$ is the dimensionless amplitude. At order ϵ, mechanical equilibrium from Eq. (15.13) is written as

$$\nabla \tilde{p}_1 = \nabla \cdot \tilde{\boldsymbol{\tau}}_1, \tag{15.79}$$

and, using the result in Eq. (15.56) for the deviatoric stress at order ϵ, we obtain the equation for the pressure as

$$\nabla \tilde{p}_1 = \frac{1 - i\mathrm{De}_2}{1 - i\mathrm{De}_1} \nabla \cdot \tilde{\boldsymbol{\gamma}}_1 = \frac{1 - i\mathrm{De}_2}{1 - i\mathrm{De}_1} \nabla^2 \tilde{\mathbf{u}}_1. \tag{15.80}$$

As expected, Eq. (15.80) is linear and therefore the boundary condition in Eq. (15.76), written in Fourier space as

$$\frac{\partial^4 \tilde{y}_1}{\partial x^4} + \mathrm{Sp}^3 \tilde{p}_1|_{(x,0)} = \tilde{f}_a(x), \tag{15.81}$$

is a linear equation relating the streamfunction to the active force. In response to a Fourier mode for the force as in Eq. (15.78), the shape is thus still of the form $\tilde{y}_1 = -ibe^{ix}$ as in Section 15.2.2 with an undetermined wave amplitude, b. Note that b is now complex to allow for a phase difference between the force and the waveform. Following this solution for the shape, the streamfunction still takes the form of a Fourier mode as in Eq. (15.51),

$$\tilde{\psi}_1 = -ib(1 + y)e^{-y}e^{ix}. \tag{15.82}$$

With this solution, we can solve for the pressure in Eq. (15.80) as

$$\tilde{p}_1 = -2b\frac{1 - i\mathrm{De}_2}{1 - i\mathrm{De}_1}e^{-y}e^{ix}, \tag{15.83}$$

and after substituting Eqs. (15.82) and (15.83) into Eq. (15.81) we obtain the wave amplitude along the e^{ix} mode as

$$ib + 2b\mathrm{Sp}^3\frac{1 - i\mathrm{De}_2}{1 - i\mathrm{De}_1} = -\hat{f}_a. \tag{15.84}$$

Note that, unsurprisingly, this equation is analogous to the linearised dynamics of an active filament in Eq. (15.34). Solving finally for the complex wave amplitude b in Eq. (15.84), we obtain

$$b = -\frac{\hat{f}_a}{\left(i + 2\mathrm{Sp}^3\dfrac{1 - i\mathrm{De}_2}{1 - i\mathrm{De}_1}\right)}. \tag{15.85}$$

Active Swimming at Order ϵ^2

With the waving amplitude obtained in Eq. (15.85), we may now revisit the calculation from the previous section on fixed-kinematics waving in order to compute the swimming velocity of the active sheet. Indeed, the only difference between the flow

problems at order ϵ^2 in Sections 15.2.2 and 15.2.3 is the fact that the waving amplitude b is a complex number. All results for the order-ϵ^2 solution in Section 15.2.2
up to Eq. (15.66) remain valid and thus, when evaluating Eq. (15.65), the product
of order-ϵ terms with the complex conjugates of other order-ϵ terms leads to the
appearance of $bb^* = |b|^2$ instead of b^2, with the rest of the algebra unchanged. The
result in Eq. (15.72) for the order-ϵ^2 swimming speed is modified therefore to

$$U_2 = \frac{|b|^2(1 + \text{De}_1\text{De}_2)}{2(1 + \text{De}_1^2)}.$$ (15.86)

Using the wave amplitude b obtained in Eq. (15.85), we obtain finally the swimming speed for the active sheet as

$$U_2 = \frac{|\hat{f}_a|^2}{2}\,\frac{1 + \text{De}_1\text{De}_2}{1 + \text{De}_1^2 + 4\text{Sp}^3\,(\text{De}_1 - \text{De}_2) + 4\text{Sp}^6\left(1 + \text{De}_2^2\right)}.$$ (15.87)

Hindered vs. Enhanced Swimming

The Newtonian limit of Eq. (15.87), $U_{2,N}$, is found by setting $\text{De}_1 = \text{De}_2 = 0$,

$$U_{2,N} = \frac{|\hat{f}_a|^2}{2}\,\frac{1}{1 + 4\text{Sp}^6},$$ (15.88)

and is the swimming velocity of the sheet driven by the same active force in a Newtonian fluid with the same long-time viscosity (i.e. same definition of the sperm
number in Eq. (12.134)). Comparing the non-Newtonian result in Eq. (15.87), written as $U_{2,NN}$, to Eq. (15.88), we have the ratio

$$\frac{U_{2,NN}}{U_{2,N}} = \frac{(1 + \text{De}_1\text{De}_2)(1 + 4\text{Sp}^6)}{1 + \text{De}_1^2 + 4\text{Sp}^6\left(1 + \text{De}_2^2\right) + 4\text{Sp}^3\,(\text{De}_1 - \text{De}_2)}.$$ (15.89)

When the sheet is rigid, i.e. in the $\text{Sp} \ll 1$ limit, Eq. (15.89) simplifies to the
fixed-kinematics result in Eq. (15.73), and thus we recover $U_{2,NN} < U_{2,N}$ for a
wave with a prescribed waveform. In contrast, when $\text{Sp} \gg 1$ and the sheet is very
flexible, the ratio in Eq. (15.89) is dominated by the Sp^6 terms and becomes approximately

$$\frac{U_{2,NN}}{U_{2,N}} \approx \frac{1 + \text{De}_1\text{De}_2}{1 + \text{De}_2^2}, \qquad \text{Sp} \gg 1.$$ (15.90)

Since $\text{De}_1 > \text{De}_2$ for nonzero wave frequencies, we have $U_{2,NN} > U_{2,N}$, and the
flexibility of the sheet allows it to experience an enhancement in swimming speed.

The difference between the result for $\text{Sp} \ll 1$ (hindered swimming) and $\text{Sp} \gg 1$
(enhanced swimming) suggests that a transition takes place at a critical value of
Sp, i.e. a threshold value of the swimmer flexibility. To see this, consider the ratio

in Eq. (15.89) in the large-De limit. When $De_1, De_2 \gg 1$, the fraction is dominated by the quadratic terms in De as

$$\frac{U_{2,NN}}{U_{2,N}} \approx \frac{De_1 De_2 (1 + 4Sp^6)}{De_1^2 + 4Sp^6 De_2^2} = \frac{\beta(1 + 4Sp^6)}{1 + 4Sp^6 \beta^2}, \qquad (15.91)$$

where $\beta = De_2/De_1 < 1$ is the ratio between the retardation and relaxation timescales, a material property of the fluid, which we assume to be constant. Imposing that the ratio in Eq. (15.91) is larger than 1, i.e. that the swimming speed in the very non-Newtonian limit is larger than that in the Newtonian case, leads to the condition

$$Sp^6 > \frac{1}{4\beta}. \qquad (15.92)$$

For the active swimmer to swim faster in a strongly non-Newtonian fluid than in a Newtonian one with the same long-time viscosity, the sperm number in Eq. (15.77) must be larger than a critical value set by the fluid. This means that, everything else being fixed, the swimmer must be sufficiently flexible and its bending modulus B sufficiently small. More rigid swimmers see their locomotion hindered by the complex fluid whereas more flexible ones swim with an enhanced speed.

These results are illustrated numerically in Fig. 15.4, where we plot the speed ratio, $U_{2,NN}/U_{2,N}$, as a function of De_1 for a fixed value of $\beta = 1/2$, corresponding to a critical value of sperm number in Eq. (15.92) of $Sp_c \approx 0.891$. For $Sp = 0.1 < Sp_c$ (solid grey line), an increase in the Deborah number leads to a systematic decrease in the swimming speed compared to the Newtonian value. For $Sp = Sp_c$, after a transient decrease the speed ratio recovers to 1 for large values of the Deborah number. In contrast, for $Sp = 10 > Sp_c$, the swimming speed is increased by non-Newtonian stresses, and it asymptotes to a value above 1 in the limit $De_1 \gg 1$.

The mechanism illustrated in Fig. 15.4 is an elasto-hydrodynamic coupling between the swimmer and the fluid. If the waving amplitude remained unchanged by the fluid, Eq. (15.73) would still apply and the swimming speed would always decrease. Instead, a flexible swimmer undergoes an increase in beating amplitude due to a non-Newtonian decrease in stress, which, for sufficiently flexible swimmers, is sufficiently large to compensate for the non-Newtonian decrease from Eq. (15.73), and so leads to an overall increase in swimming speed.

15.3 Locomotion and Transport in Heterogeneous Fluids

In the complex fluids considered in the previous sections there was no explicit modelling of the fluid microstructure. However, the heterogeneous nature of the fluid can be responsible for important new physics. For example, in polymer gels,

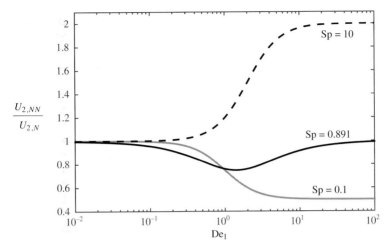

Figure 15.4 Locomotion transition for a flexible active sheet. Ratio of non-Newtonian to Newtonian swimming speeds, $U_{2,NN}/U_{2,N}$, plotted as a function of De_1 for fixed $\beta = De_2/De_1 = 1/2$. Three cases are illustrated for different values of the sperm number: speed decrease for $Sp = 0.1$ (solid grey line), transient decrease but eventually constant speed ratio for $Sp = 0.891$ (solid black line) and speed increase for $Sp = 10$ (dashed black line).

flagellated bacteria are able to create larger propulsive forces by pushing indirectly, via the fluid, against networks of entangled molecules (Berg and Turner, 1979; Magariyama and Kudo, 2002). In this section we consider two different ways to model the heterogeneity of the fluid as a continuum, and show that in both cases they lead to enhancement of locomotion and fluid transport.

15.3.1 Locomotion and Transport in a Heterogeneous Porous Medium

Brinkman Medium

The first model, termed the Brinkman medium, incorporates the multiscale nature of the fluid as a modification to the Stokes equations (Brinkman, 1947). Consider a Newtonian fluid with a random array of stationary obstacles (for example, rigid spherical particles). The fluid, in this case, may be viewed as having two separate length scales. On the length scale of the obstacles, the flow satisfies the Stokes equations in Eq. (2.9). In contrast, on much larger length scales, the fluid is subject to an effective resistance from the array of obstacles and has a behaviour akin to that of a porous medium. The Brinkman medium is a modification of the Stokes equations that allows it to encompass both behaviours at once, namely

$$\nabla \cdot \mathbf{u} = 0, \quad \nabla p = \eta \left(\nabla^2 \mathbf{u} - \kappa^2 \mathbf{u} \right). \tag{15.93}$$

Here $\kappa > 0$ is a damping coefficient due to the viscous resistance by an isotropic distribution of obstacles in the fluid (if not isotropic, κ becomes a tensor). Importantly, κ^{-1} has dimensions of length and can be interpreted as a screening length scale. On small length scales ℓ such that $\ell\kappa \ll 1$, Eq. (15.93) becomes the Stokes equations, while on large length scales with $\ell\kappa \gg 1$, it becomes the Darcy equation for flow in a porous medium (Batchelor, 1967).

Waving Sheet in Brinkman Medium

To model locomotion or transport in a Brinkman medium, we return once again to Taylor's waving sheet (Leshansky, 2009). The mathematical and geometrical setup is similar to that of Chapter 3 and Section 15.2. The only difference is that the fluid is now governed by Eq. (15.93), and therefore the sheet model allows us to address cell locomotion and fluid transport by a ciliary envelope in a fluid with a distribution of microstructure.

We consider therefore a two-dimensional waving sheet of dimensionless transverse amplitude ϵb and use the same notation as in the previous section (b is now back to being real), so the dimensionless location of material points on the surface of the sheet is given in the swimming frame by

$$x_s = x, \tag{15.94a}$$
$$y_s = \epsilon b \sin(x - t). \tag{15.94b}$$

Using k^{-1} as the relevant length scale, we define the dimensionless damping $\alpha \equiv \kappa/k > 0$, and with stresses non-dimensionalised by $\eta\omega$ and velocities by ω/k, Eq. (15.93) becomes the dimensionless Brinkman equation,

$$\nabla \cdot \mathbf{u} = 0, \quad \nabla p = \nabla^2 \mathbf{u} - \alpha^2 \mathbf{u}, \tag{15.95}$$

where we use the same symbols for convenience.

We use a streamfunction, ψ, so that the incompressibility condition is enforced automatically. In order to derive the equation satisfied by ψ, we take the curl to eliminate the pressure and obtain an equation for the vorticity, ω, as

$$\nabla^2 \omega = \alpha^2 \omega. \tag{15.96}$$

Since in two dimensions we have $\omega = -\nabla^2 \psi \mathbf{e}_z$, Eq. (15.96) becomes

$$\left(\nabla^2 - \alpha^2\right) \nabla^2 \psi = 0, \tag{15.97}$$

which we solve asymptotically in powers of ϵ in order to determine the leading-order value of the flow speed at infinity, U.

Solution at Order ϵ

The boundary conditions for ψ at order ϵ are identical to Eq. (15.50), i.e.

$$\left.\frac{\partial \psi_1}{\partial y}\right|_{(x,0)} = 0, \tag{15.98a}$$

$$\left.\frac{\partial \psi_1}{\partial x}\right|_{(x,0)} = b\cos(x-t) = \mathcal{R}\left\{be^{i(x-t)}\right\}, \tag{15.98b}$$

$$\lim_{y\to\infty} \nabla\psi_1 = U_1\mathbf{e}_y. \tag{15.98c}$$

Assuming $\alpha \neq 0$, the general unit-speed solution to Eq. (15.97) that is 2π-periodic in x is obtained by separation of variables as

$$\psi = \bar{\psi} + Ky + E_0 e^{-\alpha y} + F_0 e^{\alpha y} \tag{15.99}$$
$$+ \sum_{n\geq 1}\left(C_n e^{-ny} + D_n e^{ny} + E_n e^{-(n^2+\alpha^2)^{1/2}y} + F_n e^{(n^2+\alpha^2)^{1/2}y}\right)e^{in(x-t)},$$

where we take real parts implicitly (as allowed for linear operations). Given the boundary conditions in Eq. (15.98), the solution at order ϵ can only involve the $n = 1$ terms in the sum or terms independent of x, and we can take $\bar{\psi} = 0$ since only gradients of the streamfunction are relevant. To prevent the solution from diverging as $y \to \infty$, we need $F_0 = D_1 = F_1 = 0$ and therefore

$$\psi_1 = Ky + E_0 e^{-\alpha y} + \left(C_1 e^{-y} + E_1 e^{-(1+\alpha^2)^{1/2}y}\right)e^{i(x-t)}. \tag{15.100}$$

Since the swimmer needs to be force-free, we have $E_0 = 0$, and enforcing the three conditions in Eq. (15.98) leads to $K = 0$ and

$$C_1 = ib\frac{(1+\alpha^2)^{1/2}}{1-(1+\alpha^2)^{1/2}}, \quad E_1 = ib\frac{1}{(1+\alpha^2)^{1/2}-1}. \tag{15.101}$$

Therefore the first-order streamfunction is given finally, in real space, by

$$\psi_1 = \frac{1}{(1+\alpha^2)^{1/2}-1}\left[(1+\alpha^2)^{1/2}e^{-y} - e^{-(1+\alpha^2)^{1/2}y}\right]b\sin(x-t). \tag{15.102}$$

Once again, $U_1 = 0$ and any net fluid motion is quadratic in the amplitude.

Solution at Order ϵ^2 and Swimming Velocity

The boundary conditions for ψ_2 in real space are given by Eq. (15.69), i.e.

$$\left.\frac{\partial \psi_2}{\partial y}\right|_{(x,0)} = -b\sin(x-t)\left.\frac{\partial^2 \psi_1}{\partial y^2}\right|_{(x,0)}, \tag{15.103a}$$

$$\left.\frac{\partial \psi_2}{\partial x}\right|_{(x,0)} = -b\sin(x-t)\left.\frac{\partial^2 \psi_1}{\partial x \partial y}\right|_{(x,0)}, \tag{15.103b}$$

$$\lim_{y\to\infty} \nabla\psi_2 = U_2 \mathbf{e}_y. \tag{15.103c}$$

The solution for ψ_2 then takes the form of Eq. (15.99) with only the $n = 2$ modes and the linear solution Ky (the other terms with no x dependence are zero, as above). As in Chapter 3, and following Eq. (15.103c), the term linear in y satisfies $K = U_2$, and its value is obtained by computing the average of Eq. (15.103a) over one wavelength of the sheet. From Eq. (15.102) we evaluate

$$\left.\frac{\partial^2 \psi_1}{\partial y^2}\right|_{(x,0)} = -b(1+\alpha^2)^{1/2}\sin(x-t), \tag{15.104}$$

and then using Eq. (15.103a) we obtain

$$U_2 = (1+\alpha^2)^{1/2}\frac{b^2}{2}. \tag{15.105}$$

Discussion

Since the speed in a Newtonian Stokes flow, $U_{2,N}$, is found by setting $\alpha = 0$ in Eq. (15.105), we may use the Brinkman solution, $U_{2,B}$, and recall that $\alpha = \kappa/k$ to calculate the ratio

$$\frac{U_{2,B}}{U_{2,N}} = \left[1 + (\kappa/k)^2\right]^{1/2}. \tag{15.106}$$

It is clear that $U_{2,B} > U_{2,N}$ and the swimming of the free sheet (or the flow induced by the envelope of cilia) is always enhanced by the extra drag due to obstacles in the fluid. Note that a significant increase occurs only for long wavelengths, i.e. $k^{-1} \gtrsim O(\kappa^{-1})$.

The physical origin of the increase in flow speed can be further explored by computing the drag coefficients for slender filaments in a Brinkman medium (Leshansky, 2009). The ratio of drag coefficients is above 2, indicating an enhancement in locomotion.

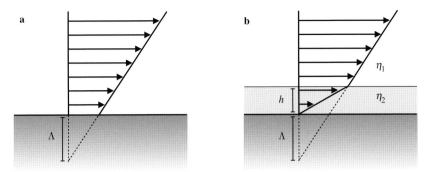

Figure 15.5 Slip and apparent slip. (a) Slip length Λ for unidirectional shear flow over a flat stationary surface. (b) Apparent slip for shear flow in a viscous fluid with a low-viscosity layer (η_2) of thickness h separating a high-viscosity fluid ($\eta_1 > \eta_2$) from a flat surface.

15.3.2 Apparent Slip-Induced Enhancement of Locomotion and Transport

Apparent Slip

The second multiscale model we consider captures the phenomenon of phase separation occurring generically in complex fluids near surfaces. In many fluidic suspensions and polymer solutions, the suspended solute interacts with boundaries, through steric interactions and also often from physicochemical potentials (such as van der Waals or electrostatic). These interactions usually lead to depletion of the solute near a surface, and thus to the presence of thin, solvent-rich fluid layers near boundaries (Barnes, 1995). As a result, the viscosity of the fluid takes the form of an approximate stratification whereby the majority of the bulk fluid has a viscosity η_1 and is separated from surfaces by a thin fluid layer of lower viscosity, $\eta_2 < \eta_1$.

To capture the impact of such depletion on the dynamics of the fluid, a classical modelling approach consists in replacing the no-slip boundary condition on the surfaces by an apparent slip (Sochi, 2011). A slip boundary condition, proposed by Navier (1823), has long been used as an empirical modelling tool to capture decreases in hydrodynamic friction near a boundary in a variety of settings (Lauga et al., 2007). For unidirectional flow over a flat surface with normal along the y direction, the slip boundary condition at $y = 0$ for the velocity $\mathbf{u} = (u_x, u_y)$ is given by $u_y = 0$ (no penetration) and

$$u_x = \Lambda \frac{\partial u_x}{\partial y}, \quad \text{at } y = 0. \tag{15.107}$$

Here $\Lambda > 0$ is the slip length, equal to the virtual length scale below the surface at which the no-slip condition would hold if extrapolated linearly (see Fig. 15.5(a)). Alternatively, Eq. (15.107) can be interpreted as a linear proportionality between surface velocity and surface stress akin to a friction condition.

The rationale behind the application of the slip boundary condition for complex fluids can be understood by taking a closer look at the dynamics near a two-fluid layer (Fig. 15.5(b)). Under a constant external shear stress, σ, the flow field is piecewise linear. If the viscosity of the fluid near the surface, η_2, is smaller than the bulk viscosity, η_1, continuity of shear stresses means that the shear rate closer to the surface, $\dot{\gamma}_2 = \sigma/\eta_2$, is larger than the bulk shear rate, $\dot{\gamma}_1 = \sigma/\eta_1$, and the flow field is as illustrated in Fig. 15.5(b). While the no-slip boundary condition does hold on the surface, if the low-viscosity layer is thin, it appears from the point of view of the bulk fluid that no-slip is violated and that the flow extrapolates linearly to zero at a distance

$$\Lambda = h \left(\frac{\eta_1}{\eta_2} - 1 \right) \tag{15.108}$$

below the surface. We call Λ the apparent slip length of the fluid, and it provides a modelling shorthand to incorporate thin low-viscosity layers in a complex fluid (Sochi, 2011).

In order to apply the slip boundary condition on the surface of a swimmer, we need to generalise Eq. (15.107) to the case of moving boundaries. Denoting by $\mathbf{n}$ the instantaneous unit normal on the surface of material points $\mathbf{r}_s$, the no-penetration boundary condition is written classically as

$$[\mathbf{n} \cdot \mathbf{u}] = 0, \quad \text{on} \quad \mathbf{r} = \mathbf{r}_s, \tag{15.109}$$

where brackets mean the difference between values in the fluid and values on the surface, while the slip condition in Eq. (15.107) becomes

$$[\mathbf{n} \times \mathbf{u}] = 2\Lambda \mathbf{n} \times (\mathbf{E} \cdot \mathbf{n}), \quad \text{on} \quad \mathbf{r} = \mathbf{r}_s, \tag{15.110}$$

where $\mathbf{E}$ is the symmetric rate of strain tensor.

Waving Sheet in Slipping Fluid

Since complex fluids have a tendency towards phase separation, we investigate the consequences of their apparent slip on locomotion. We consider the waving sheet model with the same setup and notation as in the previous section (Man and Lauga, 2015). After non-dimensionalisation, the material points on the sheet are located at

$$x_s = x, \tag{15.111a}$$
$$y_s = \epsilon b \sin(x - t), \tag{15.111b}$$

with the fluid in the domain $y \geqslant y_s$ and where we use the same symbols as the dimensional problem for convenience. We define $\phi \equiv x - t$ as in Chapter 3, in order to simplify notation, and the dimensionless slip length as $\xi = k\Lambda$. Note that a similar calculation can be carried out for slender filaments, and identical results are obtained at order ξ (Man and Lauga, 2015).

Following Eq. (15.111), the unit tangent vector along the sheet, $\mathbf{t}$, and the unit normal into the fluid, $\mathbf{n}$, are given by

$$\mathbf{t} = \frac{1}{(1 + \epsilon^2 b^2 \cos^2 \phi)^{1/2}} \begin{pmatrix} 1 \\ \epsilon b \cos \phi \end{pmatrix} \tag{15.112}$$

and

$$\mathbf{n} = \frac{1}{(1 + \epsilon^2 b^2 \cos^2 \phi)^{1/2}} \begin{pmatrix} -\epsilon b \cos \phi \\ 1 \end{pmatrix}. \tag{15.113}$$

Since the fluid is assumed to be Newtonian, the streamfunction satisfies the biharmonic equation as in Chapter 3,

$$\nabla^4 \psi = 0. \tag{15.114}$$

The first boundary condition for Eq. (15.114) is the continuity of the normal velocity, Eq. (15.109), which is written as

$$\mathbf{n} \cdot \mathbf{u} = \mathbf{n} \cdot \mathbf{u}_s, \quad \mathbf{r} = \mathbf{r}_s, \tag{15.115}$$

or, using the streamfunction,

$$\epsilon b \cos \phi \frac{\partial \psi}{\partial y}\bigg|_{(x, \epsilon b \sin \phi)} + \frac{\partial \psi}{\partial x}\bigg|_{(x, \epsilon b \sin \phi)} = \epsilon b \cos \phi. \tag{15.116}$$

The slip boundary condition, Eq. (15.110), written in a vector form is

$$\mathbf{n} \times \mathbf{u} = 2\xi \mathbf{n} \times (\mathbf{E} \cdot \mathbf{n}) + \mathbf{n} \times \mathbf{u}_s, \quad \mathbf{x} = \mathbf{x}_s. \tag{15.117}$$

In terms of the streamfunction, the components of the rate of strain tensor are

$$\mathbf{E} = \frac{1}{2} \begin{pmatrix} 2\dfrac{\partial^2 \psi}{\partial x \partial y} & \dfrac{\partial^2 \psi}{\partial y^2} - \dfrac{\partial^2 \psi}{\partial x^2} \\ \dfrac{\partial^2 \psi}{\partial y^2} - \dfrac{\partial^2 \psi}{\partial x^2} & -2\dfrac{\partial^2 \psi}{\partial x \partial y} \end{pmatrix}. \tag{15.118}$$

Using Eqs. (15.113) and (15.118), we can then evaluate the component of the slip condition, Eq. (15.117), in the $\mathbf{e}_z (= \mathbf{e}_x \times \mathbf{e}_y)$ direction and obtain the second boundary condition for ψ as

$$\epsilon b \cos \phi \frac{\partial \psi}{\partial x}\bigg|_{(x, \epsilon b \sin \phi)} - \frac{\partial \psi}{\partial y}\bigg|_{(x, \epsilon b \sin \phi)} - \epsilon^2 b^2 \cos^2 \phi = \tag{15.119}$$

$$\frac{\xi}{(1 + \epsilon^2 b^2 \cos^2 \phi)^{1/2}} \left[4\epsilon b \cos \phi \frac{\partial^2 \psi}{\partial x \partial y} + (1 - \epsilon^2 b^2 \cos^2 \phi) \left(\frac{\partial^2 \psi}{\partial x^2} - \frac{\partial^2 \psi}{\partial y^2} \right) \right]_{(x, \epsilon b \sin \phi)}.$$

Solution at Order ϵ

The boundary condition in Eq. (15.116) becomes, at order ϵ,

$$\left.\frac{\partial \psi_1}{\partial x}\right|_{(x,0)} = b \cos \phi, \tag{15.120}$$

while that in Eq. (15.119) simplifies to

$$-\left.\frac{\partial \psi_1}{\partial y}\right|_{(x,0)} = \xi \left(\frac{\partial^2 \psi_1}{\partial x^2} - \frac{\partial^2 \psi_1}{\partial y^2}\right)_{(x,0)}. \tag{15.121}$$

Noting that for the no-slip solution in Eq. (15.51) we have exactly

$$\left.\frac{\partial \psi_1}{\partial y}\right|_{(x,0)} = 0, \quad \left.\left(\frac{\partial^2 \psi_1}{\partial x^2} - \frac{\partial^2 \psi_1}{\partial y^2}\right)\right|_{(x,0)} = 0, \tag{15.122}$$

we see that Eq. (15.121) is satisfied for that solution, and therefore the streamfunction at first order with slip ($\xi \neq 0$) is the same as the no-slip one ($\xi = 0$) from Chapter 3, written in real space as

$$\psi_1 = b(1 + y)e^{-y} \sin(x - t). \tag{15.123}$$

Solution at Order ϵ^2

At order ϵ^2, the normal boundary condition in Eq. (15.116) is written as

$$\left.\frac{\partial \psi_2}{\partial x}\right|_{(x,0)} = -b \sin \phi \left.\frac{\partial^2 \psi_1}{\partial x \partial y}\right|_{(x,0)} - b \cos \phi \left.\frac{\partial \psi_1}{\partial y}\right|_{(x,0)}, \tag{15.124}$$

while the slip condition in Eq. (15.119) becomes

$$\left.\left(b \cos \phi \frac{\partial \psi_1}{\partial x} - \frac{\partial \psi_2}{\partial y} - b \sin \phi \frac{\partial^2 \psi_1}{\partial y^2}\right)\right|_{(x,0)} - b^2 \cos^2 \phi \tag{15.125}$$

$$= \xi \left[4b \cos \phi \frac{\partial^2 \psi_1}{\partial x \partial y} + \left(\frac{\partial^2 \psi_2}{\partial x^2} - \frac{\partial^2 \psi_2}{\partial y^2}\right) + b \sin \phi \left(\frac{\partial^3 \psi_1}{\partial x^2 \partial y} - \frac{\partial^3 \psi_1}{\partial y^3}\right)\right]_{(x,0)}.$$

Given these boundary conditions, the solution to the biharmonic equation in Eq. (15.114) for ψ_2 is of the general form in Eq. (3.17), i.e.

$$\psi(x, y, t) = \bar{\psi} + Ky + Gy^2 + Hy^3 \tag{15.126}$$
$$+ \sum_{n \geq 1} \left[(C_n + yD_n)e^{-ny} + (E_n + yF_n)e^{ny}\right] e^{in(x-t)},$$

and includes terms with either no dependence on x or with $n = 2$. As in Section 3.3, we can then write $\partial \psi_2/\partial y - U_2$ as a periodic function and average Eq. (15.125) over a wavelength of the waving sheet. With the solution in Eq. (15.123) at order ϵ, we get

$$\left\langle b\cos\phi \frac{\partial \psi_1}{\partial x}\bigg|_{(x,0)} \right\rangle = \frac{1}{2}b^2, \tag{15.127a}$$

$$\left\langle b\sin\phi \frac{\partial^2 \psi_1}{\partial y^2}\bigg|_{(x,0)} \right\rangle = -\frac{1}{2}b^2, \tag{15.127b}$$

$$\frac{\partial^2 \psi_1}{\partial x \partial y}\bigg|_{(x,0)} = 0, \tag{15.127c}$$

$$\left\langle \left(\frac{\partial^2 \psi_2}{\partial x^2} - \frac{\partial^2 \psi_2}{\partial y^2} \right)\bigg|_{(x,0)} \right\rangle = 0, \tag{15.127d}$$

$$\left\langle b\sin\phi \left(\frac{\partial^3 \psi_1}{\partial x^2 \partial y} - \frac{\partial^3 \psi_1}{\partial y^3} \right)\bigg|_{(x,0)} \right\rangle = -b^2, \tag{15.127e}$$

and thus obtain finally the swimming speed at leading order,

$$U_2 = \frac{1}{2}b^2 + \xi b^2. \tag{15.128}$$

Discussion

Returning to dimensional quantities, we may rewrite the result in Eq. (15.128) as a ratio between the swimming velocities in the case of slip, $U_{2,S}$, and no-slip, $U_{2,NS}$, as

$$\frac{U_{2,S}}{U_{2,NS}} = 1 + 2k\Lambda. \tag{15.129}$$

Clearly the right-hand side of this ratio is above 1, and therefore a finite slip length results in an increase in the swimming (or pumping) speed induced by the waving sheet, $U_{2,S} > U_{2,NS}$. A significant increase occurs when the slip length, Λ, is at least on the order of the wavelength, k^{-1}. Since slip lengths have been measured for polymer solutions to be on the order of 0.1–$10\,\mu m$ (Mhetar and Archer, 1998), and the wavelengths of eukaryotic flagella are on the order of tens of microns, phase separation in complex fluids can lead to large increases in flow speeds.

Further Reading

The rheology of complex fluids is a field with a long history of mathematical modelling, and many classical textbooks focusing on both the microscopic point of view and the continuum fluid mechanics are available (Bird, 1976; Bird et al., 1987a,b; Doi and Edwards, 1988; Larson, 1988; Tanner, 1988; Bird and Wiest, 1995; Larson, 1999; Morrison, 2001). Beyond the complex fluids considered in this chapter, fluids with different rheological characteristics could also be relevant to cell locomotion. For example, inelastic generalised Newtonian fluids are predicted theoretically to leave the swimming speed of inextensible swimmers unchanged at leading order in their amplitude (Vélez-Cordero and Lauga, 2013), a result consistent with experiments (Shen and Arratia, 2011; Gagnon et al., 2013). Viscoplastic fluids, which do not flow unless stresses exceed a finite yield stress, have also been shown to impact locomotion by localising flows near the surface of the swimmer (Hewitt and Balmforth, 2017, 2018). Active fluids composed of swimmer suspensions behave like non-Newtonian fluids, as shown in a series of experiments with swimming bacteria (Sokolov and Aranson, 2009; Ryan et al., 2011) and algae (Rafai et al., 2010; Mussler et al., 2013), and rationalised theoretically (Hatwalne et al., 2004; Saintillan, 2018).

Extensions to the theoretical models presented in this chapter have been proposed in the literature, giving the opportunity to address additional experiments. The role played by the amplitude of the flexible waving swimmer in allowing it to increase its locomotion speed is also relevant for helical propulsion. Experiments on rigid rotating helices showed that an increase in the force-free swimming speed was possible for De $= O(1)$ provided the helical amplitude was sufficiently large (Liu et al., 2011), which can be explained partially using numerical simulations (Spagnolie et al., 2013). For flexible swimmers, the feedback mechanism between viscoelastic stresses and kinematics derived using the waving sheet model is also relevant for the locomotion of *C. elegans* nematodes in polymeric fluids (Thomases and Guy, 2014). Similarly, polymer depletion is responsible for the increase in propulsion speeds of flagellated bacteria in macromolecular solutions (Zöttl and Yeomans, 2017), while slip plays an important role for locomotion in gels (Fu et al., 2010). Note that, using the waving sheet model, we can show that rigid swimmers can experience an increase in their swimming speed provided they combine waves travelling in both directions (Riley and Lauga, 2015). The squirmer model of Chapter 4 can also be used to probe locomotion in polymeric fluids (Zhu et al., 2011, 2012). Finally, asymptotic calculations have made it possible to compute rigorously the locomotion of finite-size three-dimensional swimmers either with small surface motion or for arbitrary motion in weakly non-Newtonian fluids (Lauga, 2009, 2014).

Beyond the biological realm, some work has gone into quantifying the impact of non-Newtonian fluids on the locomotion of artificial swimmers. Experiments show that flexible filaments actuated by magnetic fluids can swim faster in constant-viscosity viscoelastic fluids (Espinosa-Garcia et al., 2013) in a manner that depends on their deformation kinematics (Godinez et al., 2015). Nonlinearities in the relationship between stresses and deformation allow swimmers to escape the constraints of the scallop theorem seen in Chapter 2, and complex fluids can be exploited to create new types of swimmers. For example, rotated rigid helices unable to move in water can be propelled efficiently in a gel against which they can push (Schamel et al., 2014). Devices with reciprocal deformation can be made to swim by taking advantage of the nonlinear dependence of locomotion speeds on actuation frequency, as shown experimentally using Purcell's one-hinge scallop in a fluid with shear-dependent viscosity (Qiu et al., 2014). This situation is also relevant to the swimming of biological swimmers exploiting power/recovery strokes and experiencing buildup and relaxation of elastic stresses (Qin et al., 2015). Finally, propulsive forces can also result from secondary flows created by non-Newtonian normal stresses (Normand and Lauga, 2008; Pak et al., 2010, 2012; Keim et al., 2012).

Exercises

1. Consider a linear viscoelastic fluid modelled by the one-dimensional spring-dashpot system illustrated in the diagram below. Derive the relationship between the overall stress, σ, and overall shear rate, $\dot{\gamma}$, for the fluid in the form $(1 + \mathcal{D})\,\sigma = \eta\,(1 + \mathcal{L})\,\dot{\gamma}$, where $\mathcal{D}$ and $\mathcal{L}$ are two differential operators to be identified and η is a viscosity to be determined.

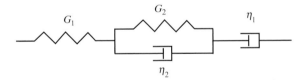

2. Calculate the relaxation modulus, $G(u)$, for the one-dimensional Jeffreys fluid with a constitutive relationship between stress, σ, and shear rate, $\dot{\gamma}$, given by

$$\left(1 + \lambda_1 \frac{\partial}{\partial t}\right)\sigma = \eta\left(1 + \lambda_2 \frac{\partial}{\partial t}\right)\dot{\gamma}.$$

3. Consider an incompressible linear viscoelastic fluid where the deviatoric stress tensor, $\boldsymbol{\tau}$, is linearly related to the shear rate tensor, $\dot{\boldsymbol{\gamma}}$, as

$$\boldsymbol{\tau} = \int_{-\infty}^{t} G(t - t')\dot{\boldsymbol{\gamma}}(t')\,dt'.$$

This fluid undergoes steady shear at rate $\dot{\gamma} > 0$ on a planar rotating table. The surface of the table is parametrised by the unit vectors $\mathbf{e}_x$ and $\mathbf{e}_y$ and rotates at constant rate Ω about the $\mathbf{e}_z$ vector such that at $t = 0$ the rotating frame coincides with the lab frame $(\mathbf{e}_1, \mathbf{e}_2, \mathbf{e}_3)$. Compute the viscosity of the fluid in the rotating frame and in the lab frame, and evaluate them at $t = 0$. Show that they are not identical in general if $\Omega \neq 0$.

4. In an upper-convected Maxwell fluid, the deviatoric stress tensor, τ, and the shear rate tensor, $\dot{\gamma}$, are related by the constitutive relationship

$$\tau + \lambda \overset{\triangledown}{\tau} = \eta \dot{\gamma},$$

where $\overset{\triangledown}{\tau}$ refers to the upper-convected derivative of τ. Its fundamental rheological behaviour can be characterised in steady shear and steady extension.

For steady shear, consider the two-dimensional shear flow $\mathbf{u} = \dot{\gamma}(y, 0, 0)$ in Cartesian coordinates. Assuming steady state, compute all components of the stress tensor and show that the shear viscosity is constant.

For steady extension, consider the three-dimensional uniform extensional flow $\mathbf{u} = \dot{\epsilon}(-x/2, -y/2, z)$. Compute the components of the stress tensor. Defining an extensional viscosity, η_{ext}, as the ratio between the extra normal stress along z and the extension rate, i.e. $\eta_{ext} = (\tau_{zz} - \tau_{xx})/\dot{\epsilon}$, show that in the Newtonian limit η_{ext} is equal to three times the shear viscosity. Show that, for the upper-convected Maxwell fluid, there is a critical extension rate $\dot{\epsilon}$ at which the extensional viscosity diverges.

5. The Gordon–Schowalter derivative is an example of an objective derivative. It is defined, for a tensor $\mathbf{T}$, as

$$\overset{\square}{\mathbf{T}} \equiv \overset{\triangledown}{\mathbf{T}} + \frac{a}{2} (\dot{\gamma} \cdot \mathbf{T} + \mathbf{T} \cdot \dot{\gamma}),$$

where $\overset{\triangledown}{\mathbf{T}}$ is the upper-convected derivative of $\mathbf{T}$, $\dot{\gamma}$ is the shear rate tensor and $a > 0$ is a small dimensionless parameter. Using this derivative, we define the Johnson–Segalman–Oldroyd fluid for which the relationship between the deviatoric stress tensor, τ, and the shear rate tensor is given by

$$\tau + \lambda_1 \overset{\square}{\tau} = \eta \left(\dot{\gamma} + \lambda_2 \overset{\square}{\dot{\gamma}} \right).$$

Calculate the steady shear viscosity for this fluid. What are the conditions on the values of λ_1 and λ_2 for the fluid to be shear-thinning (i.e. to have a steady shear viscosity that decreases with shear rate)?

6. In a Maxwell fluid, the relationship between the deviatoric stress tensor, τ, and the shear rate tensor, $\dot{\gamma}$, is given by the linear differential equation

$$\left(1 + \lambda \frac{\partial}{\partial t} \right) \tau = \eta \dot{\gamma}.$$

Show that resistive-force theory, which relates the hydrodynamic force density along a slender filament, $\mathbf{f}(s,t)$, to the velocity of the filament centreline, $\mathbf{u}(s,t)$, at arclength s and time t, becomes

$$\left(1 + \lambda \frac{\partial}{\partial t}\right) \mathbf{f}(s,t) = -\left[c_\parallel \mathbf{tt} + c_\perp (\mathbf{1} - \mathbf{tt})\right] \cdot \mathbf{u}(s,t),$$

in a Maxwell fluid.

7. We revisit the setup from Exercise 4, Chapter 5 in the case of a non-Newtonian fluid. A straight rigid rod of length L is made to oscillate in position and orientation in a linear viscoelastic fluid, due to time-varying external forces and torques, around the configuration where it is aligned with the y direction and its centre is located at the origin. The centre of the rod moves horizontally as $x(t) = \epsilon a \cos(\omega t)$, where a has dimension of length and ϵ is a dimensionless parameter. The orientation of the rod oscillates in the (x, y) plane with angle $\theta = \epsilon \cos(\omega t + \phi)$ to the y direction, where ϕ is a constant. In the viscoelastic fluid, the forces acting on the rod are captured by the modified resistive-force theory derived in Exercise 6. For small oscillating amplitude, $\epsilon \ll 1$, show that the averaged force in the y direction, at order ϵ^2, is the same as in the Newtonian case, but that the rate of work of the rod on the fluid is always smaller.

8. A waving sheet swims in a heterogeneous fluid as in Section 15.3, but now the waving motion is purely longitudinal. The dimensionless location of material points (x_s, y_s) on the waving sheet is therefore given in the swimming frame by $x_s = x + \epsilon a \cos(x - t)$, $y_s = 0$. The fluid obeys the dimensionless incompressible Brinkman equation, i.e.

$$\nabla \cdot \mathbf{u} = 0, \quad \nabla p = \nabla^2 \mathbf{u} - \alpha^2 \mathbf{u}.$$

Show that, as opposed to the transverse case studied in Section 15.3, the swimming speed of a sheet undergoing longitudinal waving motion is the same at order ϵ^2 as the Newtonian Stokes result.

9. In Chapter 12, we used a symmetry argument to show that swimmers with front–back symmetry cannot synchronise if the flow is Newtonian. Why does this symmetry argument no longer hold in general for a non-Newtonian fluid?

10. Following the result in Exercise 9, we consider the synchronisation of two front–back symmetric sheets undergoing transverse waving motion in a non-Newtonian fluid. We adopt the notation of Chapter 12 and write the instantaneous dimensionless locations of the sheets as two sinewaves, $y_1 = \epsilon b \sin(x - t)$, $y_2 = d + \epsilon b \sin(x - t + \Delta)$. The fluid between the swimmers obeys the dimensionless Oldroyd-B constitutive relationship relating the deviatoric stress tensor, τ, to the shear rate tensor, $\dot{\gamma}$, as

$$\tau + \mathrm{De}_1 \overset{\triangledown}{\tau} = \dot{\gamma} + \mathrm{De}_2 \overset{\triangledown}{\gamma},$$

where De_1 and De_2 are the two Deborah numbers. Solve for the flow between the swimmers as a power expansion in ϵ. Show that synchronisation occurs for the force-free swimmers at order ϵ^2 and always leads to phase-locking with $\Delta = 0$.

11. A model of polymeric fluid includes the phase-separated nature of the fluid near surfaces explicitly. Consider a swimmer moving below a fluid made up of two separate viscous Newtonian domains, namely a bulk fluid with viscosity η_1 and a thin solvent-rich layer near the surface of the swimmer with viscosity $\eta_2 < \eta_1$. The swimmer is a two-dimensional waving sheet undergoing transverse waving motion with dimensionless material points (x_s, y_s) located at $x_s = x$, $y_s = \epsilon b \sin(x - t)$. The bulk fluid is located in the domain $y > h$ while the solvent-rich fluid is between the swimmer and the bulk fluid, i.e. in the region $y_s < y < h$. For small-amplitude waving motion, solve for the flow in both fluids asymptotically at order ϵ, and compute the swimming speed at order ϵ^2. Show that swimming is always faster than in a homogeneous Newtonian fluid.

12. The deviatoric stress in a weakly non-Newtonian fluid, τ, is related to the shear rate tensor, $\dot{\gamma}$, as $\tau = \eta \dot{\gamma} + \epsilon \Sigma[\mathbf{u}]$, where $\Sigma[\mathbf{u}]$ is a symmetric tensor with dimension of stress, which is an arbitrary nonlinear instantaneous functional of the velocity field $\mathbf{u}$, and $\epsilon \ll 1$ is a small dimensionless number. Consider an arbitrary three-dimensional swimmer moving in such a fluid by instantaneously deforming its shape and imposing velocity boundary conditions on the fluid. Solve the swimming problem in powers of ϵ using the reciprocal theorem for Stokes flows subject to body forces derived in Exercise 3, Chapter 6. Hence show that the linear and angular velocities of the swimmer can be computed at order ϵ in an integral form using only knowledge of the solution to the Newtonian swimming problem.

13. A swimmer is located on the surface of a thin layer of non-Newtonian viscous fluid whose viscosity decreases with shear rate. The swimmer is composed of two large horizontal paddles A and B, of respective surface areas S_A and S_B, and the thin layer of fluid is located between the paddles and a rigid no-slip surface. The distance between the paddles is prescribed to oscillate in time as $\ell(t)$. Explain the physical principle allowing the paddle-swimmer to move on average, despite using reciprocal motion. How different would the result be if the fluid were shear-thickening (i.e. if its viscosity were an increasing function of the shear rate)?

References

Alberts, B., Johnson, A., Lewis, J., et al. 2007. *Molecular Biology of the Cell. 5th edn.* New York, NY: Garland Science.

Alexander, D. E. 2002. *Nature's Flyers: Birds, Insects, and the Biomechanics of Flight.* Baltimore, MD: The Johns Hopkins University Press.

Anderson, J. 1989. Colloidal transport by interfacial forces. *Annu. Rev. Fluid Mech.*, **21**, 61–99.

Audoly, B. and Pomeau, Y. 2010. *Elasticity and Geometry: From Hair Curls to the Nonlinear Response of Shells.* Oxford, UK: Oxford University Press.

Avron, J. E., Kenneth, O., and Oaknin, D. H. 2005. Pushmepullyou: An efficient microswimmer. *New J. Phys.*, **7**, 234.

Barnes, H. A. 1995. A review of the slip (wall depletion) of polymer solutions, emulsions and particle suspensions in viscometers: Its cause, character, and cure. *J. Non-Newt. Fluid Mech.*, **56**, 221–231.

Batchelor, G. K. 1967. *An Introduction to Fluid Dynamics.* Cambridge, UK: Cambridge University Press.

Batchelor, G. K. 1970a. The stress system in a suspension of force-free particles. *J. Fluid Mech.*, **41**, 545–570.

Batchelor, G. K. 1970b. Slender body theory for particles of arbitrary cross section in Stokes flow. *J. Fluid Mech.*, **44**, 419–440.

Batchelor, G. K. 1976. Developments in microhydrodynamics. In Koiter, W.T. (ed.), *Theoretical and Applied Mechanics.* North-Holland, Amsterdam. Pages 33–55.

Bennett, R. R. and Golestanian, R. 2013. Phase-dependent forcing and synchronization in the three-sphere model of *Chlamydomonas. New J. Phys.*, **15**, 075028.

Berg, H. C. 1993. *Random Walks in Biology.* Princeton, NJ: Princeton University Press.

Berg, H. C. 2003. The rotary motor of bacterial flagella. *Annu. Rev. Biochem.*, **72**, 19–54.

Berg, H. C. 2004. *E. coli in Motion.* New York, NY: Springer-Verlag.

Berg, H. C. and Anderson, R. A. 1973. Bacteria swim by rotating their flagellar filaments. *Nature*, **245**, 380–382.

Berg, H. C. and Brown, D. A. 1972. Chemotaxis in *Escherichia coli* analysed by three-dimensional tracking. *Nature*, **239**, 500–504.

Berg, H. C. and Turner, L. 1979. Movement of microorganisms in viscous environments. *Nature*, **278**, 349–351.

Berg, H. C. and Turner, L. 1990. Chemotaxis of bacteria in glass capillary arrays: *Escherichia coli*, motility, microchannel plate, and light scattering. *Biophys. J.*, **58**, 919–930.

Berke, A. P., Turner, L., Berg, H. C., and Lauga, E. 2008. Hydrodynamic attraction of swimming microorganisms by surfaces. *Phys. Rev. Lett.*, **101**, 038102.

Berman, R. S., Kenneth, O., Sznitman, J., and Leshansky, A. M. 2013. Undulatory locomotion of finite filaments: Lessons from *Caenorhabditis elegans*. *New J. Phys.*, **15**, 075022.

Bianchi, S., Saglimbeni, F., and Di Leonardo, R. 2017. Holographic imaging reveals the mechanism of wall entrapment in swimming bacteria. *Phys. Rev. X*, **7**, 011010.

Bird, R. B. 1976. Useful non-Newtonian models. *Annu. Rev. Fluid Mech.*, **8**, 13–34.

Bird, R. B. and Wiest, J. M. 1995. Constitutive equations for polymeric liquids. *Annu. Rev. Fluid Mech.*, **27**, 169–193.

Bird, R. B., Armstrong, R. C., and Hassager, O. 1987a. *Dynamics of Polymeric Liquids. Vol. 1: Fluid Mechanics*. 2nd edn. New York, NY: Wiley-Interscience.

Bird, R. B., Curtiss, C. F., Armstrong, R. C., and Hassager, O. 1987b. *Dynamics of Polymeric Liquids. Vol. 2: Kinetic Theory*. 2nd edn. New York, NY: Wiley-Interscience.

Blake, J. R. 1971a. A note on the image system for a stokeslet in a no-slip boundary. *Proc. Camb. Phil. Soc.*, **70**, 303–310.

Blake, J. R. 1971b. A spherical envelope approach to ciliary propulsion. *J. Fluid Mech.*, **46**, 199–208.

Blake, J. R. 1971c. Infinite models for ciliary propulsion. *J. Fluid Mech.*, **49**, 209–222.

Blake, J. R. 1972. A model for the micro-structure in ciliated organisms. *J. Fluid Mech.*, **55**, 1–23.

Blake, J. R. and Chwang, A. T. 1974. Fundamental singularities of viscous-flow. Part 1: Image systems in vicinity of a stationary no-slip boundary. *J. Eng. Math.*, **8**, 23–29.

Blake, J. R. and Sleigh, M. A. 1974. Mechanics of ciliary locomotion. *Biol. Rev. Camb. Phil. Soc.*, **49**, 85–125.

Blakemore, R. P. 1982. Magnetotactic bacteria. *Annu. Rev. Microbiol.*, **36**, 217–238.

Brady, J. F. and Bossis, G. 1988. Stokesian dynamics. *Annu. Rev. Fluid Mech.*, **20**, 111–157.

Bray, D. 2000. *Cell Movements*. New York, NY: Garland Publishing.

Brennen, C. 1974. Oscillating boundary layer theory for ciliary propulsion. *J. Fluid Mech.*, **65**, 799–824.

Brennen, C. and Winet, H. 1977. Fluid mechanics of propulsion by cilia and flagella. *Annu. Rev. Fluid Mech.*, **9**, 339–398.

Brenner, H. 1967. Coupling between the translational and rotational Brownian motions of rigid particles of arbitrary shape: II. General theory. *J. Colloid Interface Sci.*, **23**, 407–436.

Bretherton, F. P. 1962. The motion of rigid particles in a shear flow at low Reynolds number. *J. Fluid Mech.*, **14**, 284–304.

Brinkman, H. C. 1947. A calculation of the viscous force exerted by a flowing fluid on a dense swarm of particles. *Appl. Sci. Res. A*, **1**, 27–34.

Brokaw, C. J. 1972. Flagellar movement: A sliding filament model. *Science*, **178**, 455–462.

Brokaw, C. J. 1989. Direct measurements of sliding between outer doublet microtubules in swimming sperm flagella. *Science*, **243**, 1593–1596.

Brotto, T., Caussin, J.-B., Lauga, E., and Bartolo, D. 2013. Hydrodynamics of confined active fluids. *Phys. Rev. Lett.*, **110**, 038101.

Brumley, D. R., Polin, M., Pedley, T. J., and Goldstein, R. E. 2012. Hydrodynamic synchronization and metachronal waves on the surface of the colonial alga *Volvox carteri*. *Phys. Rev. Lett.*, **109**, 268102.

Brumley, D. R., Wan, K. Y., Polin, M., and Goldstein, R. E. 2014. Flagellar synchronization through direct hydrodynamic interactions. *eLife*, **3**, e02750.

Brumley, D. R., Polin, M., Pedley, T. J., and Goldstein, R. E. 2015. Metachronal waves in the flagellar beating of *Volvox* and their hydrodynamic origin. *J. Roy. Soc. Interface*, **12**, 20141358.

Butenko, A. V., Mogilko, E., Amitai, L., Pokroy, B., and Sloutskin, E. 2012. Coiled to diffuse: Brownian motion of a helical bacterium. *Langmuir*, **28**, 12941–7.

Calladine, C. R. 1975. Construction of bacterial flagella. *Nature*, **255**, 121–124.

Calladine, C. R. 1978. Change of waveform in bacterial flagella: The role of mechanics at the molecular level. *J. Mol. Biol.*, **118**, 457–479.

Camalet, S. and Jülicher, F. 2000. Generic aspects of axonemal beating. *New J. Phys.*, **2**, 1–23.

Campbell, N. A. and Reece, J. B. 2014. *Biology*. Boston, MA: Pearson.

Cates, M. E. and Tailleur, J. 2015. Motility-induced phase separation. *Annu. Rev. Condens. Matter Phys.*, **6**, 219–244.

Chaikin, P. M. and Lubensky, T. C. 2000. *Principles of Condensed Matter Physics*. Cambridge, UK: Cambridge University Press.

Chamolly, A., Ishikawa, T., and Lauga, E. 2017. Active particles in periodic lattices. *New J. Phys.*, **19**, 115001.

Chan, B. 2009. Bio-inspired fluid locomotion. Ph.D. thesis, Massachusetts Institute of Technology, Cambridge, MA.

Chan, B., Balmforth, N. J., and Hosoi, A. E. 2005. Building a better snail: Lubrication and adhesive locomotion. *Phys. Fluids*, **17**, 113101.

Chandrasekhar, S. 1943. Stochastic problems in physics and astronomy. *Rev. Mod. Phys.*, **15**, 1–89.

Chen, D. T. N., Heymann, M., Fraden, S., Nicastro, D., and Dogic, Z. 2015. ATP consumption of eukaryotic flagella measured at a single-cell level. *Biophys. J.*, **109**, 2562–2573.

Childress, S. 1981. *Mechanics of Swimming and Flying*. Cambridge, UK: Cambridge Universtity Press.

Childress, S. 2012. A thermodynamic efficiency for Stokesian swimming. *J. Fluid Mech.*, **705**, 77–97.

Childress, S., Levandowsky, M., and Spiegel, E. A. 1975. Pattern formation in a suspension of swimming microorganisms: Equations and stability theory. *J. Fluid Mech.*, **69**, 591–613.

Chwang, A. T. and Wu, T. Y. 1971. Helical movement of microorganisms. *Proc. Roy. Soc. Lond. B*, **178**, 327–346.

Chwang, A. T. and Wu, T. Y. 1975. Hydromechanics of low-Reynolds-number flow. Part 2: Singularity method for Stokes flows. *J. Fluid Mech.*, **99**, 411–431.

Cisneros, L., Dombrowski, C., Goldstein, R. E., and Kessler, J. O. 2006. Reversal of bacterial locomotion at an obstacle. *Phys. Rev. E*, **73**, 030901.

Cisneros, L. H., Cortez, R., Dombrowski, C., Goldstein, R. E., and Kessler, J. O. 2007. Fluid dynamics of self-propelled microorganisms, from individuals to concentrated populations. *Exp. Fluids*, **43**, 737–753.

Copeland, M. F. and Weibel, D. B. 2009. Bacterial swarming: A model system for studying dynamic self-assembly. *Soft Matter*, **5**, 1174–1187.

Cox, R. G. 1970. The motion of long slender bodies in a viscous fluid. Part 1: General Theory. *J. Fluid Mech.*, **44**, 791–810.

Darnton, N. C. and Berg, H. C. 2007. Force-extension measurements on bacterial flagella: Triggering polymorphic transformations. *Biophys. J.*, **92**, 2230–2236.

Darnton, N. C., Turner, L., Rojevsky, S., and Berg, H. C. 2007. On torque and tumbling in swimming *Escherichia coli*. *J. Bacteriol.*, **189**, 1756–1764.

Darnton, N. C., Turner, L., Rojevsky, S., and Berg, H. C. 2010. Dynamics of bacterial swarming. *Biophys. J.*, **98**, 2082–2090.

Dauparas, J. and Lauga, E. 2016. Flagellar flows around bacterial swarms. *Phys. Rev. Fluids*, **1**, 043202.

De Langre, E. 2008. Effects of wind on plants. *Annu. Rev. Fluid Mech.*, **40**, 141–168.

De Lillo, F., Cencini, M., Durham, et al. 2014. Turbulent fluid acceleration generates clusters of gyrotactic microorganisms. *Phys. Rev. Lett.*, **112**, 044502.

Denissenko, P., Kantsler, V., Smith, D. J., and Kirkman-Brown, J. 2012. Human spermatozoa migration in microchannels reveals boundary-following navigation. *Proc. Natl. Acad. Sci. U.S.A.*, **109**, 8007–8010.

Di Leonardo, R., Angelani, L., Dell'Arciprete, D., et al. 2010. Bacterial ratchet motors. *Proc. Natl. Acad. Sci. U.S.A.*, **107**, 9541–9545.

Di Leonardo, R., Dell'Arciprete, D., Angelani, L., and Iebba, V. 2011. Swimming with an image. *Phys. Rev. Lett.*, **106**, 038101.

Di Leonardo, R., Búzás, A., Kelemen, L., et al. 2012. Hydrodynamic synchronization of light driven microrotors. *Phys. Rev. Lett.*, **109**, 034104.

DiLuzio, W. R., Turner, L., Mayer, M., et al. 2005. *Escherichia coli* swim on the right-hand side. *Nature*, **435**, 1271–1274.

Doi, M. and Edwards, S. F. 1988. *The Theory of Polymer Dynamics*. Oxford, UK: Oxford University Press.

Dombrowski, C., Cisneros, L., Chatkaew, S., Goldstein, R. E., and Kessler, J. O. 2004. Self-concentration and large-scale coherence in bacterial dynamics. *Phys. Rev. Lett.*, **93**, 098103.

Doostmohammadi, A., Ignés-Mullol, J., Yeomans, J. M., and Sagués, F. 2018. Active nematics. *Nature Comm.*, **9**, 3246.

Drescher, K., Goldstein, R. E., Michel, N., Polin, M., and Tuval, I. 2010a. Direct measurement of the flow field around swimming microorganisms. *Phys. Rev. Lett.*, **105**, 168101.

Drescher, K., Goldstein, R. E., and Tuval, I. 2010b. Fidelity of adaptive phototaxis. *Proc. Natl. Acad. Sci. U.S.A.*, **107**, 11171–11176.

Drescher, K., Dunkel, J., Cisneros, L. H., Ganguly, S., and Goldstein, R. E. 2011. Fluid dynamics and noise in bacterial cell-cell and cell-surface scattering. *Proc. Natl. Acad. Sci. U.S.A.*, **108**, 10940–10945.

Dudley, R. 2002. *The Biomechanics of Insect Flight: Form, Function, Evolution*. Princeton, NJ: Princeton University Press.

Durham, W. M., Kessler, J. O., and Stocker, R. 2009. Disruption of vertical motility by shear triggers formation of thin phytoplankton layers. *Science*, **323**, 1067–1070.

Durham, W. M., Climent, E., and Stocker, R. 2011. Gyrotaxis in a steady vortical flow. *Phys. Rev. Lett.*, **106**, 238102.

Durham, W. M., Climent, E., Barry, M., et al. 2013. Turbulence drives microscale patches of motile phytoplankton. *Nature Comm.*, **4**, 2148.

Ebbens, S. J. and Howse, J. R. 2011. Direct observation of the direction of motion for spherical catalytic swimmers. *Langmuir*, **27**, 12293–12296.

Elfring, G. J. and Lauga, E. 2009. Hydrodynamic phase locking of swimming microorganisms. *Phys. Rev. Lett.*, **103**, 088101.

Elfring, G. J. and Lauga, E. 2011a. Passive hydrodynamic synchronization of two-dimensional swimming cells. *Phys. Fluids*, **23**, 011902.

Elfring, G. J. and Lauga, E. 2011b. Synchronization of flexible sheets. *J. Fluid Mech.*, **674**, 163–173.

Elgeti, J. and Gompper, G. 2013. Emergence of metachronal waves in cilia arrays. *Proc. Natl. Acad. Sci. U.S.A.*, **110**, 4470–4475.

Ellington, C. P. 1984. *The Aerodynamics of Hovering Insect Flight*. London, UK: The Royal Society.

Eloy, C. and Lauga, E. 2012. Kinematics of the most efficient cilium. *Phys. Rev. Lett.*, **109**, 038101.

Ermak, D. L. and McCammon, J. A. 1978. Brownian dynamics with hydrodynamic interactions. *J. Chem. Phys.*, **69**, 1352–1360.

Espinosa-Garcia, J., Lauga, E., and Zenit, R. 2013. Elasticity increases locomotion of flexible swimmers. *Phys. Fluids*, **25**, 031701.

Eytan, O. and Elad, D. 1999. Analysis of intra-uterine fluid motion induced by uterine contractions. *Bull. Math. Biol.*, **61**, 221–238.

Fauci, L. J. 1990. Interaction of oscillating filaments: A computational study. *J. Comput. Phys.*, **86**, 294–313.

Fauci, L. J. and Dillon, R. 2006. Biofluidmechanics of reproduction. *Annu. Rev. Fluid Mech.*, **38**, 371–394.

Fauci, L. J. and McDonald, A. 1995. Sperm motility in the presence of boundaries. *Bull. Math. Biol.*, **57**, 679–699.

Fawcett, D. W. 1975. The mammalian spermatozoon. *Dev. Biol.*, **44**, 394–436.

Fish, F. E. and Lauder, G. V. 2006. Passive and active flow control by swimming fishes and mammals. *Annu. Rev. Fluid Mech.*, **38**, 193–224.

Flores, H., Lobaton, E., Mendez-Diez, S., Tlupova, S., and Cortez, R. 2005. A study of bacterial flagellar bundling. *Bull. Math. Biol.*, **67**, 137–168.

Flory, P. J. 1969. *Statistical Mechanics of Chain Molecules*. New york: Interscience.

Friedrich, B. M., Riedel-Kruse, I. H., Howard, J., and Jülicher, F. 2010. High-precision tracking of sperm swimming fine structure provides strong test of resistive force theory. *J. Exp. Biol.*, **213**, 1226–1234.

Frymier, P. D., Ford, R. M., Berg, H. C., and Cummings, P. T. 1995. Three-dimensional tracking of motile bacteria near a solid planar surface. *Proc. Natl. Acad. Sci. U.S.A.*, **92**, 6195–6199.

Fu, H. C., Wolgemuth, C. W., and Powers, T. R. 2008. Beating patterns of filaments in viscoelastic fluids. *Phys. Rev. E*, **78**, 041913–1–12.

Fu, H. C., Wolgemuth, C. W., and Powers, T. R. 2009. Swimming speeds of filaments in nonlinearly viscoelastic fluids. *Phys. Fluids*, **21**, 033102.

Fu, H. C., Shenoy, V. B., and Powers, T. R. 2010. Low-Reynolds-number swimming in gels. *Europhys. Lett.*, **91**, 24002.

Fujii, M., Shibata, S., and Aizawa, S.-I. 2008. Polar, peritrichous, and lateral flagella belong to three distinguishable flagellar families. *J. Mol. Biol.*, **379**, 273–283.

Fulford, G. R., Katz, D. F., and Powell, R. L. 1998. Swimming of spermatozoa in a linear viscoelastic fluid. *Biorheol.*, **35**, 295–309.

Gaffney, E. A., Gadelha, H., Smith, D. J., Blake, J. R., and Kirkman-Brown, J. C. 2011. Mammalian sperm motility: Observation and theory. *Annu. Rev. Fluid Mech.*, **43**, 501–528.

Gagnon, D. A., Shen, X. N., and Arratia, P. E. 2013. Undulatory swimming in fluids with polymer networks. *Europhys. Lett.*, **104**, 14004.

Galajda, P., Keymer, J. E., Chaikin, P., and Austin, R. H. 2007. A wall of funnels concentrates swimming bacteria. *J. Bacteriol.*, **189**, 8704–8707.

Garcia, X., Rafai, S., and Peyla, P. 2013. Light control of the flow of phototactic microswimmer suspensions. *Phys. Rev. Lett.*, **110**, 138106.

Gardiner, B., Berry, P., and Moulia, B. 2016. Wind impacts on plant growth, mechanics and damage. *Plant Sci.*, **245**, 94–118.

Gaunt, J. A. 1929. The triplets of helium. *Phil. Trans. Roy. Soc. A*, **228**.

Gest, H. 2004. The discovery of microorganisms by Robert Hooke and Antoni Van Leeuwenhoek, fellows of the Royal Society. *Notes Rec. Roy. Soc. London*, **58** 187–201.

Geyer, V. F., Jülicher, F., Howard, J., and Friedrich, B. M. 2013. Cell-body rocking is a dominant mechanism for flagellar synchronization in a swimming alga. *Proc. Natl. Acad. Sci. U.S.A.*, **110**, 18058–18063.

Ghosh, A. and Fischer, P. 2009. Controlled propulsion of artificial magnetic nanostructured propellers. *Nano Lett.*, **9**, 2243–2245.

Giacché, D., Ishikawa, T., and Yamaguchi, T. 2010. Hydrodynamic entrapment of bacteria swimming near a solid surface. *Phys. Rev. E*, **82**, 056309.

Gibbons, B. H. and Gibbons, I. R. 1972. Flagellar movement and adenosine triphosphatase activity in sea urchin sperm extracted with Triton X-100. *J. Cell Bio.*, **54**, 75–97.

Gilboa, A. and Silberberg, A. 1976. In-situ rheological characterization of epithelial mucus. *Biorheol.*, **13**, 59–65.

Godinez, F. A., Koens, L., Montenegro-Johnson, T. D., Zenit, R., and Lauga, E. 2015. Complex fluids affect low-Reynolds number locomotion in a kinematic-dependent manner. *Exp. Fluids*, **56**, 97.

Goldman, A. J., Cox, R. G., and Brenner, H. 1967. Slow viscous motion of a sphere parallel to a plane wall: II Couette flow. *Chem. Eng. Sci.*, **22**, 653–660.

Goldstein, R. E. 2015. Green algae as model organisms for biological fluid dynamics. *Annu. Rev. Fluid Mech.*, **47**, 343–75.

Goldstein, R. E., Polin, M., and Tuval, I. 2009. Noise and synchronization in pairs of beating eukaryotic flagella. *Phys. Rev. Lett.*, **103**, 168103.

Golestanian, R., Liverpool, T. B., and Ajdari, A. 2007. Designing phoretic micro- and nano-swimmers. *New J. Phys.*, **9**, 126.

Götz, T. 2000. Interactions of fibers and flow: Asymptotics, theory and numerics. Ph.D. thesis, Universität Kaiserslautern, Germany.

Graham, M. D. 2018. *Microhydrodynamics, Brownian Motion, and Complex Fluids.* Cambridge, UK: Cambridge University Press.

Gray, J. 1928. *Ciliary Movement.* Cambridge, UK: Cambridge University Press.

Gray, J. and Hancock, G. J. 1955. The propulsion of sea-urchin spermatozoa. *J. Exp. Biol.*, **32**, 802–814.

Grotberg, J. B. 1994. Pulmonary flow and transport phenomena. *Annu. Rev. Fluid Mech.*, **26**, 529–571.

Guasto, J. S., Johnson, K. A., and Gollub, J. P. 2010. Oscillatory flows induced by microorganisms swimming in two dimensions. *Phys. Rev. Lett.*, **105**, 168102.

Guasto, J. S., Rusconi, R., and Stocker, R. 2012. Fluid mechanics of planktonic microorganisms. *Annu. Rev. Fluid Mech.*, **44**, 373–400.

Gueron, S. and Levit-Gurevich, K. 1999. Energetic considerations of ciliary beating and the advantage of metachronal coordination. *Proc. Natl. Acad. Sci. U.S.A.*, **96**, 12240–12245.

Gueron, S. and Liron, N. 1993. Simulations of three-dimensional ciliary beats and cilia interactions. *Biophys. J.*, **65**, 499–507.

Gueron, S., Levit-Gurevich, K., Liron, N., and Blum, J. J. 1997. Cilia internal mechanism and metachronal coordination as the result of hydrodynamical coupling. *Proc. Natl. Acad. Sci. U.S.A.*, **94**, 6001–6006.

Guirao, B. and Joanny, J. F. 2007. Spontaneous creation of macroscopic flow and metachronal waves in an array of cilia. *Biophys. J.*, **92**, 1900–1917.

Guyon, E., Hulin, J., Petit, L., and Mitescu, C. 2001. *Physical Hydrodynamics*. Oxford, UK: Oxford University Press.

Hall, W. F. and Busenberg, S. N. 1969. Viscosity of magnetic suspensions. *J. Chem. Phys.*, **51**, 137–144.

Hamel, A., Fish, C., Combettes, L., Dupuis-Williams, P., and Baroud, C. N. 2011. Transitions between three swimming gaits in *Paramecium* escape. *Proc. Natl. Acad. Sci. U.S.A.*, **108**, 7290–7295.

Hancock, G. J. 1953. The self-propulsion of microscopic organisms through liquids. *Proc. Roy. Soc. Lond. A*, **217**, 96–121.

Happel, J. and Brenner, H. 1965. *Low Reynolds Number Hydrodynamics*. Englewood Cliffs, NJ: Prentice Hall.

Hasegawa, K., Yamashita, I., and Namba, K. 1998. Quasi- and nonequivalence in the structure of bacterial flagellar filament. *Biophys. J.*, **74**, 569–575.

Hatwalne, Y., Ramaswamy, S., Rao, M., and Simha, R. A. 2004. Rheology of active-particle suspensions. *Phys. Rev. Lett.*, **92**, 118101.

Heil, M. and Hazel, A. L. 2011. Fluid-structure interaction in internal physiological flows. *Annu. Rev. Fluid Mech.*, **43**, 141–162.

Hewitt, D. R. and Balmforth, N. J. 2017. Taylor's swimming sheet in a yield-stress fluid. *J. Fluid Mech.*, **828**, 33–56.

Hewitt, D. R. and Balmforth, N. J. 2018. Viscoplastic slender-body theory. *J. Fluid Mech.*, **856**, 870–897.

Higdon, J. J. L. 1979a. Hydrodynamic analysis of flagellar propulsion. *J. Fluid Mech.*, **90**, 685–711.

Higdon, J. J. L. 1979b. Hydrodynamics of flagellar propulsion: Helical waves. *J. Fluid Mech.*, **94**, 331–351.

Hill, J., Kalkanci, O., McMurry, J. L., and Koser, H. 2007. Hydrodynamic surface interactions enable *Escherichia coli* to seek efficient routes to swim upstream. *Phys. Rev. Lett.*, **98**, 068101.

Hohenegger, C. and Shelley, M. J. 2010. Stability of active suspensions. *Phys. Rev. E*, **81**, 046311.

Hohenegger, C. and Shelley, M. J. 2011. Dynamics of complex biofluids. *New Trends in the Physics and Mechanics of Biological Systems: Lecture Notes of the Les Houches Summer School. Volume 92, July 2009*, **92**, 65.

Hotani, H. 1982. Micro-video study of moving bacterial flagellar filaments III: Cyclic transformation induced by mechanical force. *J. Mol. Biol.*, **156**, 791.

Howard, J. 2001. *Mechanics of Motor Proteins and the Cytoskeleton*. Sunderland, MA: Sinauer Associates.

Howse, J. R., Jones, R. A. L., Ryan, A. J., et al. 2007. Self-motile colloidal particles: From directed propulsion to random walk. *Phys. Rev. Lett.*, **99**, 048102.

Hu, D. L., Nirody, J., Scott, T., and Shelley, M. J. 2009. The mechanics of slithering locomotion. *Proc. Natl. Acad. Sci. U.S.A.*, **106**, 10081–10085.

Hu, J., Yang, M., Gompper, G., and Winkler, R. G. 2015. Modelling the mechanics and hydrodynamics of swimming *E. coli*. *Soft Matter*, **11**, 7867–7876.

Hyon, Y., Marcos, Powers, T. R., Stocker, R., and Fu, H. C. 2012. The wiggling trajectories of bacteria. *J. Fluid Mech.*, **705**, 58–76.

Ishijima, S., Oshio, S., and Mohri, H. 1986. Flagellar movement of human spermatozoa. *Gamete Res.*, **13**, 185–197.

Ishikawa, T., Simmonds, M. P., and Pedley, T. J. 2006. Hydrodynamic interaction of two swimming model micro-organisms. *J. Fluid Mech.*, **568**, 119–160.

Ishikawa, T., Sekiya, G., Imai, Y., and Yamaguchi, T. 2007. Hydrodynamic interaction between two swimming bacteria. *Biophys. J.*, **93**, 2217–2225.

Ishimoto, K. and Gaffney, E. A. 2015. Fluid flow and sperm guidance: A simulation study of hydrodynamic sperm rheotaxis. *J. Roy. Soc. Interface*, **12**, 20150172.

Ishimoto, K. and Yamada, M. 2012. A coordinate-based proof of the scallop theorem. *SIAM J. Applied Math.*, **72**, 1686–1694.

Izri, Z., van der Linden, M. N., Michelin, S., and Dauchot, O. 2014. Self-propulsion of pure water droplets by spontaneous Marangoni stress driven motion. *Phys. Rev. Lett.*, **113**, 248302.

Jacobs, K. 2010. *Stochastic Processes for Physicists: Understanding Noisy Systems.* Cambridge, UK: Cambridge University Press.

Jahn, T. L. and Votta, J. J. 1972. Locomotion of protozoa. *Annu. Rev. Fluid Mech.*, **4**, 93–116.

Jarrell, K. F. and McBride, M. J. 2008. The surprisingly diverse ways that prokaryotes move. *Nature Rev. Microbiol.*, **6**, 466–476.

Jeffery, G. B. 1922. The motion of ellipsoidal particles immersed in a viscous fluid. *Proc. Roy. Soc. A*, **102**, 161–179.

Jibuti, L., Qi, L., Misbah, C., et al. 2014. Self-focusing and jet instability of a microswimmer suspension. *Phys. Rev. E*, **90**, 063019.

Johnson, R. E. 1980. An improved slender body theory for Stokes flow. *J. Fluid Mech.*, **99**, 411–431.

Johnson, R. E. and Brokaw, C. J. 1979. Flagellar hydrodynamics: A comparison between resistive-force theory and slender-body theory. *Biophys. J.*, **25**, 113.

Jülicher, F., Ajdari, A., and Prost, J. 1997. Modeling molecular motors. *Rev. Mod. Phys.*, **69**, 1269.

Jülicher, F., Grill, S. W., and Salbreux, G. 2018. Hydrodynamic theory of active matter. *Rep. Prog. Phys.*, **81**, 7.

Kamiya, R., Asakura, S., and Yamaguchi, S. 1980. Formation of helical filaments by copolymerization of two types of straight flagellins. *Nature*, **286**, 628–630.

Kantsler, V., Dunkel, J., Polin, M., and Goldstein, R. E. 2013. Ciliary contact interactions dominate surface scattering of swimming eukaryotes. *Proc. Natl. Acad. Sci. U.S.A.*, **110**, 1187–1192.

Kantsler, V., Dunkel, J., Blayney, M., and Goldstein, R. E. 2014. Rheotaxis facilitates upstream navigation of mammalian sperm cells. *eLife*, **3**, e02403.

Kasyap, T. V., Koch, D. L., and Wu, M. 2014. Hydrodynamic tracer diffusion in suspensions of swimming bacteria. *Phys. Fluids*, **26**, 081901.

Katz, D. F. 1974. On the propulsion of micro-organisms near solid boundaries. *J. Fluid Mech.*, **64**, 33–49.

Katz, D. F. and Berger, S. A. 1980. Flagellar propulsion of human sperm in cervical mucus. *Biorheol.*, **17**, 169–175.

Katz, D. F., Blake, J. R., and Paveri-Fontana, S. L. 1975. On the movement of slender bodies near plane boundaries at low Reynolds number. *J. Fluid Mech.*, **72**, 529–540.

Kaya, T. and Koser, H. 2009. Characterization of hydrodynamic surface interactions of *Escherichia coli* cell bodies in shear flow. *Phys. Rev. Lett.*, **103**, 138103.

Kaya, T. and Koser, H. 2012. Direct upstream motility in *Escherichia coli*. *Biophys. J.*, **102**, 1514–1523.

Kearns, D. B. 2010. A field guide to bacterial swarming motility. *Nature Rev. Microbiol.*, **8**, 634–644.

Keim, N. C., Garcia, M., and Arratia, P. E. 2012. Fluid elasticity can enable propulsion at low Reynolds number. *Phys. Fluids*, **24**, 081703.

Keller, J. B. and Rubinow, S. I. 1976. Slender body theory for slow viscous flow. *J. Fluid Mech.*, **75**, 705–714.

Kessler, J. O. 1985. Hydrodynamic focusing of motile algal cells. *Nature*, **313**, 218–220.

Khurana, N., Blawzdziewicz, J., and Ouellette, N. T. 2011. Reduced transport of swimming particles in chaotic flow due to hydrodynamic trapping. *Phys. Rev. Lett.*, **106**, 198104.

Kim, M. and Powers, T. R. 2004. Hydrodynamic interactions between rotating helices. *Phys. Rev. E*, **69**, 061910.

Kim, M., Bird, J. C., Van Parys, A. J., Breuer, K. S., and Powers, T. R. 2003. A macroscopic scale model of bacterial flagellar bundling. *Proc. Natl. Acad. Sci. U.S.A.*, **100**, 15481–15485.

Kim, M. J. and Breuer, K. S. 2004. Enhanced diffusion due to motile bacteria. *Phys. Fluids*, **16**, L78–L81.

Kim, M. J., Kim, M. M. J., Bird, J. C., et al. 2004. Particle image velocimetry experiments on a macro-scale model for bacterial flagellar bundling. *Exp. Fluids*, **37**, 782–788.

Kim, S. and Karrila, J. S. 1991. *Microhydrodynamics: Principles and Selected Applications*. Boston, MA: Butterworth-Heinemann.

Koch, D. L. and Shaqfeh, E. S. G. 1989. The instability of a dispersion of sedimenting spheroids. *J. Fluid Mech.*, **209**, 521–542.

Koch, D. L. and Subramanian, G. 2011. Collective hydrodynamics of swimming microorganisms: Living fluids. *Annu. Rev. Fluid Mech.*, **43**, 637–659.

Koens, L. and Lauga, E. 2014. The passive diffusion of *Leptospira interrogans*. *Phys. Biol.*, **11**, 066008.

Koens, L. and Lauga, E. 2016. Slender-ribbon theory. *Phys. Fluids*, **28**, 013101.

Koens, L. and Lauga, E. 2017. Analytical solutions to slender-ribbon theory. *Phys. Rev. Fluids*, **2**, 084101.

Koens, L. and Lauga, E. 2018. The boundary integral formulation of Stokes flows includes slender-body theory. *J. Fluid Mech.*, **850**, R1.

Kotar, J., Leoni, M., Bassetti, B., Lagomarsino, M. C., and Cicuta, P. 2010. Hydrodynamic synchronization of colloidal oscillators. *Proc. Natl. Acad. Sci. U.S.A.*, **107**, 7669–7673.

Kotar, J., Debono, L., Bruot, N., et al. 2013. Optimal hydrodynamic synchronization of colloidal rotors. *Phys. Rev. Lett.*, **111**, 228103.

Koumakis, N., Lepore, A., Maggi, C., and Di Leonardo, R. 2013. Targeted delivery of colloids by swimming bacteria. *Nature Comm.*, **4**, 2588.

Kruse, K., Joanny, J.-F., Jülicher, F., Prost, J., and Sekimoto, K. 2004. Asters, vortices, and rotating spirals in active gels of polar filaments. *Phys. Rev. Lett.*, **92**, 078101.

Kruse, K., Joanny, J.-F., Jülicher, F., Prost, J., and Sekimoto, K. 2005. Generic theory of active polar gels: A paradigm for cytoskeletal dynamics. *Eur. Phys. J. E*, **16**, 5–16.

Kühn, M. J., Schmidt, F. K., Eckhardt, B., and Thormann, K. M. 2017. Bacteria exploit a polymorphic instability of the flagellar filament to escape from traps. *Proc. Natl. Acad. Sci. U.S.A.*, **114**, 6340–6345.

Lagomarsino, M. C., Bassetti, B., and Jona, P. 2002. Rowers coupled hydrodynamically: Modeling possible mechanisms for the cooperation of cilia. *Europ. Phys. J. B*, **26**, 81–88.

Lagomarsino, M. C., Jona, P., and Bassetti, B. 2003. Metachronal waves for deterministic switching two-state oscillators with hydrodynamic interaction. *Phys. Rev. E*, **68**, 021908.

Lamb, H. 1932. *Hydrodynamics*. 6th edn. Cambridge, UK: Cambridge University Press.

Landau, L. D. and Lifshitz, E. M. 1980. *Statistical Physics*. 3rd edn. Oxford, UK: Butterworth-Heinemann.

Landau, L. D. and Lifshitz, E. M. 1986. *Theory of Elasticity*. 3rd edn. Oxford, UK: Butterworth-Heinemann.

Larson, R. G. 1988. *Constitutive Equations for Polymer Melts and Solutions*. Boston, MA: Butterworth-Heinemann.

Larson, R. G. 1999. *The Structure and Rheology of Complex Fluids*. Oxford, UK: Oxford Universtity Press.

Lauga, E. 2007. Propulsion in a viscoelastic fluid. *Phys. Fluids*, **19**, 083104.

Lauga, E. 2009. Life at high Deborah number. *Europhys. Lett.*, **86**, 64001.

Lauga, E. 2011a. Enhanced diffusion by reciprocal swimming. *Phys. Rev. Lett.*, **106**, 178101.

Lauga, E. 2011b. Life around the scallop theorem. *Soft Matter*, **7**, 3060–3065.

Lauga, E. 2014. Locomotion in complex fluids: Integral theorems. *Phys. Fluids*, **26**, 081902.

Lauga, E. 2016. Bacterial hydrodynamics. *Annu. Rev. Fluid Mech.*, **48**, 105–130.

Lauga, E. and Bartolo, D. 2008. No many-scallop theorem: Collective locomotion of reciprocal swimmers. *Phys. Rev. E*, **78**, 030901.

Lauga, E. and Eloy, C. 2013. Shape of optimal active flagella. *J. Fluid Mech.*, **730**, R1.

Lauga, E. and Nadal, F. 2017. Clustering instability of focused swimmers. *Europhys. Lett.*, **116**, 64004.

Lauga, E. and Powers, T. R. 2009. The hydrodynamics of swimming microorganisms. *Rep. Prog. Phys.*, **72**, 096601.

Lauga, E., DiLuzio, W. R., Whitesides, G. M., and Stone, H. A. 2006. Swimming in circles: Motion of bacteria near solid boundaries. *Biophys. J.*, **90**, 400–412.

Lauga, E., Brenner, M. P., and Stone, H. A. 2007. Microfluidics: The no-slip boundary condition. In J. Foss, C. Tropea, and Yarin, A. (eds), *Handbook of Experimental Fluid Dynamics*. New York, NY: Springer. Pages 1219–1240.

Lazier, J. R. N. and Mann, K. H. 1989. Turbulence and the diffusive layers around small organisms. *Deep Sea Res. A*, **36**, 1721–1733.

Leal, L. G. 2007. *Advanced Transport Phenomena: Fluid Mechanics and Convective Transport Processes*. Cambridge, UK: Cambridge University Press.

Leifson, E. 1960. *Atlas of Bacterial Flagellation*. New York and London: Academic Press.

Lemelle, L., Palierne, J. F., Chatre, E., and Place, C. 2010. Counterclockwise circular motion of bacteria swimming at the air-liquid interface. *J. Bacteriol*, **192**, 6307–6308.

Lemelle, L., Palierne, J.-F., Chatre, E., Vaillant, C., and Place, C. 2013. Curvature reversal of the circular motion of swimming bacteria probes for slip at solid/liquid interfaces. *Soft Matter*, **9**, 9759–9762.

Leoni, M., Kotar, J., Bassetti, B., Cicuta, P., and Lagomarsino, M. C. 2009. A basic swimmer at low Reynolds number. *Soft Matter*, **5**, 472–476.

Leptos, K. C., Guasto, J. S., Gollub, J. P., Pesci, A. I., and Goldstein, R. E. 2009. Dynamics of enhanced tracer diffusion in suspensions of swimming eukaryotic microorganisms. *Phys. Rev. Lett.*, **103**, 198103.

Leshansky, A. M. 2009. Enhanced low-Reynolds-number propulsion in heterogeneous viscous environments. *Phys. Rev. E*, **80**, 051911.

Leshansky, A. M. and Kenneth, O. 2008. Surface tank treading: Propulsion of Purcell's toroidal swimmer. *Phys. Fluids*, **20**, 063104.

Li, G., Tam, L.-K., and Tang, J. X. 2008. Amplified effect of Brownian motion in bacterial near-surface swimming. *Proc. Natl. Acad. Sci. U.S.A.*, **105**, 18355–18359.

Liao, Q., Subramanian, G., DeLisa, M. P., Koch, D. L., and Wu, M. M. 2007. Pair velocity correlations among swimming *Escherichia coli* bacteria are determined by force-quadrupole hydrodynamic interactions. *Phys. Fluids*, **19**, 061701.

Lighthill, M. J. 1952. On the squirming motion of nearly spherical deformable bodies through liquids at very small Reynolds numbers. *Commun. Pure Appl. Math.*, **5**, 109–118.

Lighthill, M. J. 1975. *Mathematical Biofluiddynamics*. Philadelphia, PA: SIAM.

Lighthill, M. J. 1976. Flagellar hydrodynamics: The John von Neumann Lecture, 1975. *SIAM Rev.*, **18**, 161–230.

Lighthill, M. J. 1996a. Helical distributions of stokeslets. *J. Eng. Math.*, **30**, 35–78.

Lighthill, M. J. 1996b. Reinterpreting the basic theorem of flagellar hydrodynamics. *J. Eng. Math.*, **30**, 25–34.

Lin, Z., Thiffeault, J.-L., and Childress, S. 2011. Stirring by squirmers. *J. Fluid Mech.*, **669**, 167–177.

Lisicki, M., Reigh, S. Y., and Lauga, E. 2018. Autophoretic motion in three dimensions. *Soft Matter*, **14**, 3304–3314.

Liu, B., Powers, T. R., and Breuer, K. S. 2011. Force-free swimming of a model helical flagellum in viscoelastic fluids. *Proc. Natl. Acad. Sci. U.S.A.*, **108**, 19516–19520.

Lopez, D. and Lauga, E. 2014. Dynamics of swimming bacteria at complex interfaces. *Phys. Fluids*, **26**, 071902.

Lovely, P. S. and Dahlquist, F. W. 1975. Statistical measures of bacterial motility and chemotaxis. *J. Theor. Biol.*, **50**, 477–496.

Lowe, C. P. 2003. Dynamics of filaments: Modelling the dynamics of driven microfilaments. *Phil. Trans. Roy. Soc. B*, **358**, 1543–1550.

Lushi, E., Kantsler, V., and Goldstein, R. E. 2017. Scattering of biflagellate microswimmers from surfaces. *Phys. Rev. E*, **96**, 023102.

Machin, K. E. 1958. Wave propagation along flagella. *J. Exp. Biol*, **35**, 796–806.

Machin, K. E. 1963. The control and synchronization of flagellar movement. *Proc. Roy. Soc. B*, **158**, 88–104.

Macnab, R. M. 1977. Bacterial flagella rotating in bundles: A study in helical geometry. *Proc. Natl. Acad. Sci. U.S.A.*, **74**, 221–225.

Macnab, R. M. and Ornston, M. K. 1977. Normal to curly flagellar transitions and their role in bacterial tumbling: Stabilization of an alternative quaternary structure by mechanical force. *J. Mol. Biol.*, **112**, 1–30.

Magariyama, Y. and Kudo, S. 2002. A mathematical explanation of an increase in bacterial swimming speed with viscosity in linear-polymer solutions. *Biophys. J.*, **83**, 733–739.

Magariyama, Y., Ichiba, M., Nakata, K., et al. 2005. Difference in bacterial motion between forward and backward swimming caused by the wall effect. *Biophys. J.*, **88**, 3648–3658.

Makino, M. and Doi, M. 2005. Migration of twisted ribbon-like particles in simple shear flow. *Phys. Fluids*, **17**, 103605.

Man, Y. and Lauga, E. 2015. Phase-separation models for swimming enhancement in complex fluids. *Phys. Rev. E*, **92**, 023004.

Man, Y., Page, W., Poole, R. J., and Lauga, E. 2017. Bundling of elastic filaments induced by hydrodynamic interactions. *Phys. Rev. Fluids*, **2**, 123101.

Marchetti, M. C., Joanny, J.-F., Ramaswamy, S., et al. 2013. Hydrodynamics of soft active matter. *Rev. Mod. Phys.*, **85**, 1143.

Marcos, Fu, H. C., Powers, T. R., and Stocker, R. 2012. Bacterial rheotaxis. *Proc. Natl. Acad. Sci. U.S.A.*, **109**, 4780–4785.

Mathijssen, A. J. T. M., Shendruk, T. N., Yeomans, J. M., and Doostmohammadi, A. 2016. Upstream swimming in microbiological flows. *Phys. Rev. Lett.*, **116**, 028104.

Mazo, R. M. 2002. *Brownian Motion: Fluctuations, Dynamics, and Applications*. Oxford, UK: Oxford University Press.

Mendelson, N. H., Bourque, A., Wilkening, K., Anderson, K. R., and Watkins, J. C. 1999. Organized cell swimming motions in *Bacillus subtilis* colonies: Patterns of short-lived whirls and jets. *J. Bacteriol.*, **181**, 600–609.

Mettot, C. and Lauga, E. 2011. Energetics of synchronized states in three-dimensional beating flagella. *Phys. Rev. E*, **84**, 061905.

Mhetar, V. and Archer, L. A. 1998. Slip in entangled polymer solutions. *Macromolecules*, **31**, 6639–6649.

Michelin, S. and Lauga, E. 2010a. Efficiency optimization and symmetry-breaking in a model of ciliary locomotion. *Phys. Fluids*, **22**, 111901.

Michelin, S. and Lauga, E. 2010b. The long-time dynamics of two hydrodynamically-coupled swimming cells. *Bull. Math. Biol.*, **72**, 973–1005.

Michelin, S., Lauga, E., and Bartolo, D. 2013. Spontaneous autophoretic motion of isotropic particles. *Phys. Fluids*, **25**, 061701.

Miño, G. L., Dunstan, J., Rousselet, A., Clément, E., and Soto, R. 2013. Induced diffusion of tracers in a bacterial suspension: Theory and experiments. *J. Fluid Mech.*, **729**, 423–444.

Montenegro-Johnson, T. D., and Lauga, E. 2014. Optimal swimming of a sheet. *Phys. Rev. E*, **89**, 060701.

Morrison, F. A. 2001. *Understanding Rheology*. Oxford, UK: Oxford University Press.

Morse, M., Huang, A., Li, G., Maxey, M. R., and Tang, J. X. 2013. Molecular adsorption steers bacterial swimming at the air/water interface. *Biophys. J.*, **105**, 21–28.

Mussler, M., Rafai, S., Peyla, P., and Wagner, C. 2013. Effective viscosity of non-gravitactic *Chlamydomonas reinhardtii* microswimmer suspensions. *Europhys. Lett.*, **101**, 54004.

Myers, K. M. and Elad, D. 2017. Biomechanics of the human uterus. *Wiley Interdisciplinary Reviews: Systems Biology and Medicine*, **5**, e1388.

Navier, C.-L. 1823. Mémoire sur les lois du mouvement des fluides. *Mémoires de l'Académie Royale des Sciences de l'Institut de France*, **VI**, 389–440.

Niedermayer, T., Eckhardt, B., and Lenz, P. 2008. Synchronization, phase locking, and metachronal wave formation in ciliary chains. *Chaos*, **18**, 037128.

Nonaka, S., Tanaka, Y., Okada, Y., et al. 1998. Randomization of left–right asymmetry due to loss of nodal cilia generating leftward flow of extraembryonic fluid in mice lacking KIF3B motor protein. *Cell*, **95**, 829–837.

Nonaka, S., Yoshiba, S., Watanabe, D., et al. 2005. De novo formation of left–right asymmetry by posterior tilt of nodal cilia. *PLoS Biol.*, **3**, e268.

Normand, T. and Lauga, E. 2008. Flapping motion and force generation in a viscoelastic fluid. *Phys. Rev. E*, **78**, 061907.

Okada, Y., Takeda, S., Tanaka, Y., Belmonte, J.-C. I., and Hirokawa, N. 2005. Mechanism of nodal flow: A conserved symmetry breaking event in left-right axis determination. *Cell*, **121**, 633–644.

Osterman, N. and Vilfan, A. 2011. Finding the ciliary beating pattern with optimal efficiency. *Proc. Natl. Acad. Sci. U.S.A.*, **108**, 15727–15732.

Pak, O. S. and Lauga, E. 2010. The transient swimming of a waving sheet. *Proc. Roy. Soc. A*, **466**, 107–126.

Pak, O. S. and Lauga, E. 2014. Generalized squirming motion of a sphere. *J. Eng. Math.*, **88**, 1–28.

Pak, O. S., Normand, T., and Lauga, E. 2010. Pumping by flapping in a viscoelastic fluid. *Phys. Rev. E*, **81**, 036312.

Pak, O. S., Zhu, L., Brandt, L., and Lauga, E. 2012. Micropropulsion and microrheology in complex fluids via symmetry breaking. *Phys. Fluids*, **24**, 103102.

Palacci, J., Sacanna, S., Steinberg, A. P., Pine, D. J., and Chaikin, P. M. 2013. Living crystals of light-activated colloidal surfers. *Science*, **339**, 936–940.

Papavassiliou, D. and Alexander, G. P. 2017. Exact solutions for hydrodynamic interactions of two squirming spheres. *J. Fluid Mech.*, **813**, 618–646.

Pedley, T. J. 2000. Blood flow in arteries and veins. In Batchelor, G.K., Moffatt, H.K., and Worster, M.G. (eds), *Perspectives in Fluid Mechanics*. Cambridge, UK: Cambridge University Press. Pages 105–158.

Pedley, T. J. and Kessler, J. O. 1987. The orientation of spheroidal microorganisms swimming in a flow field. *Proc. Roy. Soc. B*, **231**, 47–70.

Pedley, T. J. and Kessler, J. O. 1992. Hydrodynamic phenomena in suspensions of swimming microorganisms. *Annu. Rev. Fluid Mech.*, **24**, 313–358.

Pedley, T. J., Brumley, D. R., and Goldstein, R. E. 2016. Squirmers with swirl: A model for *Volvox* swimming. *J. Fluid Mech.*, **798**, 165–186.

Pironneau, O. and Katz, D. F. 1974. Optimal swimming of flagellated microorganisms. *J. Fluid Mech.*, **66**, 391–415.

Popel, A. S. and Johnson, P. C. 2005. Microcirculation and hemorheology. *Annu. Rev. Fluid Mech.*, **37**, 43–69.

Powers, T. R. 2002. Role of body rotation in bacterial flagellar bundling. *Phys. Rev. E*, **65**, 040903.

Powers, T. R. 2010. Dynamics of filaments and membranes in a viscous fluid. *Rev. Mod. Phys.*, **82**, 1607.

Pozrikidis, C. 1992. *Boundary Integral and Singularity Methods for Linearized Viscous Flow*. Cambridge, UK: Cambridge University Press.

Purcell, E. M. 1977. Life at low Reynolds number. *Am. J. Phys.*, **45**, 3–11.

Purcell, E. M. 1997. The efficiency of propulsion by a rotating flagellum. *Proc. Natl. Acad. Soc. U.S.A.*, **94**, 11307–11311.

Pushkin, D. O., Shum, H., and Yeomans, J. M. 2013. Fluid transport by individual microswimmers. *J. Fluid Mech.*, **726**, 5–25.

Qian, B., Jiang, H., Gagnon, D. A., Breuer, K. S., and Powers, T. R. 2009. Minimal model for synchronization induced by hydrodynamic interactions. *Phys. Rev. E*, **80**, 061919.

Qin, B., Gopinath, A., Yang, J., Gollub, J. P., and Arratia, P. E. 2015. Flagellar kinematics and swimming of algal cells in viscoelastic fluids. *Sci. Rep.*, **5**, 9190.

Qiu, T., Lee, T.-C., Mark, A. G., et al. 2014. Swimming by reciprocal motion at low Reynolds number. *Nature Comm.*, **5**, 5119.

Quaranta, G., Aubin-Tam, M.-E., and Tam, D. 2015. Hydrodynamics versus intracellular coupling in the synchronization of eukaryotic flagella. *Phys. Rev. Lett,*, **115**, 238101.

Rafai, S., Jibuti, L., and Peyla, P. 2010. Effective viscosity of microswimmer suspensions. *Phys. Rev. Lett.*, **104**, 098102.

Ramia, M., Tullock, D. L., and Phan-Thien, N. 1993. The role of hydrodynamic interaction in the locomotion of microorganisms. *Biophys. J.*, **65**, 755–778.

Reichert, M. and Stark, H. 2005. Synchronization of rotating helices by hydrodynamic interactions. *Eur. Phys. J. E*, **17**, 493–500.

Reigh, S. Y., Winkler, R. G., and Gompper, G. 2012. Synchronization and bundling of anchored bacterial flagella. *Soft Matter*, **8**, 4363–4372.

Reigh, S. Y., Winkler, R. G., and Gompper, G. 2013. Synchronization, slippage, and unbundling of driven helical flagella. *PloS One*, **8**, e70868.

Reynolds, A. J. 1965. The swimming of minute organisms. *J. Fluid Mech.*, **23**, 241–260.

Riedel, I. H., Kruse, K., and Howard, J. 2005. A self-organized vortex array of hydrodynamically entrained sperm cells. *Science*, **309**, 300–303.

Riedel-Kruse, I. H., Hilfinger, A., Howard, J., and Jülicher, F. 2007. How molecular motors shape the flagellar beat. *HFSP J.*, **1**, 192–208.

Riley, E. E. and Lauga, E. 2014. Enhanced active swimming in viscoelastic fluids. *Europhys. Lett.*, **108**, 34003.

Riley, E. E. and Lauga, E. 2015. Small-amplitude swimmers can self-propel faster in viscoelastic fluids. *J. Theor. Biol.*, **382**, 345–55.

Riley, E. E., Das, D., and Lauga, E. 2018. Swimming of peritrichous bacteria is enabled by an elastohydrodynamic instability. *Sci. Rep.*, **8**, 10728.

Riley, K. F., Hobson, M. P., and Bence, S. J. 1999. *Mathematical Methods for Physics and Engineering*. Oxford, UK: Oxford University Press.

Rodenborn, B., Chen, C.-H., Swinney, H. L., Liu, B., and Zhang, H. P. 2013. Propulsion of microorganisms by a helical flagellum. *Proc. Natl. Acad. Sci. U.S.A.*, **110**, E338–E347.

Romanczuk, P., Bär, M., Ebeling, W., Lindner, B., and Schimansky-Geier, L. 2012. Active Brownian particles. *Eur. Phys. J. Spec. Top.*, **202**, 1–162.

Rothschild, L. 1963. Non-random distribution of bull spermatozoa in a drop of sperm suspension. *Nature*, **198**, 1221.

Rusconi, R. and Stocker, R. 2015. Microbes in flow. *Curr. Op. Microbiol.*, **25**, 1–8.

Rusconi, R., Guasto, J. S., and Stocker, R. 2014. Bacterial transport suppressed by fluid shear. *Nature Phys.*, **10**, 212–217.

Ryan, S. D., Haines, B. M., Berlyand, L., Ziebert, F., and Aranson, I. S. 2011. Viscosity of bacterial suspensions: Hydrodynamic interactions and self-induced noise. *Phys. Rev. E*, **83**, 050904.

Saintillan, D. 2018. Rheology of active fluids. *Annu. Rev. Fluid Mech.*, **50**, 563–592.

Saintillan, D. and Shelley, M. J. 2007. Orientational order and instabilities in suspensions of self-locomoting rods. *Phys. Rev. Lett.*, **99**, 058102.

Saintillan, D. and Shelley, M. J. 2008a. Instabilities and pattern formation in active particle suspensions: Kinetic theory and continuum simulations. *Phys. Rev. Lett.*, **100**, 178103.

Saintillan, D. and Shelley, M. J. 2008b. Instabilities, pattern formation, and mixing in active suspensions. *Phys. Fluids*, **20**, 123304.

Saintillan, D. and Shelley, M. J. 2013. Active suspensions and their nonlinear models. *C. R. Physique*, **14**, 497–517.

Sanchez, T., Welch, D., Nicastro, D., and Dogic, Z. 2011. Cilia-like beating of active microtubule bundles. *Science*, **333**, 456–459.

Sandoval, M., Marath, N. K., Subramanian, G., and Lauga, E. 2014. Stochastic dynamics of active swimmers in linear flows. *J. Fluid Mech.*, **742**, 50–70.

Saragosti, J., Silberzan, P., and Buguin, A. 2012. Modeling *E. coli* tumbles by rotational diffusion: Implications for chemotaxis. *PloS One*, **7**, e35412.

Sartori, P., Geyer, V. F., Scholich, A., Jülicher, F., and Howard, J. 2016. Dynamic curvature regulation accounts for the symmetric and asymmetric beats of *Chlamydomonas* flagella. *eLife*, **5**, e13258.

Sauzade, M., Elfring, G. J., and Lauga, E. 2011. Taylor's swimming sheet: Analysis and improvement of the perturbation series. *Physica D*, **240**, 1567–1573.

Schamel, D., Mark, A. G., Gibbs, J. G., et al. 2014. Nanopropellers and their actuation in complex viscoelastic media. *ACS Nano.*, **8**, 8794–8801.

Schnitzer, O. and Yariv, E. 2015. Osmotic self-propulsion of slender particles. *Phys. Fluids*, **27**, 031701.

Shen, X. N. and Arratia, P. E. 2011. Undulatory swimming in viscoelastic fluids. *Phys. Rev. Lett.*, **106**, 208101.

Shiratori, H. and Hamada, H. 2006. The left-right axis in the mouse: From origin to morphology. *Development*, **133**, 2095–2104.

Silverman, M. and Simon, M. 1974. Flagellar rotation and the mechanism of bacterial motility. *Nature*, **249**, 73–74.

Simha, R. A. and Ramaswamy, S. 2002. Hydrodynamic fluctuations and instabilities in ordered suspensions of self-propelled particles. *Phys. Rev. Lett.*, **89**, 058101.

Smith, D. J., Blake, J. R., and Gaffney, E. A. 2008. Fluid mechanics of nodal flow due to embryonic primary cilia. *J. Roy. Soc. Int.*, **5**, 567–573.

Smith, D. J., Smith, A. A., and Blake, J. R. 2011. Mathematical embryology: The fluid mechanics of nodal cilia. *J. Eng. Math.*, **70**, 255–279.

Sochi, T. 2011. Slip at fluid-solid interface. *Polymer Reviews*, **51**, 309344.

Sokolov, A. and Aranson, I. S. 2009. Reduction of viscosity in suspension of swimming bacteria. *Phys. Rev. Lett.*, **103**, 148101.

Sokolov, A. and Aranson, I. S. 2012. Physical properties of collective motion in suspensions of bacteria. *Phys. Rev. Lett.*, **109**, 248109.

Sokolov, A., Aranson, I. S., Kessler, J. O., and Goldstein, R. E. 2007. Concentration dependence of the collective dynamics of swimming bacteria. *Phys. Rev. Lett.*, **98**, 158102.

Sokolov, A., Apodaca, M. M., Grzybowski, B. A., and Aranson, I. S. 2010. Swimming bacteria power microscopic gears. *Proc. Natl. Acad. Sci. U.S.A.*, **107**, 969–974.

Solomentsev, Y. and Anderson, J. L. 1994. Electrophoresis of slender particles. *J. Fluid Mech.*, **279**, 197–215.

Son, K., Guasto, J. S., and Stocker, R. 2013. Bacteria can exploit a flagellar buckling instability to change direction. *Nature Phys.*, **9**, 494–498.

Soni, G. V., Jaffar Ali, B. M., Hatwalne, Y., and Shivashankar, G. V. 2003. Single particle tracking of correlated bacterial dynamics. *Biophys. J.*, **84**, 2634–2637.

Sowa, Y. and Berry, R. M. 2008. Bacterial flagellar motor. *Q. Rev. Biophys.*, **41**, 103–132.

Spagnolie, S. E. and Lauga, E. 2010. The optimal elastic flagellum. *Phys. Fluids*, **22**, 031901.

Spagnolie, S. E. and Lauga, E. 2011. Comparative hydrodynamics of bacterial polymorphism. *Phys. Rev. Lett.*, **106**, 058103.

Spagnolie, S. E. and Lauga, E. 2012. Hydrodynamics of self-propulsion near a boundary: Predictions and accuracy of far-field approximations. *J. Fluid Mech.*, **700**, 105–147.

Spagnolie, S. E., Liu, B., and Powers, T. R. 2013. Locomotion of helical bodies in viscoelastic fluids: Enhanced swimming at large helical amplitudes. *Phys. Rev. Lett.*, **111**, 068101.

Spagnolie, S. E., Moreno-Flores, G. R., Bartolo, D., and Lauga, E. 2015. Geometric capture and escape of a microswimmer colliding with an obstacle. *Soft Matter*, **11**, 3396–3411.

Srigiriraju, S. V. and Powers, T. R. 2005. Continuum model for polymorphism of bacterial flagella. *Phys. Rev. Lett.*, **94**, 248101.

Srigiriraju, S. V. and Powers, T. R. 2006. Model for polymorphic transitions in bacterial flagella. *Phys. Rev. E*, **73**, 011902.

Stone, H. A. and Samuel, A. D. T. 1996. Propulsion of microorganisms by surface distortions. *Phys. Rev. Lett.*, **77**, 4102–4104.

Subramanian, G. and Koch, D. L. 2009. Critical bacterial concentration for the onset of collective swimming. *J. Fluid Mech.*, **632**, 359–400.

Takagi, D., Palacci, J., Braunschweig, A. B., Shelley, M. J., and Zhang, J. 2014. Hydrodynamic capture of microswimmers into sphere-bound orbits. *Soft Matter*, **10**, 1784–1789.

Tam, D. and Hosoi, A. E. 2011. Optimal kinematics and morphologies for spermatozoa. *Phys. Rev. E*, **83**, 045303.

Tamm, S. L. 1972. Ciliary motion in *Paramecium*: A scanning electron microscope study. *J. Cell Biol.*, **55**, 250–255.

Tanner, R. I. 1988. *Engineering Rheology*. 2nd edn. Oxford, UK: Clarendon Press.

Tavaddod, S., Charsooghi, M. A., Abdi, F., Khalesifard, H. R., and Golestanian, R. 2011. Probing passive diffusion of flagellated and deflagellated *Escherichia coli*. *Eur. Phys. J. E*, **34**, 1–7.

Taylor, G. I. 1951. Analysis of the swimming of microscopic organisms. *Proc. Roy. Soc. A*, **209**, 447–461.

Taylor, G. I. 1952. The action of waving cylindrical tails in propelling microscopic organisms. *Proc. Roy. Soc. A*, **211**, 225–239.

Taylor, G. I. 1967. *Low-Reynolds-Number Flows*. Cambridge, MA: National Committee for Fluid Mechanics Films.

Taylor, J. R. and Stocker, R. 2012. Trade-offs of chemotactic foraging in turbulent water. *Science*, **338**, 675–679.

Ten Hagen, B., Wittkowski, R., and Löwen, H. 2011. Brownian dynamics of a self-propelled particle in shear flow. *Phys. Rev. E*, **84**, 031105.

Tennekes, H. and Lumley, J. L. 1972. *A First Course in Turbulence*. Cambridge, MA: The MIT Press.

Teran, J., Fauci, L., and Shelley, M. 2010. Viscoelastic fluid response can increase the speed and efficiency of a free swimmer. *Phys. Rev. Lett.*, **104**, 038101.

Thiffeault, J.-L. 2015. Distribution of particle displacements due to swimming microorganisms. *Phys, Rev. E*, **92**, 023023.

Thomas, D., Morgan, D. G., and DeRosier, D. J. 2001. Structures of bacterial flagellar motors from two FliF-FliG gene fusion mutants. *J. Bacteriol.*, **183**, 6404–6412.

Thomases, B. and Guy, R. D. 2014. Mechanisms of elastic enhancement and hindrance for finite-length undulatory swimmers in viscoelastic fluids. *Phys. Rev. Lett.*, **113**, 098102.

Thutupalli, S., Seemann, R., and Herminghaus, S. 2011. Swarming behavior of simple model squirmers. *New J. Phys.*, **13**, 073021.

Toner, J. and Tu, Y. 1998. Flocks, herds, and schools: A quantitative theory of flocking. *Phys. Rev. E*, **58**, 4828.

Tornberg, A. K. and Shelley, M. J. 2004. Simulating the dynamics and interactions of flexible fibers in Stokes flow. *J. Comput. Phys.*, **196**, 8–40.

Trevelyan, B. J. and Mason, S. G. 1951. Particle motions in sheared suspensions. I: Rotations. *J. Colloid Sci.*, **6**, 354–367.

Triantafyllou, M. S., Triantafyllou, G. S., and Yue, D. K. P. 2000. Hydrodynamics of fish-like swimming. *Annu. Rev. Fluid Mech.*, **32**, 33–53.

Trouilloud, R., Yu, T. S., Hosoi, A. E., and Lauga, E. 2008. Soft swimming: Exploiting deformable interfaces for low Reynolds number locomotion. *Phys. Rev. Lett.*, **101**, 048102.

Tuck, E. O. 1968. A note on a swimming problem. *J. Fluid Mech.*, **31**, 305–308.

Tung, C.-K., Ardon, F., Roy, A., et al. 2015. Emergence of upstream swimming via a hydrodynamic transition. *Phys. Rev. Lett.*, **114**, 108102.

Turner, L., Ryu, W. S., and Berg, H. C. 2000. Real-time imaging of fluorescent flagellar filaments. *J. Bacteriol.*, **182**, 2793–2801.

Uchida, N. and Golestanian, R. 2011. Generic conditions for hydrodynamic synchronization. *Phys. Rev. Lett.*, **106**, 058104.

Vélez-Cordero, J. R. and Lauga, E. 2013. Waving transport and propulsion in a generalized Newtonian fluid. *J. Non-Newt. Fluid Mech.*, **199**, 37–50.

Vicsek, T., Czirok, A., Benjacob, E., Cohen, I., and Shochet, O. 1995. Novel type of phase-transition in a system of self-driven particles. *Phys. Rev. Lett.*, **75**, 1226–1229.

Videler, J. J. 2012. *Fish Swimming*. New York, NY: Springer.

Vig, D. K. and Wolgemuth, C. W. 2012. Swimming dynamics of the lyme disease spirochete. *Phys. Rev. Lett.*, **109**, 218104.

Vogel, S. 1988. *Life's Devices: The Physical World of Animals and Plants*. Princeton, NJ: Princeton University Press.

Vogel, S. 1996. *Life in Moving Fluids*. Princeton, NJ: Princeton University Press.

Vogel, S. and Calvert, R. A. 1993. *Vital Circuits: On Pumps, Pipes, and the Workings of Circulatory Systems*. New York, NY: Oxford University Press.

Wan, K. Y. and Goldstein, R. E. 2016. Coordinated beating of algal flagella is mediated by basal coupling. *Proc. Natl. Acad. Sci. U.S.A.*, **113**, E2784–E2793.

Wang, Z. J. 2005. Dissecting insect flight. *Annu. Rev. Fluid Mech.*, **37**, 183–210.

Was, L. and Lauga, E. 2013. Optimal propulsive flapping in Stokes flows. *Bioinsp. Biomim.*, **9**, 016001.

Watari, N. and Larson, R. G. 2010. The hydrodynamics of a run-and-tumble bacterium propelled by polymorphic helical flagella. *Biophys. J.*, **98**, 12–17.

Wiggins, C. H. and Goldstein, R. E. 1998. Flexive and propulsive dynamics of elastica at low Reynolds number. *Phys. Rev. Lett.*, **80**, 3879–3882.

Williams, C. R., and Bees, M. A. 2011. Photo-gyrotactic bioconvection. *J. Fluid Mech.*, **678**, 41–86.

Wollin, C. and Stark, H. 2011. Metachronal waves in a chain of rowers with hydrodynamic interactions. *Eur. Phys. J. E*, **34**, 42.

Woolley, D. M. 2003. Motility of spermatozoa at surfaces. *Reproduction*, **126**, 259–270.

Woolley, D. M. 2010. Flagellar oscillation: A commentary on proposed mechanisms. *Biol. Rev.*, **85**, 453–470.

Woolley, D. M., and Vernon, G. G. 2001. A study of helical and planar waves on sea urchin sperm flagella, with a theory of how they are generated. *J. Exp. Biol.*, **204**, 1333–1345.

Woolley, D. M., Crockett, R. F., Groom, W. D. I., and Revell, S. G. 2009. A study of synchronisation between the flagella of bull spermatozoa, with related observations. *J. Exp. Biol.*, **212**, 2215–2223.

Wu, X. L. and Libchaber, A. 2000. Particle diffusion in a quasi-two-dimensional bacterial bath. *Phys. Rev. Lett.*, **84**, 3017–3020.

Wu, Y., Hosu, B. G., and Berg, H. C. 2011. Microbubbles reveal chiral fluid flows in bacterial swarms. *Proc. Natl. Acad. Sci. U.S.A.*, **108**, 4147–4151.

Xie, L., Altindal, T., Chattopadhyay, S., and Wu, X. L. 2011. Bacterial flagellum as a propeller and as a rudder for efficient chemotaxis. *Proc. Natl. Acad. Sci. U.S.A.*, **108**, 2246–2251.

Xu, G., Wilson, K. S., Okamoto, R. J., et al. 2016. Flexural rigidity and shear stiffness of flagella estimated from induced bends and counterbends. *Biophys. J.*, **110**, 2759–2768.

Xu, Y.-L. 1996. Fast evaluation of the Gaunt coefficients. *Math. Comp.*, **65**, 1601–1612.

Yang, Y., Elgeti, J., and Gompper, G. 2008. Cooperation of sperm in two dimensions: Synchronization, attraction, and aggregation through hydrodynamic interactions. *Phys. Rev. E*, **78**, 061903.

Yosida, K. 1949. Brownian motion on the surface of the 3-sphere. *Ann. Math. Stat.*, **20**, 292–296.

Zhang, L., Peyer, K. E., and Nelson, B. J. 2010. Artificial bacterial flagella for micromanipulation. *Lab on a Chip*, **10**, 2203–2215.

Zhu, L., Do-Quang, M., Lauga, E., and Brandt, L. 2011. Locomotion by tangential deformation in a polymeric fluid. *Phys. Rev. E*, **83**, 011901.

Zhu, L., Lauga, E., and Brandt, L. 2012. Self-propulsion in viscoelastic fluids: Pushers vs. pullers. *Phys. Fluids*, **24**, 051902.

Zhu, L., Lauga, E., and Brandt, L. 2013. Low-Reynolds-number swimming in a capillary tube. *J. Fluid Mech.*, **726**, 285–311.

Zhuang, J. and Sitti, M. 2016. Chemotaxis of bio-hybrid multiple bacteria-driven microswimmers. *Sci. Rep.*, **6**, 32135.

Zhuang, J., Park, B.-W., and Sitti, M. 2017. Propulsion and chemotaxis in bacteria-driven microswimmers. *Adv. Sci.*, **4**, 1700109.

Zöttl, A. and Stark, H. 2012. Nonlinear dynamics of a microswimmer in Poiseuille flow. *Phys. Rev. Lett.*, **108**, 218104.

Zöttl, A. and Stark, H. 2013. Periodic and quasiperiodic motion of an elongated microswimmer in Poiseuille flow. *Eur. Phys. J. E*, **36**, 4.

Zöttl, A. and Yeomans, J. M. 2017. Enhanced bacterial swimming speeds in macromolecular polymer solutions. *Nature Phys.*, **15**, 554–558.

Index